DR. L. W. SHEMILT
DEAN, FACULTY OF ENGINEERING
McMASTER UNIVERSITY
HAMILTON, ONTARIO, CANADA

ADVANCES IN ELECTROCHEMISTRY AND
ELECTROCHEMICAL ENGINEERING

VOLUME 10

ADVANCES IN ELECTROCHEMISTRY AND ELECTROCHEMICAL ENGINEERING

•

EDITORS

HEINZ GERISCHER, *Fritz Haber Institute,*
Berlin-Dahlem, Germany
CHARLES W. TOBIAS, *University of California,*
Berkeley, California

ADVISORY BOARD

E. J. CAIRNS, *General Motors Research Laboratory,*
Warren, Michigan
P. DELAHAYE, *New York University,*
New York, New York
M. FLEISCHMANN, *University of Southampton,*
Southampton, Great Britain
A. N. FRUMKIN, *Academy of Sciences,*
Moscow, U.S.S.R
N. IBL, *Federal Institute of Technology,*
Zurich, Switzerland
J. A. A. KETELAAR, *University of Amsterdam,*
The Netherlands
J. KORYTA, *Academie of Sciences,*
Praha, Czechoslovakia
R. PARSONS, *University of Bristol,*
Bristol, England
C. WAGNER, *Max-Planck-Institut,*
Göttingen, Germany
E. B. YEAGER, *Case-Western Reserve University,*
Cleveland, Ohio

ADVANCES IN ELECTROCHEMISTRY AND ELECTROCHEMICAL ENGINEERING

•

VOLUME 10

Edited by

HEINZ GERISCHER

Fritz-Haber Institute
Berlin-Dahlem, Germany

and

CHARLES W. TOBIAS

University of California, Berkeley

A Wiley-Interscience Publication

JOHN WILEY & SONS, New York • London • Sydney • Toronto

Copyright © 1977, by John Wiley & Sons, Inc.

All rights reserved. Published simultaneously in Canada.

No part of this book may be reproduced by any means, nor transmitted, nor translated into a machine language without the written permission of the publisher.

Library of Congress Catalog Card Number: 61-15021

ISBN 0-471-87527-9

Printed in the United States of America

10 9 8 7 6 5 4 3 2 1

Introduction to the Series

This series was planned to make available authoritative reviews in the area of electrochemical phenomena, and to bridge the gap between electrochemistry as a part of physical chemistry and electrochemical engineering. Increased interest in electrochemistry and its applications renders this publication timely.

Chapters will vary considerably in their method of approach and length. Some chapters will be definitive while others, dealing with unsettled problems in rapidly evolving areas, will have to be somewhat tentative. Coverage over a number of volumes should be quite comprehensive. Emphasis will be placed in successive volumes either on the physicochemical or the engineering approach, although some volumes may cover both aspects. Specific problems in electroanalytical chemistry will not be included, except for certain theoretical questions not usually considered by analytical chemists.

1961

PAUL DELAHAY
CHARLES W. TOBIAS

We regret very much that our colleague Paul Delahay, because of interests and obligations in other areas, has decided to resign after fifteen years as co-editor of the Advances. The favorable acceptance of the first nine volumes and an increasing interest in electrochemical science and technology, in our opinion, well justify continuation of the series. We expect to maintain the high standards set in the past and plan no departure from the purpose stated in the Introduction to the Series.

1976

HEINZ GERISCHER
CHARLES W. TOBIAS

Preface

Unlike earlier volumes of this series which were devoted either to topics in electrochemistry or in electrochemical engineering, the present volume contains a combination of chapters that should appeal to both fundamental and applied interests.

The functions and performance of electrochemical sensors used in the analysis and control of gas compositions are critically discussed by Dietz, Haecker, and Jahnke. The direct involvement of the authors in the development of these devices is well reflected in their treatment of one of the more important practical advancements of recent years.

Jaenicke's authoritative review of the electrochemical aspects of photographic processes treats an area of electrochemistry that is terra incognita to many electrochemists. For this reason, and also because some of the problems are of general fundamental interest, readers should find Jaenicke's perspective of this interdisciplinary field very valuable.

The chapter by Trasatti on the role of work function in electrochemistry deals with the general question: to what degree can solid-state properties of metals be used for the characterization of their electrochemical behavior? The author provides in a very stimulating manner his personal view of this somewhat controversial subject.

Because of its wealth of interesting questions and many potential applications, the science and technology of ionically conducting solid-state materials is attracting increasing attention. Huggins, in addition to the characterization of these materials, also considers analogous behavior in metals and semiconductors, and comments on theoretical models relating to fast ionic conduction.

The urgent need for more efficient and lighter storage batteries has focused attention on sulfur electrodes that have already been shown to be capable of superior performance. Tischer and Ludwig, in their comprehensive review of the thermodynamics and kinetics of electrode systems involving sulfur and sulfides, present a careful analysis of available experimental data and relevant theory.

HEINZ GERISCHER
CHARLES W. TOBIAS

Berlin, Germany
Berkeley, California
August 1976

Contributors to Volume 10

HERMANN DIETZ
  Robert Bosch GmbH, Research Center, Gerlingen/Stuttgart, Germany

WOLF-DIETER HAECKER
  Robert Bosch GmbH, Research Center, Gerlingen/Stuttgart, Germany

ROBERT A. HUGGINS
  Center for Materials Research Stanford University, Stanford, California

WALTHER JAENICKE
  Institute of Physical Chemistry, University of Erlangen-Nürnberg, Erlangen, Germany

HORST JAHNKE
  Robert Bosch GmbH, Research Center, Gerlingen/Stuttgart, Germany

FRANK A. LUDWIG
  Research Staff, Ford Motor Company, Dearborn, Michigan

RAGNAR P. TISCHER
  Research Staff, Ford Motor Company, Dearborn, Michigan

SERGIO TRASATTI
  Laboratory of Electrochemistry, University of Milan, Milan, Italy

Contents

Electrochemical Sensors for the Analysis of Gases, *by* H. Dietz, W. Haecker, and H. Jahnke . . . 1

Electrochemical Aspects of the Photographic Processes, *by* W. Jaenicke 91

The Work Function in Electrochemistry, *by* S. Trasatti . 213

Ionically Conducting Solid State Membranes, *by* Robert A. Huggins 323

The Sulfur Electrode in Nonaqueous Media, *by* Ragnar P. Tischer, and Frank A. Ludwig 391

Index 483

ADVANCES IN ELECTROCHEMISTRY AND ELECTROCHEMICAL ENGINEERING

VOLUME 10

Electrochemical Sensors for the Analysis of Gases
H. DIETZ, W. HAECKER, and H. JAHNKE
Robert Bosch GmbH, Research Center, Gerlingen/Stuttgart, Germany

1	Introduction	3
2	Thermodynamic Treatment of Electrodes Involving a Gas Phase	4
3	Influence of Mass Transport on Electrode Processes Involving a Gas Phase	7
4	Preceding Reactions in the Gas Phase	9
	4.1 Electrode Reactions with Gases under Equilibrium Conditions	9
	4.2 Nonequilibrium with the Gas Phase	11
5	Ionic and Electronic Conductivity in Solids	12
6	Methods Based on Voltage Measurement	16
	6.1 Systems at Equilibrium Potential	16
	6.1.1 Electrochemical Cells with Solid Electrolytes	19
	6.1.2 Electrochemical Cells with Liquid Electrolyte	44
	6.2 Systems with Mixed Potentials	46
	6.2.1 Determination of Nitrogen Dioxide in Gas Mixtures	47
	6.2.2 Determination of Oxygen in Water	48

7	Methods Based on Current Measurement		50
	7.1	Measurement of Limiting Current	50
		7.1.1 Principle of Measurement	50
		7.1.2 Criteria for the Selection of Suitable Electrode Materials and Electrode Potentials	54
		7.1.3 Electrolyte and Membrane	57
		7.1.4 Survey of Published Work on Limiting Current Sensors	60
	7.2	Coulometric Measurements	63
	7.3	Determination of Gas Concentration by Selective Inhibition of an Electrode Reaction	65
8	Methods Based on Measurement of Electrical Conductivity		67
9	Electrochemical Oxygen Sensors in Motor Vehicles		73
10	Future Prospects		83
	References		84

1 INTRODUCTION

The demand for clean air and surface water and the legal regulations to this end call for devices for testing and supervising the composition of these media. In such work, the analysis of gaseous components is particularly important. In the past, such demands have led to the development of sensors for determining the concentration of a gaseous component in a gaseous mixture or in a solution in a liquid. Apart from this, however, sensors for gases find many applications in science and technology. One type of such sensors operates on electrochemical principles. In consequence, many electrochemical processes for the determination of gaseous components have been described. The present paper presents a summary of this work.

The term "sensor" denotes a device with which the concentration of certain gaseous components can be measured. In the broad sense of the term, it includes the electronic transducer, to which usually an electronic evaluating device is connected. In our description, the emphasis is on the transducer. Peripheral devices necessary to ensure proper functioning of the measuring system, such as regulating devices for maintaining constancy of temperature and velocity of flow, are only mentioned in passing.

Indirect processes, in which the electrochemical reaction is only used to indicate the endpoint of a chemical titration or the like, are not considered.

A sensor must meet the following requirements:

    High accuracy,
    Wide range of measurement,
    Good reproducibility and constancy of the indicated value,
    Low sensitivity to external disturbances,

High selectivity for the components to be determined,
Rapid response, particularly when the sensor is
to initiate a control function.

The preceding requirements differ in relative importance according to the purpose of the application.

A further important point to be borne in mind in considering the applications of sensors is the costs, which are principally determined by the peripheral devices. These costs are usually difficult to estimate, for which reason they are not considered here.

2 THERMODYNAMIC TREATMENT OF ELECTRODES INVOLVING A GAS PHASE

Electrochemical reactions are characterized by the passage of electric charges (ions or electrons) through a metal-electrolyte phase boundary (or in some cases semiconductor-electrolyte). This produces electrical potential differences between the conducting phases. The same holds when a gas phase participates in the electrode reaction: electrons are exchanged between the electrode material and the gas molecules, and an oxidation or reduction reaction product of the gas passes into the electrolyte. It is characteristic for the electrode reactions of gases that the electrode material is not attacked, but only effects the electron exchange. Such electrodes are termed redox electrodes. The electrode reaction with a gas phase can proceed either at a metal liquid boundary, e.g.

$$\tfrac{1}{2}H_2 \text{ (gas)} \rightleftharpoons H^+(aq) + e^-(metal) \qquad (1)$$

or at a metal-solid electrolyte boundary, e.g.

$$\tfrac{1}{2}O_2 \text{ (gas)} + 2e^-(metal) \rightleftharpoons O^{--}(ZrO_2) \qquad (2)$$

For the thermodynamic treatment of gas electrodes, the individual intermediate stages of the overall reactions play no part, because in equilibrium a definite concentration of a gas in the dissolved and adsorbed states corresponds to its partial pressure in the gaseous phase. The potential of the electrode at the metal-electrolyte interface measured as cell voltage E against a definite reference electrode depends on the concentrations (activities) a_i and partial pressures p_i of the components i of the overall equation. This is expressed in the Nernst equation,

$$E = E^o + \frac{RT}{zF} (\Sigma \nu_i \ln a_i + \Sigma \nu_i \ln p_i) \qquad (3)$$

In this equation E^o is the standard equilibrium potential, ν_i are the stoichiometric coefficients of the substances involved in the overall reaction, and z is the number of the positive charges moved through the interphase in the chosen direction of the electrode reaction.

For nonaqueous solid electrolytes with oxygen-ion conduction, the usual reference electrode is an oxygen electrode with a pressure of 1 atm. In this case the pressure dependence of the potential of the oxygen electrode (2) is given by

$$E = -4.96 \cdot 10^{-5} \, T \log p_{O_2} \qquad (4)$$

3 INFLUENCE OF MASS TRANSPORT ON ELECTRODE PROCESSES INVOLVING A GAS PHASE

As in many other electrochemical reactions, polarization effects play an important part also in electrochemical sensors. One effect, namely the influence of transport processes on the electrode kinetics, is used in the diffusion limiting current sensors and has to be discussed therefore in detail. If the transport is impeded in any way, so that the velocity of the transfer reaction can be greater than that of the material transport, the latter determines the velocity of the electrochemical reaction. The transfer reaction produces concentration changes at the electrode, until in the stationary state a sufficient force for the transport to and from the reaction site of the particles taking part is built up. There are particularly simple conditions, if the effects of migration and convection are largely eliminated. This may be realized by selection of proper experimental conditions, which are described later on. In this case, a convection-free liquid layer exists immediately adjoining the electrode surface, so that, when a current flows, material transport is effected solely by diffusion as a result of the built-up concentration gradient.

The stationary diffusion current determined by the concentration gradient is given by Fick's first law of diffusion as

$$i_{diff} = D_i \cdot zF \cdot \frac{C_i - C_o}{\delta} \tag{5}$$

Here, i_{diff} is the diffusion current per unit surface area and D_i is the diffusion coefficient of the reacting substance. δ is the thickness of the diffusion layer, C_i and C_o are the concentrations of the reacting substance remote from the electrode and immediately adjoining its surface, respectively.

If a substance is removed from the system by the electrode reaction, then, if the polarization is strong enough, the concentration C_o at the electrode surface can become zero, so that every particle reaching the electrode reacts at once. The diffusion current of Eq. (5) then attains the limiting value:

$$i_L = D_i \cdot zF \frac{C_i}{\delta} \qquad (6)$$

which is termed the limiting current density. This is proportional to the concentration of the diffusing reacting substance in the bulk of the solution, a fact widely used for analytical purposes, for example, for the polarographic analysis from the gas phase.

If the electrode consists of the three-phase system: solid electronic conductor-liquid electrolyte-gas phase, the electrode reaction can proceed by solution of gas molecules in the liquid electrolyte and diffusion to the electrode surface, where the electrochemical reaction proper, namely, charge transfer, occurs. The conditions are then the same as for dissolved solid substances. For the gas concentration in the electrolyte, Henry's law can be applied, according to which the solubility is proportional to the partial pressure.

The problem becomes more complicated when additional diffusion impedance in the gas phase or by a membrane occurs. In this case, various diffusion and permeation resistances may additively determine the velocity of the overall reaction. For example, partial pressure gradients can occur in the gas phase, if the gas involved in the electrode reaction is mixed with an excess of an unreactive gas. This is termed gas-side concentration polarization (1). Above all, when material transport takes place through the pores of a solid, which is the case with all porous electrodes and in particular with gas-diffusion electrodes, such effects play an important part. The theory of these phenomena has been developed chiefly in connection with work on fuel cells (2,3). Equally important is diffusion through interposed membranes, as in membrane-covered polarographic gas detectors. In such cases, the material transport to the electrode can be treated as linear diffusion through a two-layer system (4).

For practical use, the so-called gas diffusion electrode, in which a porous matrix of solid electron-conducting material (metal, carbon) is partly penetrated by electrolyte and partly filled with gas, is important. Partial hydrophobization or slight excess pressure of the gas fixes the three-phase boundary in the electrode, the gas standing in capillary equilibrium with the electrolyte and the solid electrode material. According to the thin-film model, a thin wetting film of the electrolyte covers the walls of the pores in the neighborhood of the meniscus, and the electrochemical reaction can occur only in a restricted zone in which the diffusion path through this wetting film is short and the electrolyte resistance low (see Fig. 1).

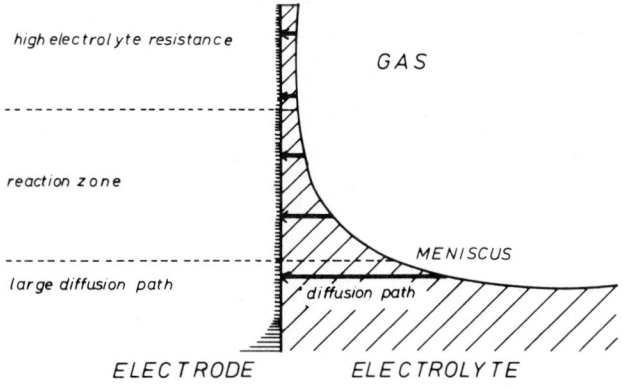

Fig. 1. Schematic representation of the gas/electrolyte interface on a partially immersed electrode.

Even at gas electrodes with two solid phases (metal and solid electrolyte), in which there is naturally no wetting film, it is often the case that transport processes determine the reaction velocity. So, for the oxygen reduction at porous noble metal films on a zirconia electrolyte, diffusion limiting currents have been found under particular conditions (5-7).

4 PRECEDING REACTIONS IN THE GAS PHASE

In the cases of electrode reactions of a gas hitherto considered, it was assumed that the gas acts directly as reaction partner. This means that the Nernst equation for the overall reaction contains the partial pressure directly present in the gas phase, which is not produced by a chemical reaction. Such an electrode is termed an electrode of the first kind, in contrast to an electrode of the second kind, in which a preceding chemical equilibrium determines the activity of the components affecting the potential.

The participation of the gas phase in preceding chemical reactions is particularly important for electrochemical gas sensors. Above all, when no equilibrium with the gas phase exists, potential deviations occur. The two possibilities, of equilibrium and nonequilibrium with the gas phase, are discussed in the following sections.

4.1 Electrode Reaction with Gases under Equilibrium Conditions

If the partial pressure of a gas that determines the potential is determined by a chemical equilibrium, an electrode of the second kind exists. The equilibrium can be homogeneous, for example,

$$CO_2 \rightleftharpoons CO + \frac{1}{2} O_2 \qquad (7)$$

or heterogeneous, for example,

$$NiO \rightleftharpoons Ni + \frac{1}{2} O_2 \qquad (8)$$

Such equilibria - particularly involving oxygen - are particularly important for gas chains with solid electrolytes (8) at high temperatures, because equilibrium at such temperatures is readily attained. These equilibria can be used to produce extremely small partial pressures very exactly at an electrode. Thus, for example, oxygen partial pressures from 10^{-5} to 10^{-21} atm. can easily be produced through the CO_2-CO equilibrium, and H_2-H_2O mixtures give oxygen partial pressures in the range 10^{-20} to 10^{-23} atm. (9). For reference potentials in oxygen measuring cells, heterogeneous equilibria involving a metal/metal-oxide mixture are commonly used, for example, Cu-Cu_2O, Ni-NiO, Co-CoO, Fe-FeO, Mo-MoO_2, and others (10).

Occasionally, a gas mixture contains a number of components, all involved in the attainment of equilibrium. An important example is the exhaust gases of a motor vehicle, in which the reaction of hydrocarbons with air may be expected to produce a wide variety of products. At thermodynamic equilibrium, an oxygen partial pressure dependent on the air/fuel ratio is attained, which determines the potential of the oxygen electrode. This potential can be measured and used for the control of an optimal air/fuel ratio (see Chapter 9).

4.2 Nonequilibrium with the Gas Phase

If the system gas-phase/electrode is not in thermodynamic equilibrium and if a chemical reaction involving the gas phase proceeds at the electrode surface, a reaction-dependent potential is attained, differing from the single electrode potential of the individual components. In considering this, it must be assumed that the reaction at the electrode surface proceeds by adsorbed species, occurring as intermediate products and effectively determining the potential.

Basically, two different mechanisms for the heterogeneous chemical reaction are possible (11). On the one hand, there is an electrochemical mechanism, in which the components react with one another in consecutive anodic and cathodic reactions with zero net current at the phase boundary, and on the other hand there is a nonelectrochemical mechanism, in which a purely chemical transfer reaction proceeds in the adsorbed state. In the case of the electrochemical mechanism, a mixed potential is formed at the electrode.

The nonelectrochemical mechanism (12) which is valid for many catalytic processes, proceeds through a purely chemical transfer reaction. The reactants are adsorbed and dissociate at the electrode surface, where a transfer of atomic species occurs. In stationary reaction conditions, the velocities of surrendering and accepting the transmitted particles are equal, and a definite activity of the species determining the potential is attained at the surface and is different from the equilibrium activity.

Many catalytic hydrogen and oxygen transport reactions proceed according to this mechanism. For example, in the water-gas reaction at a Wüstite surface (13) adsorbed oxygen is produced as intermediate product, and in the stationary state of the reaction,

$$CO_2 + H_2 \longrightarrow CO + H_2O \tag{9}$$

the velocity of oxygen transport to the catalyst,

$$CO_2 \longrightarrow CO + O_{ad} \tag{10}$$

is equal to the velocity of withdrawal of oxygen from the catalyst,

$$O_{ad} + H_2 \longrightarrow H_2O. \tag{11}$$

The same effect arises in high-temperature oxygen sensors with solid electrolyte when they are exposed to a reactible gas mixture not in equilibrium. The oxygen activity attained at the electrode during the course of the reaction then does not correspond to the thermodynamic equilibrium, so that false results can be obtained.

5 IONIC AND ELECTRONIC CONDUCTIVITY IN SOLIDS

The interaction between a gas and a solid, involving heterogeneous chemical or electrochemical reactions, has been used in many different ways for sensors. In particular, the conductivity properties of a solid play an important part and must be considered if the phenomena are to be fully understood. Thus, for example, a number of gas sensors are solid electrolyte systems with electrode reactions at solids of predominantly ionic conductivity. In such cases, the utility of a given substance as electrolyte requires practically pure ionic conductivity, with only a slight electronic conductivity.

In ionic conduction, not only are electric charges transported, but also chemical species. In solid materials this proceeds through disorders in the crystal lattice. Basically, three sorts of charged atomic defects responsible for ionic conduction in solids are distinguished, namely,

1. Charged defects, when individual particles are lacking in the lattice;
2. Ions in interstitial positions, when individual particles are present in excess;
3. Foreign ions, when lattice positions are occupied by ions of other substances. These can cause, under suitable conditions, an ionic conduction.

The charge condition is always related to the undisturbed lattice.

In electronic conduction, the mobile particles are quasifree electrons or positive holes. The concentrations of defect order centers in solids depend on the partial pressure of a component in the gas phase.

In recent work, doped zirconium dioxide and doped thorium dioxide have become particularly important as solid electrolytes with predominantly oxygen-ion conduction. On adding a few percent of CaO, MgO, or Y_2O_3 to ZrO_2 or ThO_2, a cubic lattice of the fluorite type is stabilized, in which, furthermore, the ratio of cations to anions is displaced in the direction of metal ions. The lattice contains vacant oxygen sites, which are responsible for the conduction at high temperatures.

This concentration of vacancies produced by doping, and hence the ionic conduction, changes hardly at all on alteration of the oxygen partial pressure:

$$(V_o) \approx \text{const} \qquad (12)$$

although oxygen reacts sparingly with the lattice defects in accordance with the equation

$$\frac{1}{2} O_2 \text{ (gas)} + Vo^{++} + 2e^- = O_o \qquad (13)$$

The law of mass action can be applied to this equation as follows:

$$P_{O_2}^{1/2} (Vo^{++})(e^-)^2 = K_1 \qquad (14)$$

which gives, taking the constant concentration of oxygen defects into account,

$$(e^-) \propto P_{O_2}^{-1/4} \qquad (15)$$

That is, the concentration of quasifree electrons is proportional to the inverse of the fourth root of the oxygen partial pressure with which the oxide is in equilibrium.

Analogously, the concentration of holes is given by the equation

$$\frac{1}{2} O_2 \text{ (gas)} + Vo^{++} = O_o + 2h^+ \qquad (16)$$

from which the following relationship for the concentration of holes is obtained:

$$(h^+) \propto P_{O_2}^{1/4} \qquad (17)$$

The electrical conductivity of ZrO_2 containing 10 mole % Y_2O_3 as a function of the oxygen partial pressure has been studied at various temperatures by Burke, Rickert, and Steiner (14) using the Hebb-Wagner technique (15,16) with an inert blocking electrode (Fig. 2).

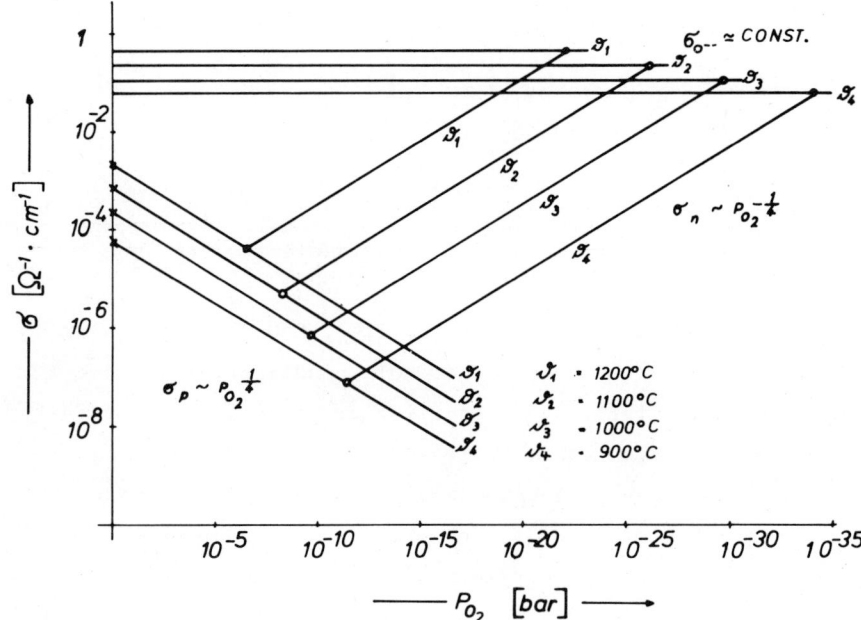

Fig. 2. Partial conductivity in ZrO_2 + 10 mole % Y_2O_3 at various temperatures as a function of the oxygen partial pressure.

It can be seen that at oxygen partial pressures below 10^{-20} bar the electronic conductivity attains the same order of magnitude as the ionic conductivity of the stabilized zirconium dioxide.

The oxygen partial pressure at which the ionic transfer number attains the value 0.5 is a function of the temperature and differs for different materials. If an electrochemical cell is set up using a mixed conductor as solid electrolyte, and if it is operated with various oxygen partial pressures, with chemical potentials of the oxygen μ'_{O_2} and μ''_{O_2}, the emf attained is in any case smaller than the thermodynamic cell voltage. Wagner (16) has calculated the reduction of the cell voltage in the general case and obtained the following relationship:

$$E = \frac{1}{4F} \int_{\mu_{O_2}''}^{\mu_{O_2}'} t_{ion} \cdot d\mu_{O_2} = \frac{RT}{4F} \int_{p_{O_2}''}^{p_{O_2}'} t_{ion} \cdot d \ln p_{O_2} \qquad (18)$$

This relationship between t_{ion} and E makes it possible to calculate the ionic transfer number of a solid electrolyte on a theoretical basis.

Other oxide phases exist, such as TiO_2, CoO, ZnO, or Cu_2O, that are predominantly electron or hole conductors and whose electrical conductivity under equilibrium conditions accordingly depends on the oxygen partial pressure in the surrounding gas phase. In such cases, the oxygen partial pressure determines the metal excess or deficiency in the oxide phase and hence the concentration of the charge carriers. The dependence of conductivity on oxygen partial pressure makes it possible to establish the conduction mechanism in the oxide.

6 METHODS BASED ON VOLTAGE MEASUREMENT

6.1 Systems at Equilibrium Potentials

Nernst's equation (3) gives the potential of an electrode as a function of the concentration or partial pressure of the gas involved. The partial pressure of such a gas, and hence its concentration, can in principle be determined by combining such an electrode, the measuring electrode, with a reference electrode giving a known potential and measuring the equilibrium potential E of the electrochemical cell thus formed. Knowledge and constancy of the other parameters in the Nernst equation are, however, necessary conditions.

If the same gas is involved in the reactions determining the potentials at the two electrodes, the electrochemical cell is of the type whose equilibrium voltage is given by Eq. (19):

$$E = \frac{RT}{zF} \cdot \ln \frac{p'}{p''} \tag{19}$$

where p" and p' are the partial pressures of the gas determining the potential at the measuring electrode and at the reference electrode, respectively.

To obtain unambigious results, it must be ensured that, apart from the gas to be determined, which is regarded as determining the potential, no other components of the mixture participate in electrode reactions; otherwise, the electrode potential, and hence the voltage of the cell, is falsified. In addition, it must be known whether the gas to be determined, which determines the potential, reacts with other components of the mixture. Particularly clear conditions are given if the other components of the mixture are inert gases, with which no reactions are possible. If, on the other hand, a component reacts with the gas to be determined before the equilibrium is reached, thereby altering its concentration, unambiguous results are only obtained if the concentration of the component and the equilibrium constant of the reaction are known. Conversely, this allows the possibility of determining the concentration of a gas not involved in the electrode reaction, if it influences the concentration of the potential-determining gas in a known way through a chemical reaction equilibrium.

The equilibrium potential can only be measured exactly when no current is flowing in the circuit. Because small currents always do flow, even when voltage-measuring devices of very high input resistance are used, therefore the measuring electrode and the reference electrode are both polarized according to their current/voltage characteristic curves. This leads to erroneous results which can in principle be reduced by carrying out the measurement at a higher temperature, since under such conditions the transfer overvoltage, the diffusion overvoltage, and, usually, the reaction overvoltage are reduced. At an early stage of the work the attempt was made to use the high reversibility of electrodes at high temperatures for measurement of the equilibrium potential in cells with a high-temperature solid electrolyte (17,18). These and further measurements (19,20) led to the possibility of using electrochemical cells with solid electrolytes for the analysis of gases (21).

Numerous investigations in this field have greatly extended the possibilities of gas analysis. On the other hand, the importance of electrochemical cells with liquid electrolytes has fallen off. The latter type is therefore only briefly discussed at the end of this chapter.

6.1.1 ELECTROCHEMICAL CELLS WITH SOLID ELECTROLYTES Only ionic conductors whose ionic partial conductivity at the working temperature is at least 1000 times as large as their electronic partial conductivity are suitable as solid electrolytes. The overall electrical conductivity should be as high as possible. In addition, the starting material for the solid electrolyte must be capable of being formed into a gas-tight electrolyte body.

Two groups of solid ionic conductors are particularly suitable for application as solid electrolytes: ionic conductors with CaF_2 structure on ZrO_2 and ThO_2 basis, and β-aluminium oxide electrolytes. The members of the first group become anionic conductors even at temperatures above about 300 to 400°C, whereas the second group shows cationic conductivity at room temperature. β-aluminium oxides, however, can only with difficulty be formed into gastight electrolyte bodies, so that they have not yet been applied on the large scale (22). Oxygen-ion-conducting solid electrolytes based on ZrO_2, however, have already been extensively used in practice. The mode of action and range of application of cells with solid electrolytes as sensors for gases are accordingly discussed here taking these materials as samples.

Determination of Gases in Gas Mixtures

The oxygen concentration cell,

$$O_2 \, (p_{O_2}'') \, Me/ZrO_2, \, stab./Me, \, O_2(p_{O_2}')$$

with the equilibrium potential

$$E = \frac{RT}{4F} \cdot \ln \frac{p_{O_2}'}{p_{O_2}''} \qquad (20)$$

was suggested for the measurement of oxygen partial pressures or concentrations by Weissbart and Ruka (23) and by Schmalzried (24, 25). The electrolyte was ZrO_2 doped with oxides such as CaO and Y_2O_3. Doping with more than 15 mole % CaO or 7 mole % Y_2O_3 stabilizes the cubic phase of ZrO_2 (26) which shows a high electrical conductivity principally due to oxygen ions. Fig. 3 shows the temperature dependence of the electrical conductivity of stabilized ZrO_2 (26).

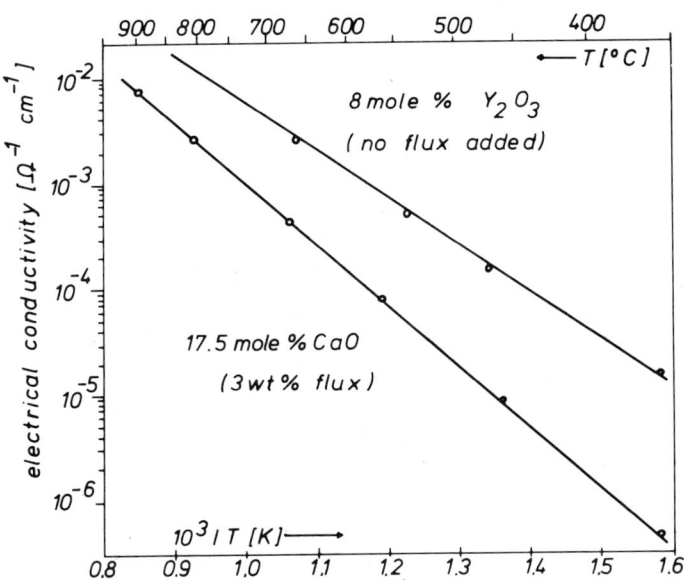

Fig. 3. Temperature dependence of the electrical conductivity of ZrO_2 (17.5 mole % CaO) and ZrO_2 (8 mole % Y_2O_3).

The measuring cell usually consists of a tube of stabilized ZrO_2, sintered at temperatures above 1600°C to a dense body, carrying platinum electrodes on its inner and outer surfaces. The electrodes are usually applied as so-called platinum paste fired at temperatures around 1000°C to give sufficient adhesion, thus producing porous platinum electrodes.

The gas whose partial pressure of oxygen is to be determined is passed through the interior of the tube at a constant flow rate (Fig. 4). The reference gas is usually air from the surrounding atmosphere, whose oxygen partial pressure under normal condition is 0.2095 atm. = 0.20903 bar.

Because oxygen is present mixed with inert gases, so that the oxygen partial pressure cannot be influenced by reaction

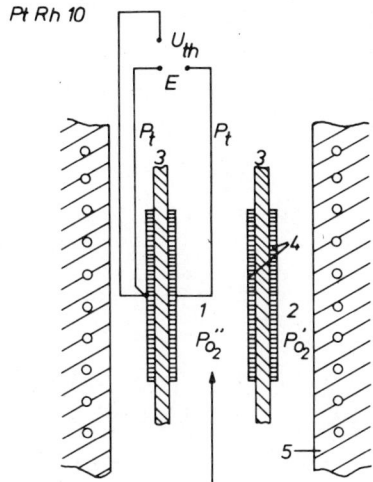

Fig. 4. Schematic view of an oxygen measuring cell with stabilized ZrO_2 as solid electrolyte. 1, gas mixture; 2, reference gas; 3, solid electrolyte; 4, porous electrodes; 5, furnace.

equilibria, Eq. 20 is valid without reservation. With air as reference gas, it takes the form

$$E = -0.03369T - 0.04960T \cdot \log p''_{O_2} \qquad (21)$$

To eliminate the temperature dependence of the equilibrium potential, the temperature of measurement must be kept constant. This is achieved by means of a controllable heating system surrounding the measuring cell. The system is controlled by the voltage of a thermocouple situated directly at one of the two platinum electrodes. For this purpose, one of the platinum leads of the electrodes can be used as one limb of a Pt/PtRh thermocouple. The measuring cell and the heating system must be carefully designed to avoid temperature gradients in the solid electrolyte, because otherwise the result may be in error by a thermovoltage amounting to several millivolts. In place of a ZrO_2 tube open on both sides, a tube closed at one end may be used, its interior being irrigated with air as reference gas and the measuring electrode being applied to the outside. This arrangement has the advantage of greater flexibility of design, since metal oxides producing definite partial pressures of oxygen may be introduced into the reference gas space in place of oxygen.

The value of p_{O_2}'' at a known temperature can be calculated from the measured voltage by Eq. 21. To state the accuracy of measurement and range of applicability of ZrO_2 solid electrolyte cells, various factors must be taken into account, involving the properties of the electrolyte, the working conditions of the device (temperature of measurement, flow rate of the gas) and the characteristics of the voltage measurement system.

Figure 5 shows the dependence of the equilibrium potential on temperature and oxygen partial pressure, calculated from Eq. 21 (27).

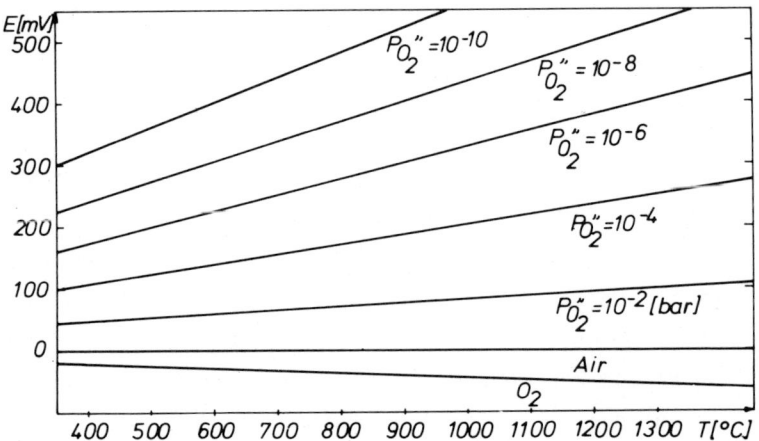

Fig. 5. Equilibrium potential E, calculated by Eq. 21, as a function of temperature and oxygen partial pressure.

In measuring oxygen partial pressures over the temperature range shown, considerable deviations from the calculated equilibrium potential values occur at high and very low temperatures. The reasons for them are discussed below.

Figure 2 shows the partial conductivity of stabilized ZrO_2 (10 mole % Y_2O_3) as a function of temperature and oxygen partial pressure. It is apparent that, at low partial pressures, the ratio of oxygen ionic conductivity to electronic conductivity falls off with increasing temperature and decreasing oxygen partial pressure. The oxygen partial pressure p'_e at which ionic and electronic excess conductivity are equal (24,25,28) is thus displaced towards larger values with rising temperature. p'_e is related to the ionic transfer number t_{ion}, which is equal to 1 for pure ionic conductivity, by Eq. 22 (28):

$$t_{ion} = \left[1 + \left(\frac{p_{O_2}}{p'_e}\right)^{-1/4}\right]^{-1} \qquad (22)$$

The measured voltage differs from the calculated equilibrium potential more strongly the more t_{ion} differs from 1, that is the larger p'_e is. For Y_2O_3-stabilized ZrO_2 (10 mole % Y_2O_3), p'_e is about 10^{-35} bar at 900°C and 10^{-30} bar at 1000°C, so that the measured voltage in both cases has only about half its calculated value. At 900°C and a partial pressure of 10^{-30} bar the electronic conductivity is about an order of magnitude smaller than the ionic conductivity. Under these conditions, t_{ion} is nearly equal to 1. Equation 20 is then valid when $t_{ion} > 0.99$, that is, when the electronic conductivity component does not exceed 1%. The upper temperature limit, which may not be exceeded if accurate partial pressure measurements are required, is shown as a function of the oxygen partial pressure in Fig. 6 (29):

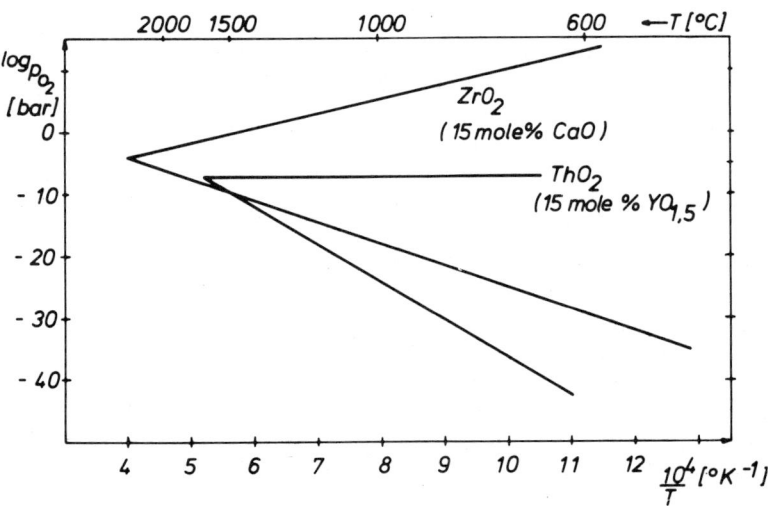

Fig. 6. Ranges of temperature and partial pressure in which the electronic partial conductivity amounts to less than 1% of the total conductivity.

It can be seen that the range of conditions within which oxygen partial pressure can be measured accurately by this means becomes smaller with increasing temperature and decreasing partial pressure. The width of the usable range depends strongly on the purity of the electrolyte. Thus, for example, traces of transition metal ions, which donate excess electrons simultaneously with the oxidative change of their valency, can restrict the range (30). Nevertheless, with relatively pure ZrO_2 solid electrolyte and high temperatures, the range of oxygen partial pressures that can be measured accurately extends down to 10^{-30} bar (at 600°C) or 10^{-20} bar (at 1000°C).

Higher degrees of electronic partial conductivity produce

errors of measurement in two ways: by invalidating the assumptions underlying the simplified Nernst equation, and by making the solid electrolytes permeable to oxygen. This produces an internal electric short circuit, with the consequence that oxygen is transported from the high to the low partial pressure side (27,28,32). This changes the partial pressures or concentrations at both electrodes, the effect being greater for larger electronic conductivity components and for smaller velocities of gas exchange with the gas space surrounding the electrode. The concentration overvoltage thus produced, which falsifies the measuring value, can be reduced by using electrodes with small film thickness and large pore diameter, that is, with low diffusion resistance, and by increasing the irrigation velocity of the gas. Similar measuring errors are produced by using a voltage measuring device with too low input resistance, even when the electronic partial conductivity is zero. When the cell is connected to the resistance, oxygen is transported through the electrolyte in the manner described previously.

Measuring errors can also arise from microscopic fissures and pores in the electrolyte, through which gas exchange between the two electrode compartments can occur. Experience shows that solid electrolytes with a density of more than 95% of the theoretical value are suitable for exact equilibrium potential measurements. This value can be attained without difficulty in producing electrolytes based on ZrO_2.

Although the p'_e value of stabilized ZrO_2 falls off with decreasing temperature, the range over which exact oxygen partial pressure measurement is possible is limited toward lower temperatures by various factors. Figure 7 shows the concentrations of oxygen calculated and measured in Ar/O_2 mixtures, as a function of temperature (33).

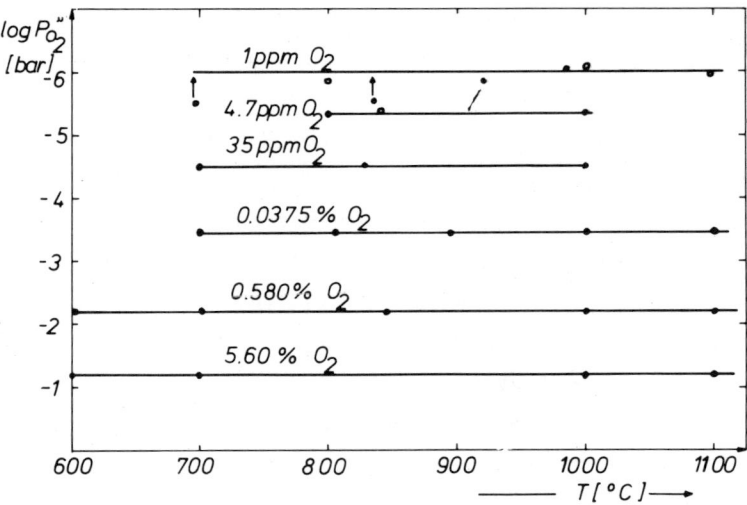

Fig. 7. Oxygen partial pressures calculated and measured for Ar/O_2 mixtures, as a function of temperature.

At partial pressures of 10^{-6} bar, corresponding to about 1 ppm O_2, large deviations from the theoretical value occur even at temperatures of about 800°C. At 500°C, exact measurements are possible only at partial pressures above 10^{-3} bar. Various reasons have been given for this behavior. Hartung and Möbius (31) have postulated the occurrence of diffusion overvoltage at the electrodes or interference through reaction of oxygen with the platinum electrode material. Ullmann (27) ascribes the deviations to contamination of the ZrO_2 ceramic material, which he supposed to produce an increase of the electronic transfer number at low temperatures. The response time, which is a fraction of a second at high temperatures, becomes longer at lower temperatures. At the same time, the reproducibility of the measured values is impaired. Because, in addition, the specific electrical resistance of the ZrO_2 electrolyte becomes very high at low temperatures (Fig. 3), an exact equilibrium potential measurement is technically quite difficult to obtain.

In order to carry out oxygen partial pressure measurements at low pressures with the greatest possible accuracy, the distance between the high- and low-temperature regions must be kept large. In choosing the measuring temperature, it must be borne in mind that greater sensitivity, that is, a greater change of equilibrium potential for each decade change in oxygen partial pressure, is attained at higher temperatures (Fig. 7). The measuring temperature is usually in the range from 700 to 1000°C. For example, in the range of partial pressures down to 10^{-3} bar an error of measurement of less than ± 1% or down to $10^{-5.5}$ bar of less than ± 8% can be attained (33).

Following a suggestion of Beekmans and Heyne (34), a ZrO_2 measuring cell can be used, not only to determine oxygen, but also, by a very simple modification, to control oxygen partial pressures in gas mixtures (34). The ZrO_2 tube through which the gas flows is provided with a measuring electrode pair and, at an upstream position, with a dosage electrode pair. A change in the oxygen partial pressure is indicated by the cell as a change in voltage. This is led to a control unit where it is compared with a voltage corresponding to the nominal partial pressure. The control unit causes a current to flow through the dosage electrodes, which pump oxygen through the electrolyte, either from the measuring gas to the reference gas or in the opposite direction. The current continues to flow until the cell voltage is equal to the nominal voltage, that is, until the original oxygen concentration is regained.

Cells with ZrO_2 solid electrolyte have been incorporated in commercially available oxygen measuring devices for some time

(35,36). The short response time makes a practically delay-free measurement possible, providing ideal conditions for the connection of a controlled system. A further advantage over most of the other methods of determining oxygen is the fact that the oxygen concentration to be measured is not altered by the measurement. The working life of the ZrO_2 measuring cell is considerable. In one case, a working life in a corrosive industrial atmosphere of 12 to 18 months at $1150°C$ has been attained (36).

A further promising electrolyte with CaF_2 structure is stabilized thorium oxide. ThO_2 stabilized with Y_2O_3 makes possible the exact measurement of still smaller oxygen partial pressures (Fig. 6), because the p'_e value for the electronic excess conductivity is smaller than that of stabilized ZrO_2. However, at pressures greater than 10^{-6} bar, defect electron conductivity occurs, making exact equilibrium potential measurement impossible.

Other solid electrolytes, such as Al_2O_3 (25), aluminium silicate with a mullite phase (32,37,38), or magnesium orthosilicate (39), are less important than oxides with CaF_2 structure. Although mullite solid electrolytes show a wider range of temperatures or partial pressures with pure ionic conductivity than stabilized ZrO_2 solid electrolytes, their application in oxygen-measuring cells has up to now been considered only at very high temperatures, for the determination of oxygen in molten metals (37,38). These measurements are discussed later.

For low temperatures, Yuan and Kröger (40) recommended Pyrex, a borosilicate glass, as solid electrolyte. Using the cell

$$O_2 \ (p'_{O_2}), \ Pt/Pyrex/Pt, \ O_2 \ (p''_{O_2})$$

whose equilibrium potential follows the Nernst equation even at low temperatures, reliable values can be obtained at about $200°C$. The mechanism of ionic conduction in Pyrex electrolytes has, however, not been completely explained. A variant of this measuring cell, in which oxygen reacts with the electrolyte through combination with sodium,

$$\text{Na (liquid), W/Pyrex/Pt, } O_2 \text{ } (p_{O_2})$$

is particularly interesting, because it can be used for measuring the partial pressure of sulfur vapor (40). The measuring cell for oxygen is shown schematically in Fig. 8.

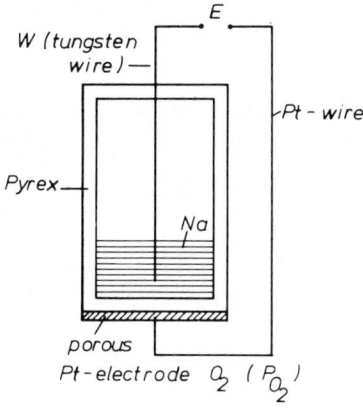

Fig. 8. Schematic representation of a Pyrex cell for determination of oxygen partial pressure.

The cell is, as usually, kept at constant temperature by a controlled heating system. The equilibrium potential of the cell is determined by the reaction

$$2 \text{ Na (liquid)} + \frac{1}{2} O_2 \rightleftharpoons Na_2O \text{ (glass)}$$

but depends, apart from the oxygen partial pressure and the temperature, also on the arrangement of the Na_2O or its ions in the glass.

In the mathematical relationship between equilibrium potential and oxygen partial pressure, in addition to the activities of the reacting substances in equilibrium with oxygen, there are involved enthalpy and entropy terms describing the interaction between Na_2O and the crystal lattice of the solid electrolyte. At constant temperature the activities and the entropy and enthalpy terms are all constant, so that the equilibrium potential is a linear function of the oxygen partial pressure. The measured oxygen partial pressures correspond to the calculated values down to oxygen partial pressures of about 10^{-30} atm. and are reproducible. The temperature range over which reliable measurement is possible extends down to $150°C$.

The mechanism and the equation for the sodium/oxygen cell can also be applied to the sodium/sulfur cell (40), in which Na_2S is formed in the glass. The measuring cell contains a sulfur-resisting graphite electrode on the gas side instead of a platinum electrode. Here, too, the equilibrium potential is a linear function of the sulfur partial pressure over the temperature range 250 to $450°C$, so that sulfur partial pressures can be measured with such a cell. According to a suggestion by Yuan and Kröger (40), the cell

$$O_2\ (p'_{O_2}),\ Pt/Pyrex/C,\ S(p''_S)$$

also appears promising. Up to now, however, no sensor on this basis has been described.

Determination of oxygen in mixtures containing oxidizable gases

If measuring cells based on ZrO_2 or ThO_2 with porous platinum electrodes are used for the determination of oxygen in mixtures containing oxidizable gases, the measurement is only exact if the oxygen is present in large excess. The free oxygen reacts with the oxidizable components at the catalytically active platinum, until the thermodynamic equilibrium valid for the temperature of measurement is reached. In this case, therefore, the three-phase boundary is offered not the free oxygen but the oxygen present in equilibrium, which is termed the equilibrium oxygen. The electrode of the first type, used for oxygen/inert gas mixtures, then becomes an electrode of the second type, whose potential depends on the equilibrium constants of the preliminary redox reaction.

Figure 9 shows the dependence of the oxygen partial pressure on the mole ratio of the mixture CO/O_2, in each case after the attainment of thermodynamic equilibrium. Substitution of the calculated values of oxygen partial pressure in the Nernst equation gives the relationship between the equilibrium potential of a solid electrolyte cell and the mole ratio CO/O_2 shown in Fig. 10.

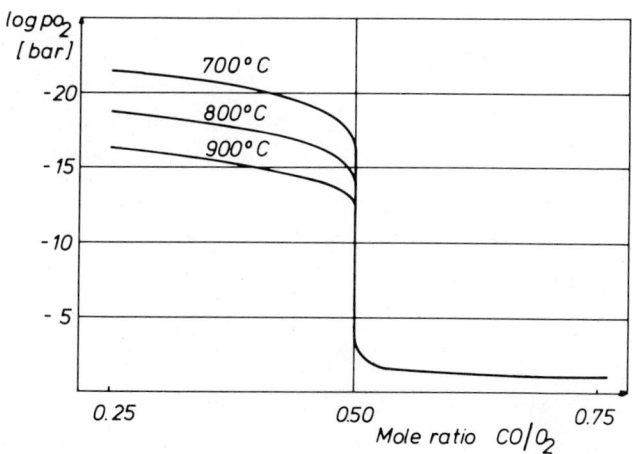

Fig. 9. Dependence of the oxygen partial pressure from the mole ratio of the mixture CO/O_2 at thermodynamic equilibrium and various temperatures.

Because platinum is a very good catalyst for the oxidation of

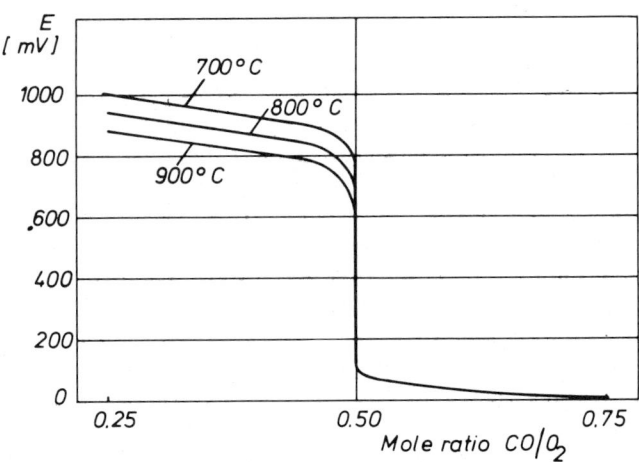

Fig. 10. Dependence of the equilibrium potential of an oxygen concentration cell from the mole ratio of the mixture CO/O_2 at thermodynamic equilibrium and various temperatures.

CO at the high temperatures used for measurement, thermodynamic equilibrium is always obtained at the measuring electrode, so that curves as shown in Fig. 10 are obtained. The oxygen partial pressure required can be calculated from the observed equilibrium potential if the equilibrium constant of the reaction and the concentrations of the reactants at equilibrium are known. This process can, however, only simply be applied in well-understood binary systems, but not in complex gas mixtures.

The direct determination of free oxygen is nevertheless possible, if in place of platinum other electrode materials are used that show no catalytic effect even at the high temperatures. The dependence of equilibrium potential on oxygen partial pressure is then given by the straight line of Nernst's equation over the whole range of partial pressure. Thus, for example, oxygen in mixtures with methane and inert gases can be determined very exactly at temperatures of 600 to 750°C using ZrO_2 measuring cells with silver electrodes, the latter being catalytically inactive for the oxidation of methane (41). In every case, however, it must be established that the measuring electrode is inactive for all possible oxidation reactions; otherwise, oxygen will be withdrawn from the mixture above the three-phase boundary and deviations from the Nernst's linear relationship will occur.

Figure 10 shows that a sharp potential step occurs on passing from an oxidising to a reducing atmosphere at the stoichiometric composition, or in the reverse direction. This property makes possible the application of measuring cells with solid electrolyte and catalytically active measuring electrodes in many different ways, for example, for the control and regulation of redox reactions in chemical processing technology or of combustion processes in industrial heating systems and internal combustion engines (26,35,36,42,43). Cells with stabilized ZrO_2 have recently become particularly important for the last-named application.

Determination of oxidizable gases in mixtures containing oxygen

Concentration cells with solid electrolyte and catalytically active measuring electrodes can be used for determining the concentration of oxidizable gases in the presence of oxygen. To show the versatility of solid electrolyte cells in this field, the determination of hydrogen, carbon monoxide, and methane is discussed.

As shown above, the potential of the measuring electrode is determined by the partial pressure of the excess oxygen, if the system is over-stoichiometric with respect to oxygen, or to the equilibrium oxygen partial pressure, if the system is under-stoichiometric with respect to oxygen (cf. Figs. 9 and 10). According to Hartung and Möbius (31), hydrogen concentration can be determined in the system $H_2/H_2O/(N_2)$ through the dependence of the equilibrium potential of the ZrO_2 measuring cell:

$$(N_2), H_2, H_2O, O_2\ (p''_{O_2})/ZrO_2 \cdot \text{stab.}/\text{air}\ (p'_{O_2})$$

on the equilibrium partial pressure p''_{O_2} of the oxygen.

For the formation of water at the catalyst, the equations

$$K = \frac{p_{H_2} \cdot p''^{1/2}_{O_2}}{p_{H_2O}} \qquad (23)$$

and

$$p''_{O_2} = K^2 \left(\frac{p_{H_2O}}{p_{H_2}}\right)^2$$

are valid. The oxygen partial pressure at the reference electrode is

$$p'_{O_2} = X_{O_2} \cdot P \qquad (24)$$

where X_{O_2} is the mole fraction of oxygen in the air and P is the atmospheric pressure. The hydrogen partial pressure in H_2/inert gas mixtures, in which at room temperature a definite water vapor partial pressure is produced, is given by

$$p_{H_2} = X_{H_2}(P - p_{H_2O}) \qquad (25)$$

where X_{H_2} is the mole fraction of hydrogen in the water free mixture with inert gas and P is the total pressure. At low H_2 concentrations, p_{H_2O} is approximately constant, even in the hot ZrO_2 measuring cell, so that the p_{H_2O} value produced at room temperature can be used. The Nernst equation, with Eqs. 24 and 25 substituted, then takes the following form:

Determination of H_2 and CO

$$E = \frac{RT}{4F} \ln \frac{X_{O_2}}{K^2} + \frac{RT}{4F} \ln P^{1/2} \left(P - P_{H_2O}\right) \left(\frac{X_{H_2}}{P_{H_2O}}\right) \qquad (26)$$

On substituting the known numerical values, including the value for K at the temperature of measurement, and with knowledge of the partial pressure of the water, the hydrogen concentration can be determined by measuring E. The measurement is conducted in such a manner that the dry hydrogen/oxygen/inert gas mixture is passed through water or sulfuric acid at a particular temperature, producing a definite and known water vapor partial pressure. The values of hydrogen found using a ZrO_2 measuring cell in the concentration range from 0.001 % to about 1 % and the temperature range 600 to 1000°C agree with the values calculated from Eq. 26. This method shows itself superior to amperometric hydrogen titration particularly at low concentrations. Conversely, if the hydrogen concentration is known, the water vapor partial pressure can be determined.

Similarly, the CO content of exhaust gases from the incomplete combustion of fuel can be determined using the ZrO_2 measuring cell (34). A part of the flow of the exhaust gases is fed to the measuring cell, which is preferably operated at temperatures in the range 600 to 1000°C. From the relationship

$$P''_{O_2} = K \left(\frac{P_{CO_2}}{P_{CO}}\right)^2 \qquad (27)$$

and the Nernst equation, there follows

$$E = \frac{RT}{zF} \left(2 \ln \frac{P_{CO_2}}{P_{CO}} + \ln \frac{k}{P'_{O_2}}\right) \qquad (28)$$

where p'_{O_2} is the oxygen partial pressure in the reference air electrode. At low CO concentrations, p_{CO_2} is approximately constant and equal to the value that can be calculated for complete combustion of the fuel. P_{CO} can then be calculated from the values of K, E, T, and P'_{O_2}. The method gives exact results over the range from 10^{-4} to 10 vol % CO and is thus suitable for the supervision of combustion processes.

Sandler (41) has suggested a ZrO_2 concentration cell for the determination of oxidizable gases in the presence of an excess of oxygen. One electrode of the first and one of the second type are combined.

The internal electrode, fitted with an upstream catalyst, receives a gas mixture practically in thermodynamic equilibrium, so that it determines the oxygen remaining after reaction with the oxidizable gas. The external electrode, however, receives the unreacted gas mixture, so that it shows the true oxygen content of the mixture. The capabilities of such a measuring cell have been demonstrated using methane/air mixtures (Fig. 11). The oxygen concentration of the measuring gas must be known. If the mixture initially contains a moles of CH_4 and b moles of O_2, the inner electrode receives the equilibrium mixture

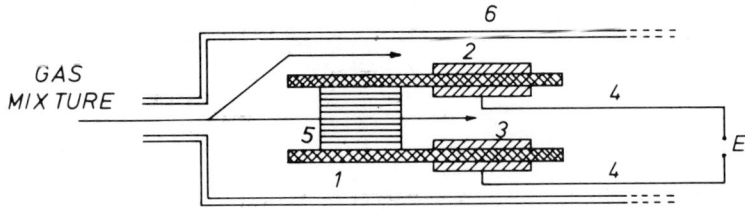

Fig. 11. Principle view of the measuring cell for the determination of methane in air. 1, solid electrolyte; 2 and 3, porous silver electrodes; 4, electrical leads; 5, catalyst.

$$aCO_2 + 2aH_2O + (b-2a)O_2$$

The voltage of the cell is given by Nernst's equation, substituting the proportional number of moles in place of the partial pressures:

$$E = \frac{RT}{4F} 2.30 \log \left(\frac{b}{b-2a}\right) \tag{29}$$

b can be calculated from a, E, and T.

The sensor has been suggested for use as a detector for methane in coal mines and for locating leaks.

The accuracy of the determination of oxidizable gases with oxygen concentration cells is naturally limited by the same factors as in the determination of oxygen. In addition, however, the constancy of the cell temperature is much more critical, since the potentials of electrodes of the second kind are more strongly temperature dependent than those of cells of the first kind.

Determination of Gases in Liquid Phases

Determination of oxygen in molten metals

The concentration of oxygen in molten metals is of great importance for the properties of the melt and of the material subsequently prepared from it. The dissolved oxygen interacts through redox equilibria with the components of the melt that affect the material properties. Thus, for example, in the manufacture of steel by blasting crude iron in a blast furnace, the impurities in the iron, such as C, P, and Si, are oxidized, and the carbon content, which determines the properties of the steel, depends on the oxygen content of the melt. The concentration of dissolved oxygen is thus an important criterion for assessing the condition of the melt. Rapid and accurate determination of oxygen concentration over a wide range of values therefore opens up possibilities of testing the melt and undertaking corrections in the event of departure from nominal values, in the ideal case automatically (44-46).

Conventional methods for determining the oxygen concentration of a melt, such as hot extraction, require taking a sample. A disadvantage of such methods is that a certain time elapses from sampling to measurement, so that changes in the properties of the melt occurring in this time can not be registered. Furthermore, hot extraction only gives the total oxygen concentration, including the oxygen combined in the suspended oxides, instead of the dissolved free oxygen, which is primarily of interest. The latter can be determined by means of oxygen concentration cells with ionic-conducting solid electrolyte,

$$\text{reactants} \rightleftharpoons O_2 \; (p''_{O_2}), \; \text{Me/solid electrolyte/Me}, \; O_2 \; (p'_{O_2})$$
$$\text{(melt)} \hspace{4cm} \text{(ref. electrode)}$$

using the Nernst equation.

The cell, a tube closed at one end consists, at least partly, of the solid electrolyte. The reference electrode is situated inside the tube and is supplied either with oxygen from the air or with equilibrium oxygen of metal/oxide mixtures. The outside of the tube does not need to carry a metal electrode, since the melt itself is conducting and functions as an electrode.

In determining oxygen concentration particularly in molten steel, involving temperatures of 1500 to 1700°C and concentrations of 1 to 1000 ppm (28), the requirements made on the cell are even more severe than in the determination of oxygen concentration in gases. The cell is directly immersed in the melt and attains its operating temperature within a few seconds, so that the solid electrolyte must be able to withstand sudden extreme changes of temperature. In addition, it must be stable enough, at such temperatures, to permit determinations over a long period.

The best solid electrolyte for determination of oxygen in gases, fully stabilized ZrO_2, is unsuitable as single material for cells for determination in molten metals, because it fractures under the temperature conditions involved. This difficulty is overcome mainly in two ways. The one method is to use partially stabilized ZrO_2, its electrical conductivity is less than that of the fully stabilized material, although its temperature stability is better. Alternatively the principle of the quartz tube cell (Fig. 12) can be used.

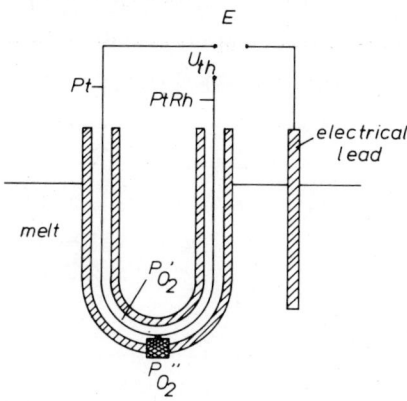

Fig. 12. Principle of the quartz tube cell for determination of oxygen in molten metals.

In this arrangement, the ZrO_2 solid electrolyte consists of a pellet fused into a U-shaped quartz tube of adequate temperature stability. The latter serves both as carrier for the pellet and as separating wall between the reference gas compartment and the melt (28,47).

The oxygen partial pressure in the melt calculated from the Nernst equation is converted into oxygen activity, $a_{O,Me}$ and oxygen concentration in percent or ppm. The relation between oxygen activity and concentration involves the activity coefficient of the oxygen in the melt:

$$a_{O,Me} = f_{O,Me} \; (wt \; \% \; O)_{Me} \qquad (30)$$

which is in its turn dependent on the nature and concentration of the other elements present. The effect of these other elements in the oxygen activity coefficient can be calculated and taken into account (38). In practice, the cell is usually calibrated, using samples of known composition of melts of the type to be investigated.

The necessity of measuring very low oxygen concentrations at very high temperatures makes severe demands on the conductivity properties of the solid electrolyte. As already mentioned in connection with cells for determination of oxygen partial pressures in gases, the degree of electronic partial conductivity in stabilized ZrO_2 increases with falling oxygen partial pressure and rising temperature, leading to false results. This is clearly shown in Fig. 13. The pe' value at $1600^\circ C$ is $10^{-15.8}$ bar (28,48).

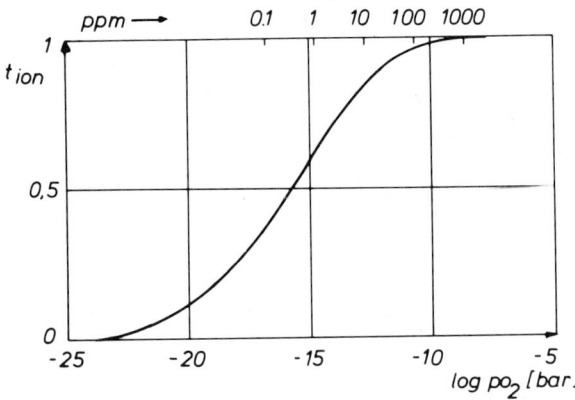

Fig. 13. Relationship between ionic transfer number t_{ion} and oxygen concentration in molten steel in ppm, at $1600^\circ C$, using stabilized ZrO_2.

It can be seen that significant deviations from $t_{ion} = 1$ occur even at oxygen contents of 10 to 100 ppm. This leads to false values of concentration if the measurements are calculated by the simplified Nernst equation. It is therefore necessary, in this region of very low oxygen concentrations, to modify the Nernst equation by introducing the pe' value, as suggested by Schmalzried (24,25),

$$E = \frac{RT}{F} \ln \frac{(p'_e)^{1/4} + (p'_{O_2})^{1/4}}{(p'_e)^{1/4} + (p''_{O_2})^{1/4}} \tag{31}$$

If the process of dissolution of oxygen in the melt is governed by

$$\tfrac{1}{2} O_2 \text{ (gas)} \rightleftharpoons O_{melt}$$

and the partial pressure of the dissolved oxygen by

$$p''_{O_2} = (a_{O,Me} \cdot k^{-1})^2 \tag{32}$$

with $a_{O,Me}$ = activity of oxygen in the melt, and if the reference gas pressure is taken as that over a Mo/MoO$_2$ mixture, then

$$a_{O,Me} = K \left[\left(p_{Mo,MoO_2} \right)^{1/4} \cdot \exp\left(-\frac{E}{RT} \right) - \left(p'_e \right)^{1/4} \right]^2 \tag{33}$$

This relationship serves for calculation of oxygen activity and concentration from the measured cell voltage at oxygen concentrations below 100 ppm (28). The oxygen permeability of the solid electrolyte, because of the electronic partial conductivity, leads to polarization of the measuring electrode and hence to an error in the value indicated, as in the measurement of oxygen concentrations of about 10 ppm. In general, the above mentioned causes of erroreous values can be determined by reducing the electronic conductivity component, and hence the pe' value, and by increasing the thickness of the tube wall, reducing the oxygen permeability. The pe' value can also be lowered by using purer material, containing a lower proportion of electronically conducting impurities, for the ZrO_2. The wall thickness is restricted to at most about 1 mm by the requirement of quicker and more uniform heat conduction.

The other factors influencing the accuracy of the measurement, which have already been discussed in connection with cells for measuring oxygen concentration in gases, have also to be borne in mind for cells for oxygen determination in melts.

Stabilized thorium oxide, which has a vanishingly small electronic partial conductivity at high temperatures and low oxygen partial pressures, is in this respect preferable to ZrO_2, but it is difficult to produce stable electrolyte bodies from it by sintering, and it is also costly. It is therefore less important than ZrO_2 in this field of application. Solid electrolytes based on aluminium silicates have also been considered (38,39). On the one hand, they contain the mullite phase, $3Al_2O_3 \cdot 2SiO_2$, which is a good conductor and shows only slight electronic conductivity components in the range of oxygen partial pressures of interest, but on the other hand they usually contain corundum or SiO_2. The glass component present in the structure with excess of SiO_2 admittedly increases the temperature stability of the material but lowers its working life at high temperature. The electrolyte can be used at temperatures up to $1500^{o}C$, for example, in cast-iron melts (38), but at higher temperatures under low oxygen partial pressures it shows a too high electronic conductivity component.

<u>Determination of oxygen in water</u>   The oxygen content of liquid

phases with a relatively low boiling point cannot be determined with solid electrolyte galvanic cells directly immersed in the liquid phase, because the operating temperatures of the cells are too high. Instead, either the oxygen partial pressure in the gas atmosphere over the liquid phase, which is uniquely determined by the oxygen concentration in the liquid at a given temperature, or the oxygen content of the vapor mixture produced by evaporating a sample of the liquid can be determined.

In one particular field of application, however, namely that of watersupplies, not only the oxygen content of surface water is of interest, but also the quantity of dissolved oxygen that would be used for complete biological decomposition of the organic substances contained in the water. In practice, the sum of all the organic components is determined <u>by oxidation with $K_2Cr_2O_7$ or $KMnO_4$</u>, in a standard process. The value obtained is a measure of the quantity of oxygen that would have to be supplied to the water to compensate for oxygen loss through decomposition of the organic substances, to avoid anaerobic, harmful fermentation processes.

According to Butzelaar and Hoogeveen (49), the combination of a measuring and dosage device with a ZrO_2 solid electrolyte described previously offers the possibility of determining rapidly and reliably the quantity of oxygen used by the sum of all the organic substances contained in the water. A constant flow of nitrogen passes through a controlled ZrO_2 measuring and dosage cell and takes up there a constant, known oxygen partial pressure. It then passes through a combustion chamber containing catalytically active Pt mesh, and a second measuring and dosage cell, with whose help changes in the oxygen partial pressure can be established at any time.

A small quantity of the water to be examined is sprayed into the combustion chamber and the organic components are oxidized on the Pt mesh. The reduction of the oxygen partial pressure thereby produced causes the second ZrO_2 cell to add to the gas stream a quantity of oxygen sufficient to restore the original oxygen partial pressure. This quantity of oxygen is equal to that used by the water. The method affords an accuracy of measurement of 2 to 3 % and is in this respect comparable with the usual standard $K_2Cr_2O_7$ method, but has the advantage of giving a result much more quickly. Such a cell is for this reason usable as a part of a regulating circuit, for example, for the automatic regulation of sewage disposal units.

6.1.2 ELECTROCHEMICAL CELLS WITH LIQUID ELECTROLYTE   As mentioned previously, exact gas determinations from the measured voltage of a galvanic cell, using the Nernst equation, are only possible if the polarization at the measuring and reference electrodes is negligible and if, in addition, further conditions are fulfilled, for example, that the gas components have reacted to the thermodynamic equilibrium. Polarization-free measurement is possible at low temperatures, at which liquid electrolytes can be used, only with few systems. Charge-transfer polarization in aqueous solutions, even with very active catalysts, is for most of the potential determining gases so large that no reversible electrode potentials can be measured. An exception is hydrogen, which gives reversible potentials at platinum electrodes even at room temperature. A very exact method of determining hydrogen in inert gas mixtures on this bases has been described by Bianchi et al. (50). According to Nernst, for the hydrogen concentration cell

$$H_2(p_{H_2}''), \; Pt \; / \; H_2SO_4 \; / \; Pt, H_2(p_{H_2}')$$

the equation

$$E = \frac{RT}{2F} \ln \frac{P'_{H_2}}{P''_{H_2}} \qquad (34)$$

is valid.

The cell is symmetrically constructed and contains two hydrogen diffusion electrodes of the type shown in Fig. 14.

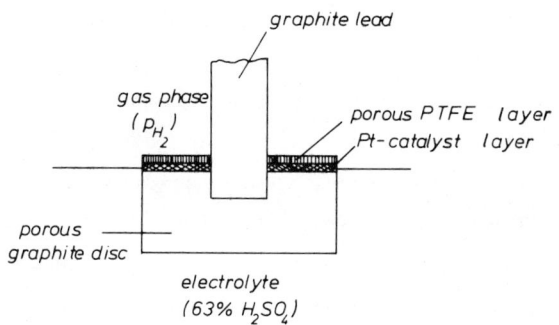

Fig. 14. H_2 gas diffusion electrode for determination of hydrogen partial pressures (63).

The measuring device is so constructed that

> the same total gas pressure and temperature ($25^\circ C$) exist in the anode and cathode compartments;
> no oxygen from the surrounding atmosphere can penetrate into the apparatus;
> the electrodes are connected with only a small quantity of electrolyte. In this case solution equilibria between gas and liquid phases are rapidly reached, so that the response time of the device is short.

The high concentration of sulfuric acid prevents rapid drying out of the electrolyte. The exchange current density at the Pt/C catalyst is high, so that not only a surprisingly short response time, of a few seconds, but also a high accuracy of measurement is attained. The accuracy amounts to ± 0.2 % down to small hydrogen partial pressures, and the device responds to concentrations of even 10 ppm.

Measuring cells of this kind have the advantage over ZrO_2 high temperature cells of more easily attaining a constant temperature. For this benefit, however, serious disadvantages must be accepted: the Pt electrode responds to many gases, for example, oxygen, and is in addition susceptible to poisoning at low temperatures, so that even traces of other gases or substances affect the result. Furthermore, cells with liquid electrolyte cannot attain the extremely short response times of high-temperature cells and are therefore not usable in so many applications. For this reason it is not possible to determine oxygen with such cells, since the reversibility of the electrodes is not sufficient for exact measurement.

6.2 Systems with Mixed Potentials

In the methods for determination of gas partial pressure or concentrations on the basis of a voltage measurement, electrochemical cells were described in which the potential of the measuring electrode depends directly on the concentration or the partial pressure of the potential determining component in the gas phase. This concentration at the three-phase boundary can be influenced by superimposed gas-phase reactions. In cells whose equilibrium potential serves as a measure of the concentration of the material determining the voltage, this material interacts through a redox equilibrium with its reduced or oxidized form in the liquid or solid phase, involving electron transfer through the phase boundary. If further potential determining processes are superimposed on this direct redox reaction, a mixed potential is measured. There are but few examples of electrochemical cells with electrodes whose mixed potential is a reliable measure of the concentration of gases. The potential determining gas interacts with its reduced or oxidized form not directly, but, for example, through a reaction with the electrode material or after displacing another potential determining material in accordance with a definite adsorption equilibrium. As a consequence of such potential determining mechanisms, slopes for the dependence of the electrode potential and the cell voltage on the concentration of the potential determing material are measured that deviate strongly from the values to be expected for direct and exclusive redox equilibria.

6.2.1 DETERMINATION OF NITROGEN DIOXIDE IN GAS MIXTURES Barna and Jasinski (51,52) describe a sensor specific for NO_2. The measuring electrode material is an oxide chalcogenide glass with the composition $Se_{60}Ge_{28}Sb_{12}Fe_n$ (n = 1.3 - 2). To secure selectivity for NO_2, short response time, and good reproducibility of the voltage, the substance must be subjected to oxidative heat treatment, whose influence on the properties of the solid is not clear. The active electronic-conducting electrode material is used in the form of a disk. The gas under investigation passes through a porous PTFE membrane, which protects the system from drying out, to the electrode surface, which is on the other side wetted with electrolyte (salt solutions in the pH range 3.7 to 8) by means of a wick. The reference electrode is a Beckman-Perma probe electrode. The cell operates at room temperature and shows a response time of a few minutes. Gases such as NO, SO_2, CO, CH_4, and O_2 do not disturb the measurement. With gas mixtures of various NO_2 concentrations, linear dependence of the electrode potential or cell voltage on the logarithm of the NO_2 concentration is obtained down to 1 ppm. The slope is 70 mV per concentration decade and is the same for different samples of the electrode material. Obviously, however, the sensor must be calibrated for various electrolytes, since the intercepts of the straight lines in the plot of mV against log ppm depend on the nature and concentration of the electrolyte. The unusually large decade value makes it probable that no simple redox reaction is responsible for the potential, but a mixed potential mechanism occurs. Whether the measured value is actually produced by a mixed potential cannot be directly determined from the experimental results at present available. The mechanism of a reversible chemisorption

$$NO_2 + G \rightleftharpoons NO_2G \longrightarrow NO_2^- + G^+$$

(gas) (electrode surface) (solid)

suggested by the authors (51,52) cannot be reconciled with the suggestion of a mixed potential mechanism.

If the electrode material is set in PTFE-bonded gas diffusion electrodes, it can be used in cells for amperometric determination of NO_2, in which the current at a definite voltage is proportional to the NO_2 concentration (see Section 7.1).

6.2.2 DETERMINATION OF OXYGEN IN WATER According to a suggestion of Hahn et al. (53), sodium tungsten bronzes can be used as electrode materials for the determination of oxygen in aqueous solutions. The most suitable materials are crystals with the composition Na_xWO_3 (x = 0.6 to 0.65) with a smooth surface, which are heat-treated in an inert gas atmosphere. In the cell

$$O_2, Na_xWO_3 \; / \; KOH \; / \; sat. \; cal. \; el.$$

(solution) (pH 12)

using aqueous KOH solutions at room temperature, unusually large decade values of 120 mV per decade oxygen concentration are obtained in the range of concentrations 0.2 to 8 ppm O_2. With a response time of the order of minutes, an accuracy of ± 5 % is attained. The conditions for measurements of such accuracy are temperature constancy of the cell, thorough stirring, and removal of heavy metal ions by complex formation with EDTA solution. Because the slope depends on the nature of the specimen of electrode material, the measuring cell must be calibrated.

The strong pH dependence of the electrode potential makes a potential determining mechanism probable involving an OH^- adsorption step at the electrode surface. It is supposed that, in oxygen-free solutions, an OH^- adsorption step determines the potential and that, when oxygen is added, an adsorption equilibrium between OH^- and oxygen, dependent on the oxygen concentration, is attained, its magnitude influencing the mixed potential. The oxygen content of surface water and drainage water can be determined with galvanic cells of this type.

The definite and reproducible dependence of the mixed potential of a measuring electrode on the concentration of a gas has been observed in alcohol sensors, which operate on the amperometric principle and can be used as detectors in gas chromatography (54, 55). A definite dependence on the open-circuit electrode potential of a gas diffusion electrode in acid electrolyte on the alcohol content of air stream is found. The electrode potential is a mixed potential produced by displacement of chemisorbed oxygen at the platinum surface by, for example, ethanol, corresponding to its concentration in the gas, and attainment of chemical equilibrium, with oxidation of the ethanol and reduction of oxygen. The electrode is nevertheless not operated on open circuit, but normally in a closed circuit, the current delivered being taken as a measure of the alcohol concentration. In the literature, there is no information about the rate determining step.

7 METHODS BASED ON CURRENT MEASUREMENT

7.1 Measurement of Limiting Current

7.1.1 PRINCIPLE OF MEASUREMENT In Section 3 the limiting current density i_L was introduced and defined as the current density at which, because of the drop in concentration, every particle reaching the electrode is at once removed by the electrode reaction, so that the concentration C_o at the electrode surface becomes zero. A linear relationship between limiting current and concentration C_i is then obtained:

$$I_L = \frac{D \cdot q \cdot z \cdot F}{\delta} C_i \tag{35}$$

Here, $\frac{\delta}{D \, q}$ is the diffusion resistance. Equation 35 forms the basis of all limiting current measuring processes. Such measurements accordingly involve a polarographic process from the gas phase. If the impedance of the reaction results from diffusion, the term "diffusion limited current" is used.

The diffusion resistance can arise either through an interposed membrane, through a diffusion layer in the gas or liquid or filling the pores of the electrode, or through the liquid electrolyte, in which the gas molecules dissolve and through which they must diffuse to the electrode surface. It is sometimes difficult to separate diffusion polarization from superimposed reaction polarization.

By choice of suitable membranes, electrolytes, and electrode materials and by operating at optimum electrode potentials, the sensor can be made specific for measurement of any desired gas component, but there are limits to this. The problems involved are discussed in the following sections.

The mechanism by which the sensor functions can best be explained with reference to the schematic diagram, Fig. 15.

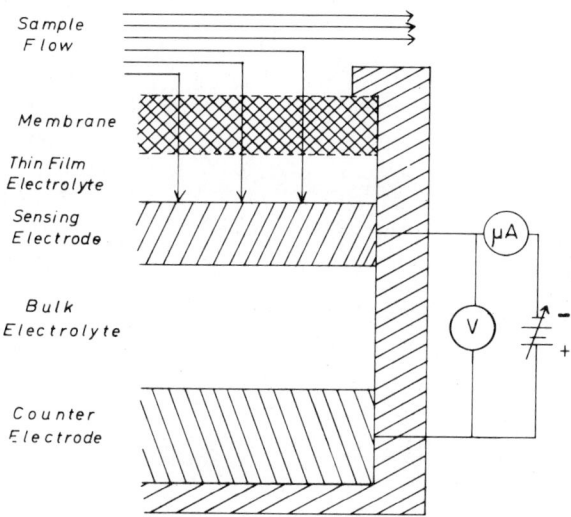

Fig. 15. Schematic construction of a limiting current sensor.

The sample flow passes over the sensor membrane. A small part of the gas component to be measured diffuses through the membrane and the usually very thin electrolyte film on the surface of the electrode. There it reacts completely, either by anodic oxidation or by cathodic reduction. The reaction products either pass into the electrolyte or diffuse back into the gas stream.

By suitable choice of the quantity of electrolyte and of the dimensions of its bulk, the concentration of the reaction products in the electrolyte can, if necessary, be kept low. In some cases the reaction products can also react with the other electrode, forming sparingly soluble compounds.

The counter electrode should be as little polarizable as possible, since it also serves as a reference electrode. Suitable materials for the counter electrode are redox systems, depending on the nature of the electrolyte, for example, $Pb/PbSO_4$, $PbO_2/PbSO_4$, Ag/Ag_2O, or quinone/hydroquinone.

A constant voltage is maintained between the sensor electrode and the counter electrode. The value chosen depends on the current/voltage characteristics of the gas components to be determined. When the electrochemical reactible component is present, a limiting current flows between the electrodes, its magnitude being proportional to the concentration of the component in the gas stream.

At different gas concentrations, such a device yields various current/voltage curves, as shown in Fig. 16. These curves were

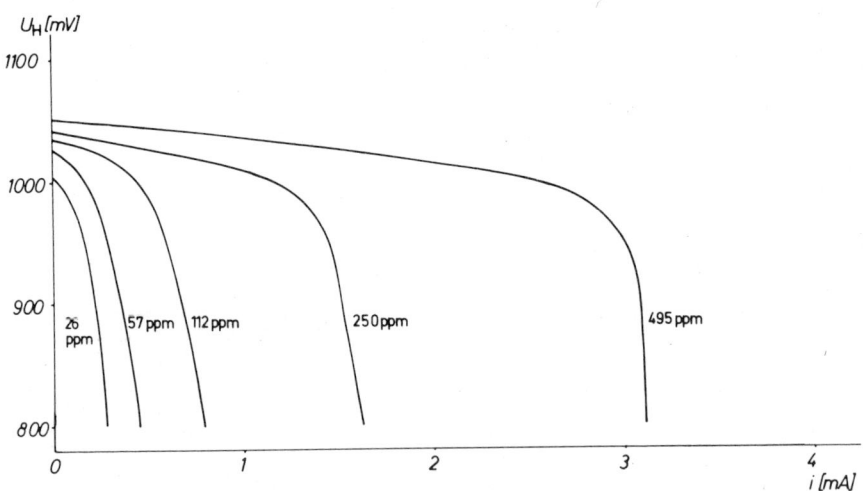

Fig. 16. Current/voltage curves for the cathodic reduction of NO_2 in 4.5 N H_2SO_4 at various NO_2 contents.

obtained in the cathodic reduction of NO_2 on cobalt phthalocyanine in 4.5 N H_2SO_4 at room temperature (56). The area of the electrodes was 6 cm^2, and measurements were made under potentiostatic conditions. The rest potential lies between 1000 and 1040 mV, depending on the concentration of NO_2 and measured against a hydrogen electrode in the same solution. If the electrode potential is reduced stepwise, the cathodic current first rises until a value is reached at which further reduction of the potential effects no further increase of the current. This is the limiting current. Below 900 mV the electrode operates in the limiting current region at all NO_2 concentrations. A plot of the limiting current as a function of the NO_2 concentration shows that the theoretically derived equation, Eq. 35, is also valid for this case. Figure 17 shows the corresponding diagram, based on the limiting current values at 900 mV. Because, in general, at least one diffusion step occurs before the electrochemical

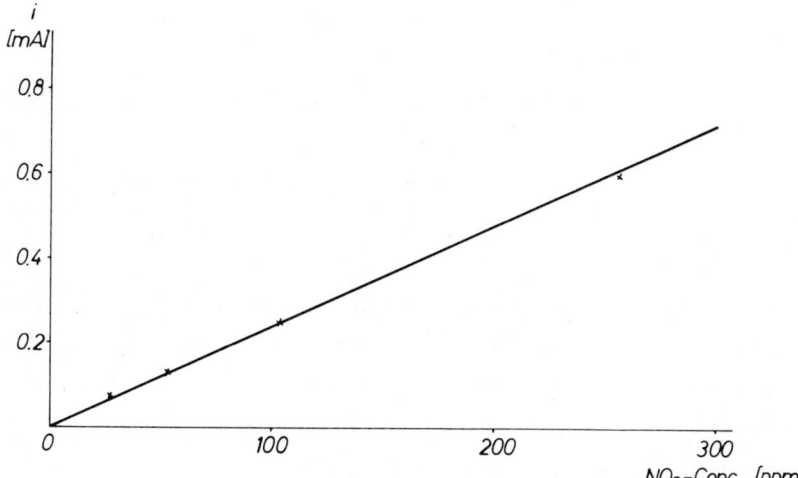

Fig. 17. Limiting currents in the cathodic reduction of NO_2 on cobalt phthalocyanine as a function of the concentration (E_H = 900 mV, electrolyte 4.5 N H_2SO_4).

reaction, the response time of limiting-current sensors is usually longer than that of the potential sensors. It lies between seconds and some minutes, depending on the design of the sensor.

7.1.2 CRITERIA FOR THE SELECTION OF SUITABLE ELECTRODE MATERIALS AND ELECTRODE POTENTIALS In principle, all gases that can be electrochemically oxidized or reduced can be determined by limiting current measurement. A potential must be applied to the electrode at which only the desired reaction proceeds. An approximate indication of the suitable range of potential is provided by the reversible potentials of the reactions involved. Some values are given in Table 1. This includes gases for which either sensors have been described in the literature or whose determination is urgently required.

More exact information as to a suitable range of potentials is afforded by the kinetics of the oxidation or reduction reaction, which can only be discussed in connection with the electrode material used.

If the gas to be determined occurs in a mixture containing several reactible components, the concentration of the desired component can be determined selectively by choice of a suitable electrode material and potential.

As an example, we consider the determination of NO_2 concentration by cathodic reduction in the exhaust gases of motor vehicles containing, as reducible components, not only NO_2 but also O_2, and in addition CO, H_2, hydrocarbons, and partially oxidized hydrocarbons as oxidizable components. Platinum is inadmissible as electrode material for this application because it catalyzes the reactions of all the substances named and because, over the whole range of available potentials, mixed potentials, depending on all the exhaust gas components, occur on the platinum.

Suitable Electrode Reactions

TABLE I

Equations for the Reactions of Gaseous Substances with Thermodynamic Potentials

Type of Reaction	E_H (mV)
Oxidation reactions	
$H_2 \longrightarrow 2 H^+ + 2 e^-$	0
$C_2H_5OH + 3 H_2O \longrightarrow 2 CO_2 + 12 H^+ + 12 e^-$	87
$HCHO + H_2O \longrightarrow CO_2 + 4 H^+ + 4e^-$	-123
$CH_4 + 2 H_2O \longrightarrow CO_2 + 8 H^+ + 8e^-$	169
$CO + H_2O \longrightarrow CO_2 + 2 H^+ + 2e^-$	-103
$SO_2 + 2 H_2O \longrightarrow SO_4^{--} + 4 H^+ + 2e^-$	170
$NO + 2 H_2O \longrightarrow NO_3^- + 4 H^+ + 3e^-$	957
$NO_2 + H_2O \longrightarrow NO_3^- + 2 H^+ + e^-$	775
Reduction reactions	
$O_2 + 4 H^+ + 4e^- \longrightarrow 2 H_2O$	1230
$O_3 + 2 H^+ + 2e^- \longrightarrow O_2 + H_2O$	2076
$NO_2 + H^+ + e^- \longrightarrow HNO_2$	1093

The situation with cobalt phthalocyanine is more favorable. It is known that none of the oxidizable substances named can react at CoPc (56,57). An interfering sensitivity to these substances is therefore not to be expected. CoPc is, however, active for the cathodic reduction of oxygen (58). Sweep voltammetry makes it possible to decide whether NO_2 can be selectively used in the presence of O_2 and, if so, in what potential range. Figure 18 shows such a diagram, plotted over the potential range from 500 to 1050 mV at a sweep rate of 60 mV/min. The curves show that NO_2

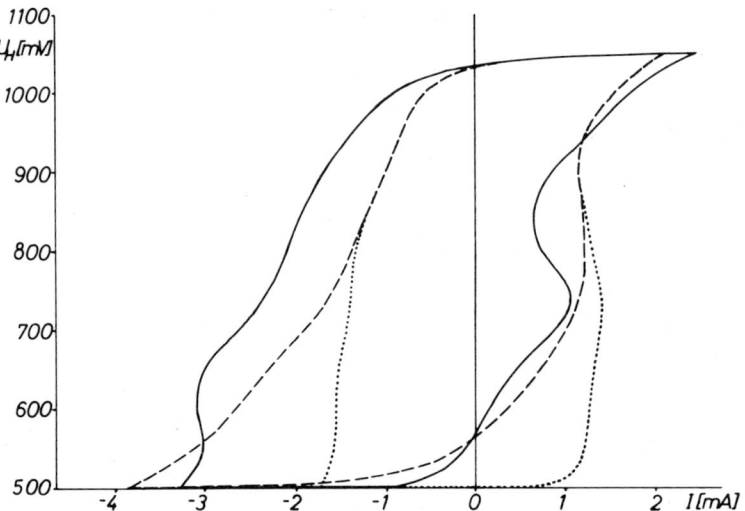

Fig. 18. Sweep voltage diagram for the cathodic reduction of NO_2 (——) and O_2 (---) on CoPc. For comparison, N_2 (···). Electrolyte, 4.5 N H_2SO_4, sweep rate 60 mV/min, electrode surface area 6 cm^2.

is reduced on CoPc at higher potentials than O_2. Hence in the potential range from 800 to 1050 mV, NO_2 can be selectively determined in the presence of an excess of O_2. Below 800 mV, O_2 is increasingly reduced along with NO_2, so that the exact determination of NO_2 is no longer possible.

Similar arguments and investigations for the choice of suitable electrode materials and electrode potentials are applicable for all gases whose concentrations have to be determined selectively by the limiting current method. If the gas to be determined has to be reduced, selective determination is only possible if the accompanying gases react at higher potentials than that to be determined. Analogously, components that have to be oxidized can only be selectively determined in the presence of other components that are oxidized at lower potentials than that to be determined.

Compounds containing elements in an intermediate oxidation state (e.g., NO_2) can, in principle, be determined either by anodic oxidation or by cathodic reduction. Necessary conditions are, however, that electrode materials at which reactions of both kinds can proceed be known, and that these materials be stable in the necessary potential range. If aqueous electrolytes are to be used, the range of electrode potentials available is further limited by the decomposition potential of water.

7.1.3 ELECTROLYTE AND MEMBRANE The electrolyte is the conducting medium between the sensor electrode and the counter electrode. Additionally, it must dissolve the reaction products formed at the sensor electrode. A thin electrolyte film between the sensor electrode and the membrane acts as a diffusion barrier. The great diversity of geometrical forms used for electrodes makes it impossible to generalize as to the thickness and extent of the electrolyte film.

The most commonly used electrolytes are aqueous solutions, usually of sulfuric acid, because of its insensitivity to the

CO_2 frequently occurring in the gas mixtures. The disadvantage of these electrolytes is the restriction of the applicable range of potentials by the decomposition potential of water, and the high vapor pressure of water. This can lead to drying up of the electrolyte when used with gas mixtures not saturated with water vapor.

These difficulties can be overcome by the use of solid electrolytes such as silver iodide (59), ceramics such as ZrO_2/CaO (60), mixtures of benzoic acid and propylene carbonate (61) or organic solvents of low vapor pressure to which a conducting electrolyte has been added (62). In all cases, however, there are attendant disadvantages.

Chand and Cunningham (62) investigated the aniodic behavior of NO, NO_2, and SO_2 on gold in various low-vapor pressure organic electrolytes with KPF_6 as conducting electrolyte, using the technique of sweep voltammetry. According to their results, β-butyrolacetone, N,N'-dimethyl formamide, and formamide are suitable for the aniodic oxidation of SO_2, while 1,2-propane-di-ol cyclic carbonate is suitable for the oxidation of NO and NO_2. No information has been published on the diffusion rates of different gases in electrolytes.

The membranes are often porous hydrophobic films, for example, of PTFE, such as have been developed for the air-breathing cathodes of fuel cells (63). Such films prevent penetration of the electrolyte into the gas phase but permit high diffusion and permeation rates of the gases to the electrode surface. Naturally, such membranes are not selective to different gases. The situation is different with non porous membranes. The selectivity ratios determined by Chand and Marcote (64) are summarized in Table 2. Membranes of different selectivities, suitable for

TABLE II

Selectivity Ratios of Selective Membranes

Types of membranes	Thickness (μ)	Selectivity ratios		
		SO_2	NO	NO_2
Tetrafluorethylene	6.3	0.83	0.73	1.00
MEM 213	30.5	1.00	0.08	0.39
Polyethylene	30.5	1.00	0.08	0.03
Polyethylene	7.6	1.00	0.14	0.0
Polyvinylchloride	38.1	1.00	0.0	0.0
Pellicon	12.7	0.94	0.0	1.00
Zitex 12 - 137B	50.8	1.00	0.30	0.69
Cellophane	25.4	1.00	0.0	0.0
Polyethylene + Pellicon Double membrane	38.1	1.00	0.0	0.0

various problems, are availabe. Lucero (65) has shown that diffusion through non porous membranes, rather than through the electrolyte in such systems, determines the diffusion limiting current.

7.1.4 SURVEY OF PUBLISHED WORK ON LIMITING CURRENT SENSORS

Limiting current sensors for the determination of O_2, NO, NO_2, SO_2, H_2S and CO are at present commercially available. The nature of the electrode material or of the electrolyte or the range of applied potentials is often not stated in the manufacturers' literature, so that this survey refers, as far as possible, to the scientific publications or patents relating to the devices concerned, as shown in Table 3. Halpert and Foley (71) investigated the polarographic behavior of oxygen on platinum at low temperatures. Sawyer et al. (72) determined the half-wave potentials and diffusion limiting currents of O_2, SO_2, Cl_2, Br_2, and NO_2 on platinum, using PE and PTFE membranes. Fabjan (73), using the rotation disk/ring electrode, showed that O_3 can be determined selectively, even in the presence of O_2, in the range of the diffusion limiting current, if potentials in the range 900 to 1100 mV (against H_2) are used. La Conti and Maget (78) showed that a linear relation between gas connection and sensor current is obtained with H_2, CO, or C_3H_8, even in the presence of O_2. According to Blurton and Sedlack (77), the oxidation of CO on Pt proceeds preferentially at the PtOH sites, while PtO has a lower activity and uncovered Pt practically none at all. Dutta and Landolt (80) have made a considerable contribution to the understanding of the reaction of NO, explaining the anodic and cathodic behavior of NO on Pt. At a comparatively early date, results were published (82) that made possible the determination of the concentration of oxygen in dissolved water. The first breakthrough in this field, however, was achieved by Clark et al. (83) who interposed membranes between the electrode and the fluid to be analyzed, thus providing a definite diffusion resistance. Limiting current sensors have now been extensively developed for measurement of the concentration of oxygen dissolved in water

TABLE III

Survey on Published Work on Limiting Current Sensors

Type of gas	Electrode material	Electrolyte	Potential range mV versus H_2	Ref.
O_2	Kohle	30% KOH	390	66
O_2 in hydrocarbons	Ag	24% KOH	?	67
O_2	Au	KCl-Gel	530 to 800	68,69,70
O_2	Pt	KCl	390 to 840	71,72
O_2	Pt	ZrO_2/CaO	650	60
O_2 in water	Pt	KCl or NH_4Cl	565 / 300	84 / 85
O_2 in human tissues and blood	Pt	KCl	300	83,86,87
O_3	Pt	H_2SO_4	900 to 1100	73
CO	MoS_2 or WS_2	H_2SO_4	?	74
CO	Pt	H_2SO_4	900 to 1500	75,76,77
CO	Pt	H_2SO_4	710	78
C_3H_8, C_8H_{18}	Pt	H_3PO_4	1060	78
HCHO	Pt	H_2SO_4	890	89
H_2	Pt	cation exchange membrane	250 to 650	78
NO_2 in air	Pt	AgJ	310 to 510	59
NO_2	CoPc	H_2SO_4	550 to 1050	56
NO	Au,Pt,Pd	H_2SO_4	960 to 1030	79
NO_2	Au,Pt,Pd	H_2SO_4	800	79
NO	Pt	H_2SO_4	750	80
SO_2	Au,Pt,Pd	H_2SO_4	170	81
SO_2	quinone	H_3PO_4	redox potential quinone/hydroquinone	98

(64,84,85) and are commercially available (69,70).

After Clark et al. (83) had used such sensors for determination of the concentration of oxygen in blood, they were extensively used in biomedical technology, among other purposes for transcutaneous determination of oxygen concentration in blood (86,87). The sensor is applied to the surface of the skin, so that the latter forms an additional diffusion resistance. Such sensors have also been used to determine the oxygen concentration in urine (88).

In considering very high working temperatures, in the range of 600 to 1000°C, the ZrO_2/CaO sensor deserves particular attention (60). As shown in Fig. 19, it consists of a tubular electrolyte

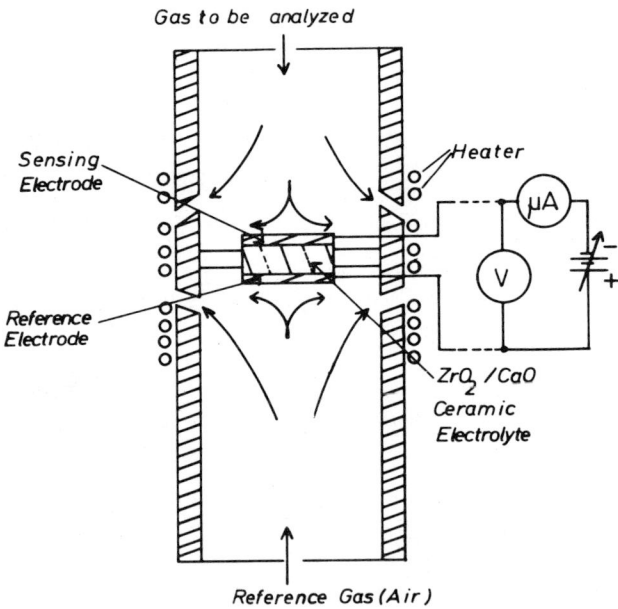

Fig. 19. Schematic construction of a limiting current sensor with ZrO_2/CaO solid electrolyte.

body, at whose upper end the gas to be analyzed flows in through
a small hole, while the reference gas (air) flows in through the
lower opening. The sensor is devided into two parts, an upper and
a lower, by a partition, and both gases can flow out into the
atmosphere through openings in the sides of the device above and
below this partition, respectively. The partition also serves as
support for the ceramic solid electrolyte, to which platinum electrodes are applied on both sides. A constant voltage of 500 mV
is applied between the electrodes and the current passing is measured. The sensor is maintained by external heating at the operating temperature, between 600 and 1000°C. In this arrangement,
the diffusion resistance is determined by the diffusion coefficient, by the distance from the upper opening to the electrode,
and by the cross section of the diffusion path. The latter two
criteria can easily be accomodated to the necessary dimensions.

7.2 Coulometric Measurements

The coulometric analysis of the gases may be regarded as the
reverse of electrolysis. As is well known, in electrolysis each
quantity of charge liberates a quantity of gas, for example, of
hydrogen and oxygen, exactly determined by Faraday's law. In
coulometry, the gas to be determined is quantitatively converted,
so that the current flowing is a measure of its concentration.
The counter electrode is usually not a gas electrode but a metal
electrode. For the measurement of oxygen concentrations by cathodic reduction, metals undergoing corrosion as the anodic process
are often used for counter electrodes.

Figure 20 shows the schematic construction of a coulometric
cell for flowing gas. The current produced by the electrochemical
reaction is proportional to the concentrations of the substances

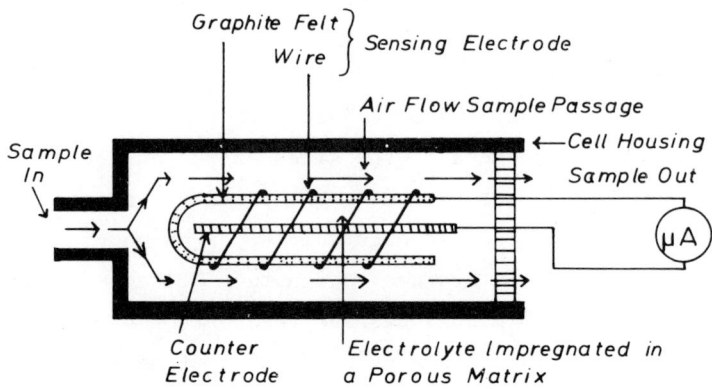

Fig. 20. Schematic structure of a coulometric cell.

to be determined and to the velocity of flow, that is, to the quantity of gas passing per unit time. This velocity must therefore be kept very accurately constant. The sensor electrode is a porous material, for example, an impregnated graphite felt, which is contacted with a metal wire to increase its conductivity. The counter electrode is made of a corrosible metal. Suitable anode materials are lead amalgam or cadmium electrolytically deposited on a nickel mesh. The electrolyte, which in the previously mentioned cells is potassium hydroxide solution, wets the anode and also part of the cathode, as a thin film on the cathode surface. In this way rapid response is secured, of the same order as that of limiting current sensors (see Section 7.1). The cell can be made operable in any position if the electrolyte be adsorbed in a porous matrix, such as asbestos. The current in an external circuit is measured. The selectivity of the process is subject to the same considerations as disscussed in Section 7.1.2.

Hersch (90) has carried out pioneer work on the coulometric analysis of gas components in the liquid or the gaseous phase.

He was the first to show that the method can be practically used for the determination of oxygen. He did so by reducing O_2 at a platinum wire cathode in 24 % KOH solution, using a counter electrode of lead amalgam.

This cell was capable of improvement. In a later publication (91) a silver mesh was used instead of platinum as cathode material and was contacted with silver wire. The cell was calibrated by electrolysis in situ, enabling an accuracy in oxygen determination of a few ppm to be attained. If the concentration changed, 90 % of the final value was attained within 30 sec. The temperature coefficient of the cell, at a flow of 100 ml per minute, was 2.5 % per degree centigrade. Hersch has patented practical designs of coulometric oxygen sensors (94,95).

Keidel (92), using a porous silver cathode, was able to increase the accuracy over a range of concentrations from 1 ppm to 1 vol %. Using his device with a flow velocity of 100 ml/min, he obtained the theoretical value of 26.3 mA per ppm O_2.

Hersch has also described a coulometric sensor in which NO_2 is cathodically reduced to NO in KCl solution or in phosphate buffers (96). A device identical in principle can be used to measure the concentration of oxygen dissolved in water (97).

The development of coulometric analysis seems recently to have come to a standstill. In practice, it is used only for determining oxygen. It is in this way possible to determine very small quantities, down to ppm, but the conditions of measurement must be kept very exactly constant, in particular the velocity of flow.

7.3 Determination of Gas Concentrations by Selective Inhibition of an Electrode Reaction

The suggestion was recently made (93) that selective inhibition of an electrode reaction by a gas phase should be used to determine the concentrations of the inhibiting components. Thus the cathodic reduction of oxygen at platinum or palladium electrodes is inhibited by very low concentrations of CO. The potentiodynamic triangular voltage diagrams (Fig. 21) in 2 N H_2SO_4 with CO-free and CO-containing air show clearly the reduction of the cathodic current due to oxygen reduction in the presence of carbon monoxide.

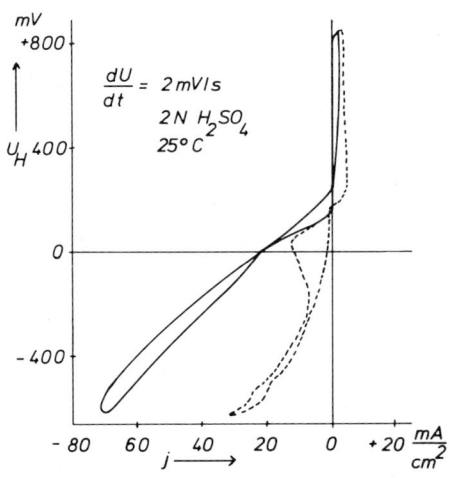

Fig. 21. Sweep voltage diagram on Palladium black with CO_2-free (——) and CO_2-containing (---) air in 2 N H_2SO_4.

With suitably constructed electrodes, this effect can be used as the basis of a method of measurement of CO concentrations, as shown in Fig. 22. This shows the cathodic currents due to oxygen reduction, at first with CO-free air, using black platinum electrodes irrigated with gas at a potential of 275 mV. When 0.1 ppm CO_2 is added to the mixture, the current slowly drops as a result of inhibition. This effect is, however, reversible, as is shown by the curve for subsequent operation with CO-free air. The limit

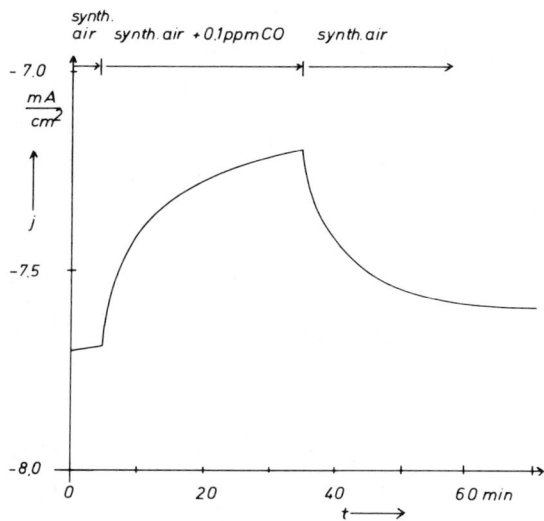

Fig. 22. Potentiostatic measurement on Palladium black (10 mg/cm^2) with air containing 0.1 ppm CO in 2 N H_2SO_4 at 275 mV versus H_2.

of applicability of this process is reached at 0.01 ppm CO. No information about the reproducibility of this method is available. Although technical development of this process is far from complete, the present state shows that it is in principle possible to develop new methods of electrochemical gas analysis.

8 METHODS BASED ON MEASUREMENT OF ELECTRICAL CONDUCTIVITY

If a gas dissolved in liquid phase alters the specific electric conductivity of the latter, its concentration can in principle be determined by a conductivity measurement. Thus, for example, it is in principle possible to bring a sample of gas of known volume, whose SO_2 content is to be measured, into contact with an alkaline liquid in which the gas dissolves. The change in electrical conductivity of the liquid phase can, after calibration with samples of known SO_2 content, be used to determine the SO_2 concentration in the test sample. Such discontinuous processes are, however, laborious, and the smallness of the changes observed make it difficult to attain accuracy.

On the other hand, the dependence of the electrical conductivity of semiconductors on the concentration of certain compounds in the gas phase, discussed in Section 5, is the basis of gas sensors of comparable utility, in many cases, with the measuring cells based on voltage measurement already described.

Semiconductor sensors for the determination of gases, using mainly oxides of the transition metals, can be divided into two groups according to the temperature range within which they are best suited to measurement. The first group is used for determining oxygen at temperatures in the range 700 to $1400°C$, the second for determining a large number of reducing gases at temperatures up to approximately $500°C$.

At temperatures of about $1000°C$, various transition metal oxides show a definite, reproducible dependence of electrical conductivity on the partial pressure of oxygen in the gas phase. As examples, titanium dioxide and cobalt oxide are considered.

Titanium dioxide, TiO_2, is an excess electron conductor at high temperatures and in low oxygen partial pressures. Its interaction with the oxygen in the gas phase is given (101-105) as

$$TiO_2 \rightleftharpoons Ti^{3+}_{int} + 3e^- + O_2$$

where Ti^{3+}_{int} represents interstitial titanium ions. This equation leads to the relationship

$$\sigma = K \cdot p_{O_2}^{-1/4} \quad \text{at } T = \text{const} \qquad (36)$$

between electrical conductivity and oxygen partial pressure, which also holds for stabilized ZrO_2 over similar ranges of temperature and oxygen partial pressure.

The electrical conductivity of cobalt oxide, which has cobalt defects in the lattice and is a defect electron conductor, follows the equation

$$\sigma = K \cdot p_{O_2}^{1/4} \qquad (37)$$

at similar temperatures and relatively high oxygen partial pressures. At relatively low oxygen partial pressures the index ranges from 1/4 to 1/6 (100, 106-108).

Figure 23 shows the dependence of the electrical resistivity on $\log p_{O_2}$ for TiO_2 and CoO at various temperatures. The curves

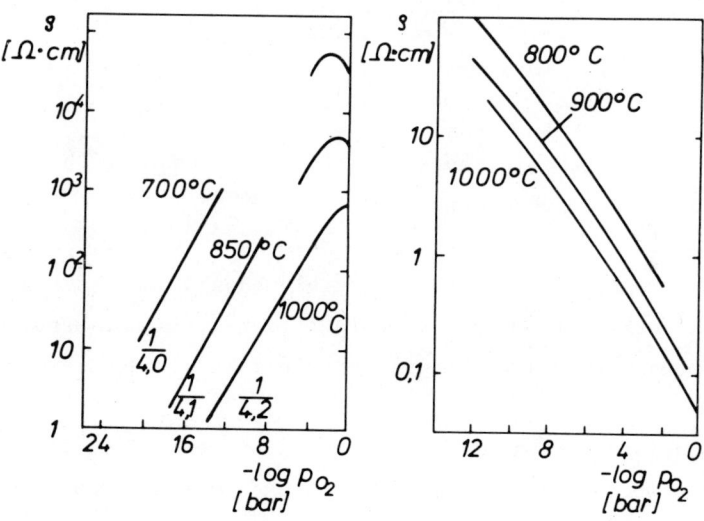

Fig. 23. Dependence of the electrical resistivity σ for TiO_2 (left) and CoO (right) on the oxygen partial pressure in the gas phase (99, 100).

were obtained by measurement in the stationary state on test pieces produced by sintering the oxide powders, and are reproducible above 700°C. The curve for TiO_2 shows deviations from Eq. 36 at high oxygen partial pressures, which do not occur with CoO.

Figure 24 shows a suggested design for a sensor with a measuring element of TiO_2 or CoO (100).

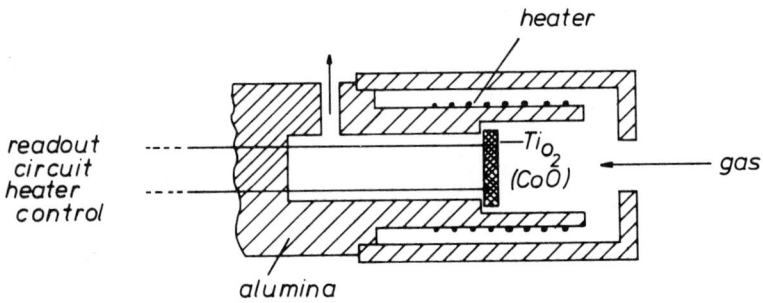

Fig. 24. Schematic representation of a sensor with a semiconductor measuring element.

Sensors operating on this principle should be so designated that the current flowing at constant voltage is not used directly as the measuring value, but rather the voltage produced by this current.

The resistive element is a sintered plate of TiO_2 or CoO powder, into whose interior the gas to be determined can penetrate. The diameter of the oxide particles is 1 to 5 µ. The curves of resistance against partial pressure measured on such oxide bodies correspond fairly exactly to Eqs. 36 and 37. The response time of such sensors with particle size in the above specified range is about 0.1 to 0.3 sec and increases with particle size. This fact and the conformity to Eqs. 36 and 37 make it probable that free oxygen interacts with the oxide lattice by incorporation or removing in the temperature range considered. The incorporation and removing process may be expected to extend into the interior of the grains.

If sensors with a TiO_2 measuring element are used in the gas mixtures containing oxygen together with easily oxidizable components, such as H_2 and CO, a sharp step in the measured value is observed at transition from an oxidizing to a reducing atmosphere, as in concentration cells with a catalytically active measuring electrode (see Section 6.1). The reason is that TiO_2 can catalyze the reaction of H_2 or CO with oxygen (see Figs. 9 and 10). Such sensors are therefore chiefly suitable for regulating combustion processes, particularly in internal combustion engines. Such applications are discussed in Section 9.

Sensors for use in relatively low temperatures use a different mechanism to produce the measuring value. It is well known that the electrical conductivity of many semiconductors is changed by chemisorption processes from the gas phase. This change is probably due to the action of reducing and oxidizing gases, respectively, as electron donors and acceptors on adsorption at the solid surface, with consequent provision of electrons to the solid or withdrawal of electrons from it. The type of semiconductor, that is, whether p or n, has an effect on the magnitude of the conductivity change. Changes in the following cases have been described: adsorption of alcohols, esters, SO_2, O_2, CO, NH_3, H_2S, NO_2, NO, N_2O, ammonium ions, HCl, I_2, benzene groups, and water on organic substances, such as β-carotin, at 20 to 50°C (109-112), H_2, hydrocarbons, CO, SO_2, and NH_3 on inorganic compounds such as ZnO, CdO, Fe_2O_3, SnO_2, CuO, Al_2O_3, TiO_2, NiO, and CoO at 400 to 500°C (113,114).

In chemisorption processes, the electron density is probably altered only in the immediate neighborhood of the solid surface and not throughout the whole lattice. In measuring the conductivity of sintered oxides, the contribution of the surface and its immediate neighborhood is greater than that of the interior of the oxide particles, which is but little influenced by the chemisorption. The relatively strong dependence of conductivity on the chemisorption of even small quantities of material is then easily explained. At high temperatures, such effects play no part, as practically no chemisorption is possible above about 500°C.

Sensors using inorganic semiconductors have been developed and used as detectors for reducing gases. One such sensor (115), with a measuring unit consisting of a sintered mixture of SnO, ZnO, and Fe_2O_3, operates at temperatures between 60 and 340°C and shows a reproducible dependence of electrical resistance on concentration of CO and other reducing gases over an extended working life.

Figure 25 shows the characteristics of such a sensor for various gases. The sensor resistance is proportional to the voltage across the load resistance.

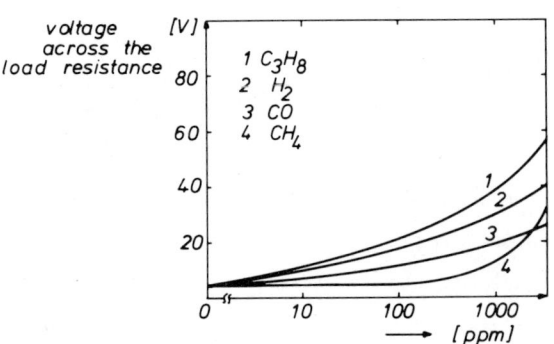

Fig. 25. Characteristics of a semiconductor sensor for various gases (115).

Sensors of this kind are hardly specific for individual gases. Hence they only give an unambiguous value in a gas mixture containing only a single reducing component or if only the sum of all the reducing components is interesting. Their chief field of application is therefore warning or controlling the sealing of gas pipes.

9 ELECTROCHEMICAL OXYGEN SENSORS IN MOTOR VEHICLES

Upper limits for the emission of harmful components by motor vehicles are set in several countries. The values at present in force can be maintained by methods affecting the operation of the engine - for example, more exact mixture dosage, retarded ignition, exhaust gas recirculation - or by incorporating oxidation catalysts. For the values intended in the United States for 1978 (0.4 g/mile hydrocarbons "HC", 3.4 g/mile carbon monoxide CO, and 2.0 g/mile nitrogen oxides NO_x), however, such measures are inadequate. The most promising method of reaching the required values, and the soonest realizable, is that of intensive decontamination of the exhaust gases.

Figure 26 shows the CO, HC, and NO_x emission of a gasoline engine as a function of the equivalence ratio:

$$\lambda = \frac{\text{actual air / fuel ratio}}{\text{stoichiometric air / fuel ratio}} \quad (38)$$

At λ-values below 1, there is an excess of fuel; at values above 1, an excess of air. At $\lambda = 1$, a stoichiometric mixture exists. The solid lines show the dependence of the emission values before conversion, the broken lines after conversion by a catalytic converter.

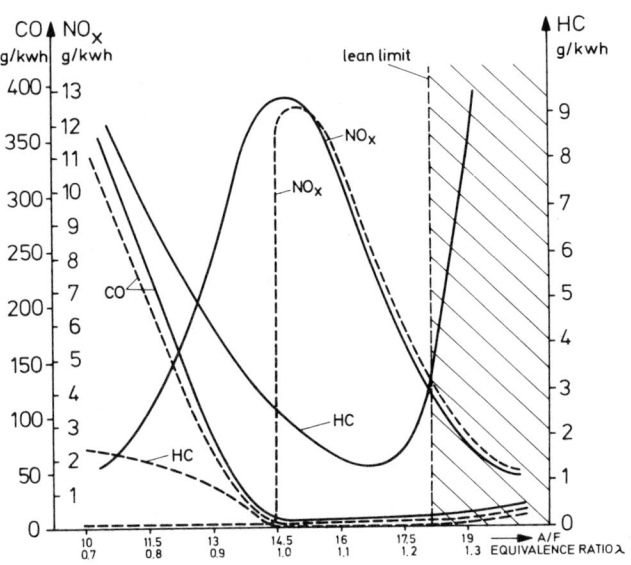

Fig. 26. Exhaust emissions with (---) and without (——) catalyst.

Three principles of exhaust gas decontamination can be derived from this diagram:

1. The dual-bed catalyst system. Two exhaust gas catalysts are provided in the exhaust system. Between them, additional air for combustion is injected by a pump. In the first catalyst, at λ-values in the range 0.7 to 0.95, the nitrogen oxides emitted by the engine are reduced to nitrogen with the aid of the CO, which is present in large quantities, while the CO is oxidized to CO_2. In the second catalyst, at λ values above 1.0, the CO still present and the uncombusted hydrocarbons are oxidized by the added air. The limits intended

in the USA for 1978 can be held with such systems. A serious disadvantage of such systems, however, is the fuel consumption, which is higher than that of present-day engines by about 10-20 %.

2. The three-way catalyst system. From Fig. 26 it can be seen that the lowest emission of all three harmful components is attained at $\lambda = 1$, the stoichiometric mixture, and that this can be attained with only one catalyst. If the stoichiometric mixture can be maintained under all engine operating conditions to a fraction of 1 %, the required emission limits could be met. An advantage of this method, in principle, is that the fuel consumption is not greater than that without catalyst. However, as in the dual-bed system, lead-free fuel must be used to avoid lead poisoning of the catalyst.

3. The concept of fuel-lean mixture accomodation. As shown in Fig. 26 very low emission values can be attained at λ-values of about 1.2, that is, close to the lean limit of the engine, even without catalyst. The emission can be further lowered by after-burning, in which hydrocarbons and CO are reacted, and by double ignition, that is, fitting two spark plugs to each cylinder, which lowers the NO_x content. The advantage of the method are the relatively low fuel consumption, the elimination of a catalyst, and the possibility of using lead-containing fuels. A serious disadvantage, on the other hand, is the difficulty of attaining a sufficiently good compromise between NO_x emission and drivability.

An exhaust gas decontamination system has been developed on the basis of the three-way catalyst system (26,43,116,117). The prospects for its use in motor vehicles, assuming continued validity of the present emission limits, are good (Fig. 27).

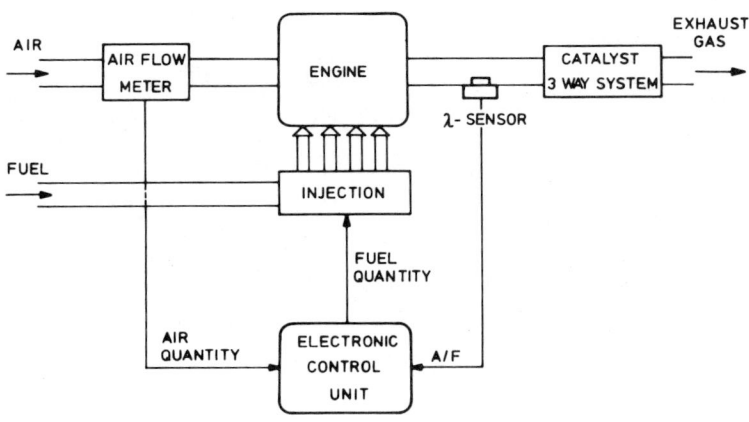

Fig. 27. Simplified principle of the emission control system (116,117).

In Section 6.2 the use of oxygen concentration cells with solid electrolyte for measurement of the concentration of oxygen in mixtures with oxidizable gases was described. Cells with a catalytically active measuring electrode show a pronounced voltage step on passing from an oxidizing to a reducing atmosphere or back (Figs. 9 and 10). Such an oxygen concentration cell with ZrO_2 solid electrolyte is the most important element in an emission control system such as that shown in Fig. 27 (26,43,116,119). The system represents a closed control loop. The air flow is controlled by a throttle valve and is measured by an air flow meter. The required amount of fuel is determined as a function of the measured air flow and other operating parameters, such as engine speed and temperature, by the electronic control unit and is injected by a magnetically operated valve. The stoichiometric air/

fuel ratio $\lambda = 1$ can be set approximately correctly by this part of the system. The oxygen concentration cell fitted in the exhaust gas flow, the so-called λ-sensor, provides the control signal for the electronic control unit through the voltage step occurring at $\lambda = 1$. The closed control loop arising through this possibility of superposition makes it possible to set the mixture very accurately at $\lambda = 1$ and hence provides the basis for optimal conversion of the harmful exhaust components in the three-way catalyst system. The solid electrolyte of the λ-sensor is ZrO_2 stabilized with CaO or Y_2O_3. The reference and the measuring electrode are porous and consist of sintered platinum. The reference gas is the surrounding atmosphere. The construction of the sensor is shown in principle in Fig. 28.

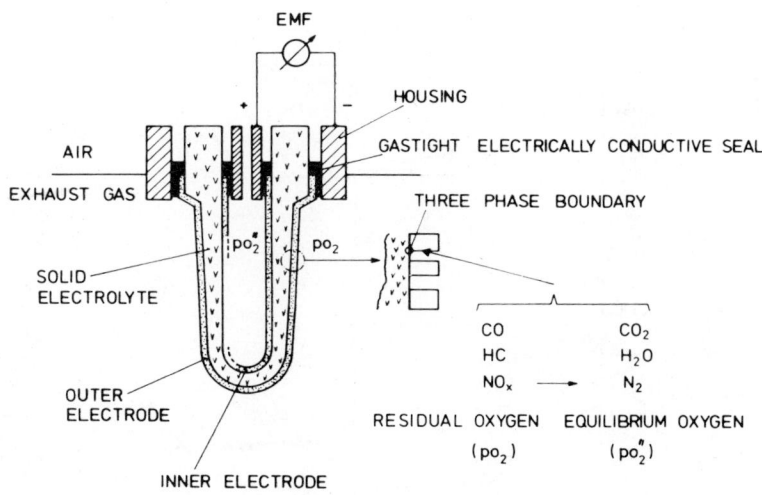

Fig. 28. Schematic view of the λ-sensor.

Even when a stoichiometric mixture is supplied to the motor, the exhaust gases are not completely in chemical equilibrium. They therefore contain definite quantities of unreacted oxygen and reducing components. The sensor shows the voltage step correctly only if the catalytic activity of the platinum electrode is sufficient for the redox reactions possible in the exhaust gases to proceed rapidly to at least approximate completion. This is indicated by at least approximate attainment of the thermodynamically calculable equilibrium oxygen partial pressure at the three-phase boundary. For example, partial pressures at 10^{-19} to 10^{-28} bar are attained with $\lambda = 0.7$ at temperatures of 500 to 950°C. If the equilibrium oxygen partial pressures are computed as a function of value and temperature (118), assuming the series of reaction equilibria shown in Fig. 29, then, substituting the values obtained in the simplified Nernst equation, various curves of the equilibrium voltage of the sensor are obtained, as shown in Fig. 29. $\lambda = 1$ corresponds to an equilibrium voltage

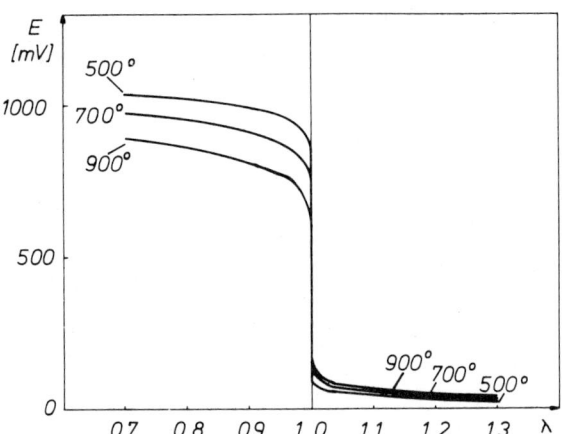

Fig. 29. Exhaust gas reactions taken as a basis for calculation. Dependence of equilibrium voltage on λ at various temperatures.

of 177 mV at 550°C or 268 mV at 900°C. The λ-sensor produces the calculated curves without sufficient accuracy.

The requirements of ZrO_2 cells to be used in the exhaust system of motor vehicles are different from those of normal cells for measuring oxygen concentration. This is shown in Table 4.

TABLE IV

Requirements of the λ-Sensor;
Maximum Permissible Limits of Operating Conditions

Requirements		Operating condition limits
Height of the voltage signal	$\geq$ 500 to 600 mV, measured between λ = 0.95 and λ = 1.05	Temperature at the ceramic tip: max. 900°C
Response time	$\leq$ 50 msec at 700°C	Thermal shocks at the ceramic tip: max. 20°C/sec
Starting temperature	300 to 400°C	
Internal resistance	$\leq 10^5$ (400°C)	Vibration acceleration of the ceramic body: max. 60 g
Lifetime	$\geq$ 15,000 miles	

Important above all is attainment of a definite minimum voltage signal height. On the other hand, it is less important to measure accurately oxygen partial pressure at λ-values larger or smaller than 1.

A short response time of the sensor is more important than in measuring cells for normal applications, because it enables practically inertialess control to be attained. The ceramic solid electrolyte must be developed with great care in order to fulfil the requirements given in Table 4 under the severe operating conditions encountered. A particular problem is that the height of voltage step departs from the calculated values at low exhaust gas temperatures (below 500°C), tending towards zero, while on the other hand the sensor is heated only by the hot exhaust gases and must deliver a usable signal as soon as possible after starting the engine. The problem described in Section 6.2 of the deviation of the voltage of high-temperature measuring cells from the calculated values of the equilibrium voltage at these temperatures is solved by using ZrO_2 ceramic stabilized with Y_2O_3 and containing only a very low proportion of electronically conducting impurities. This is shown in Fig. 30, in which the temperature dependence of the measured sensor voltage for a λ-value of 0.75 (s-shape curves) is compared with the curves calculated for λ = 0.7 and 0.9.

The sensor shown in Fig. 31 (119) is capable of fulfilling the requirements listed in Table 4 and can therefore be used as a reliable controlling sensor for the three-way catalyst system.

Sensors designed on similar priciples, which have in some cases been extensively tested in practice, have been described by several authors (120 to 122). Recently, attempts have been made to control fuel dosage with a carburetor instead of by injection (123,124). Promising results have been attained in this way.

Control of the basis of the three-way catalyst system can also be effected with the TiO_2 sensor described in Section 8 (99,125). At λ = 1, corresponding to an air/fuel ratio of about

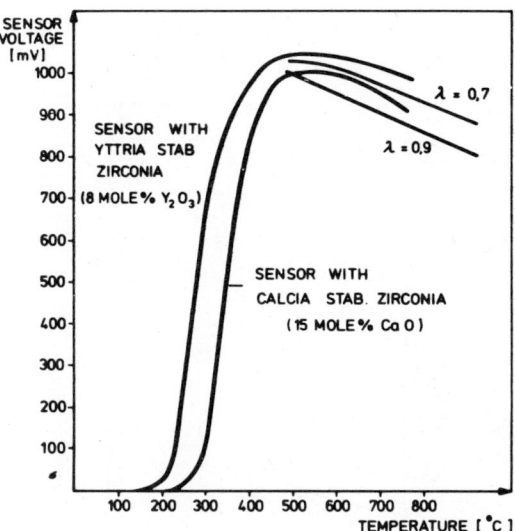

Fig. 30. Temperature dependence of the sensor voltage.

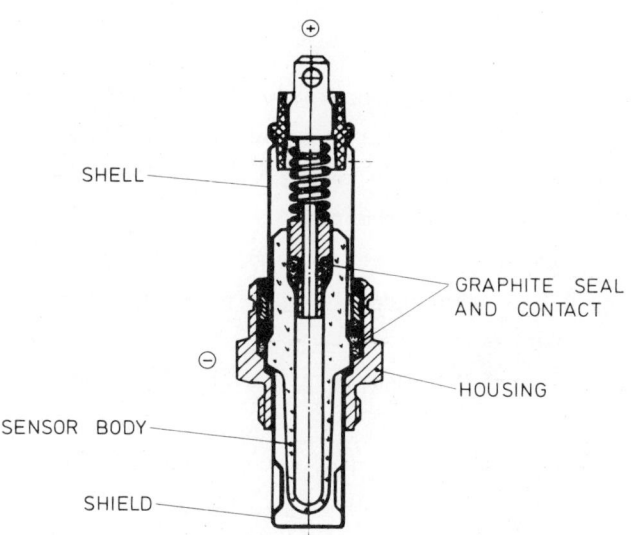

Fig. 31. Section of the λ-sensor.

14.5, a sharp change in the resistance of the TiO_2 is observed (Fig. 32), which can be transformed into a voltage change and used as control signal for the electronic control unit.

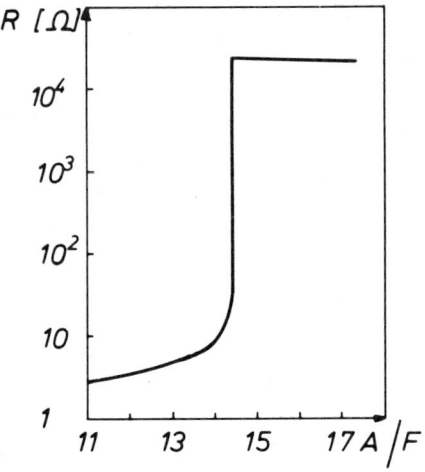

Fig. 32. Resistance as a function of air/fuel ratio for a TiO_2 sensor at $700°C$ (99).

The sensor, which is shown in priciple in Fig. 24, is not heated by the exhaust gas stream, as are the ZrO_2 sensors, but by a controlled heating system to ensure that the active TiO_2 plate is brought to the nominal temperature of $700°C$ and kept there. The response time is 200 msec, considerable longer than that of the very rapid ZrO_2 sensor. As far as present results show, the TiO_2 sensor (99) has not been as extensively studied as have some of the ZrO_2 sensors. A comparison between TiO_2 and ZrO_2 sensors is given in (127).

For the principle of lean-mixture accomodation, a control sensor indicating a λ-value of about 1.2 is possible. In principle, the combination of a controlled measuring and dosage cell based on ZrO_2, which has already been described in several publications

(34), is suitable for this application. The sensor, which is placed in a current of part of the exhaust gas, indicates departures from the nominal λ value through the measuring unit. The control unit then causes a steadily increasing electric current to flow through the dosage part of the sensor until the nominal λ value is regained. The magnitude of the current is a measure of the departure of the λ value of the exhaust gas from its nominal value. With this device, not only can the stoichiometric point be detected, but values of λ other than 1 can be maintained (126). The response time of the system is, however, very long. Up to now, no results of practical trials have been reported.

The suitability in priciple of a semiconductor sensor with CoO as active material (100) as a control sensor for λ-values greater than 1 could not be confirmed in experiments (100). The material shows a definite and reproducible dependence of electric resistance on λ in this range, but the material properties of the CoO ceramic caused difficulties.

10 FUTURE PROSPECTS

In many cases success has been attained in the replacement of conventional methods of analysis by directed development of electrochemical sensors. One reason for this is the advantages of the electrochemical measuring process, including simple construction, simple operation, and low price. The simple provision of the measured value and short response time make many of these sensors particularly suitable as active elements in control circuits. In many cases, however, the range of applicability on the conventional methods of analysis is not attained. This provides a wide field for future development. Two directions of investigation are particularly important: improvement of existing sensors in respect of durability, accuracy, and shortness of response time, and the development of new sensors showing, in addition to the properties mentioned above, better selectivity. Many of these processes described have been developed empirically, without exact knowledge of the special electrochemical mechanism involved. It may be expected that considerable advances can be attained in this field by more intensive fundamental research. This should include work towards a better understanding of the electrode processes, development of new electrolytes, and improved electrode constructions, as well as technologic advances.

REFERENCES

1. M. Eisenberg, Electrochim. Acta 6, 93 (1962).
2. C. Berger, Handbook of Fuel Cell Technology, Prentice Hall, Hemel Hempstead (1968).
3. M.W. Breiter, Electrochemical Processes in Fuel Cells, Springer Verlag, Berlin, 1969, p. 14..
4. K.H. Many, D.A. Okun, and C.N. Reilley, J. Electroanal. Chem. 4, 65 (1962).
5. F.A. Kröger, J. Electrochem. Soc. 120, 75 (1973).
6. R.J. Brook and T.L. Markin, in van Gool (Ed.), Fast Ion Transport in Solids, North-Holland Publ. Comp., Amsterdam, 1973, p. 533.
7. S. Pizzini, in van Gool (Ed.), Fast Ion Transport in Solids, North-Holland Publ. Comp., Amsterdam, 1973, p. 461.
8. H.H. Möbius, Z. Phys. Chem. 230, 396 (1965).
9. T.H. Etsell and S.N. Flengas, J. Electrochem. Soc. 118, 1890 (1971).

References

10. W.A. Fischer and G. Pateisky, Arch. Eisenhüttenwes. 41, 661 (1970).
11. C. Wagner, Electrochim. Acta 15, 987 (1970).
12. C. Wagner, Advan. Catalysis 21, 323 (1970).
13. S. Stotz, Ber. Bunsenges. Phys. Chem. 70, 37 (1966).
14. L.D. Burke, H. Rickert, and R. Steiner, Z. Phys. Chem. N.F. 74, 146 (1971).
15. M. Hebb, J. Chem. Phys. 20, 185 (1952).
16. C. Wagner, Proc. of the 7th CITCE Meeting, Butterworths Scientific Publications, London, 1957, 361.
17. F. Haber, Z. Electrochem. 11, 593 (1905).
18. F. Haber, Z. Electrochem. 12, 415 (1906).
19. H.H. Möbius, Dissertation, Rostock (1958).
20. H. Peters and H. Möbius, Z. Phys. Chem. 209, 298 (1958).
21. H. Peters and H. Möbius, DDR Patent 21673, 7.8.1961.
22. N.S. Choudhury, J. Electrochem. Soc. 120, 1663 (1974).
23. J. Weissbart and R. Ruka, Rev. Sci. Instr. 32, 593 (1961).
24. H. Schmalzried, Z. Electrochem. 66, 572 (1966).
25. H. Schmalzried, Z. Phys. Chem. N.F. 38, 87 (1963).
26. H. Düker, K.H. Friese, and W.D. Haecker, SAE-Paper 750223 (1975).
27. H. Ullmann, D. Naumann, and W. Burke, Z. Phys. Chem. 237, 337 (1968).
28. W. Pluschkell, Arch. Eisenhüttenwes. 46, 11 (1975).
29. J.W. Patterson, J. Electrochem. Soc. 118, 1033 (1971).
30. W.L. Worrell, Ceram. Bull. 53, 425 (1974).
31. R. Hartung and H.H. Möbius, Chem. Ing. Tech. 40, 592 (1968).
32. P.A. Cerkasov, W.A. Fischer, and C. Pieper, Arch. Eisenhüttenwes. 42, 873 (1971).
33. T.H. Etsell and S.N. Flengas, Metallogr. Trans. 3, 27 (1972).
34. N.M. Beekmans and L. Heyne, Philips Tech. Rundschau 31, 120 (1970/71).

35. K. Petanides, B. Marincek, and G. Heimke, Chemie-Anlagen & Verfahren Heft 6, 167 (1973).
36. R.G.H. Record, Instr. Pract. 24, 161 (1970).
37. W. Löscher, Hoesch Berichte Heft 1, 30 (1970).
38. B. Schuh and B. Marincek, Schweizer Archiv 36, 379 (1970).
39. W. Pluschkell and H.J. Engell, Ber. Deutsche Ker. Ges. 45, 388 (1968).
40. D. Yuan and F.A. Kröger, J. Electrochem. Soc. 118, 841 (1971).
41. Y.L. Sandler, J. Electrochem. Soc. 118, 1378 (1971).
42. C.H. Loos, Deutsche Offenlegungsschrift No. 2010793, 8.10.70.
43. H. Düker and H. Neidhard, Sec. Autom. Emmission Conf., Ann Arbor (1973).
44. W.A. Fischer and A. Hoffmann, Arch. Eisenhüttenwes. 28, 739, 771 (1957).
45. W.A. Fischer and W. Ackermann, Arch. Eisenhüttenwes. 36, 643, 695 (1965).
46. K.H. Ulrich and K. Borowski, Arch. Eisenhüttenwes. 39, 259 (1968).
47. G. Heimke and E. Gugel, Deutsche Offenlegungsschrift No. 2306551, 10.2.1973.
48. D.A.J. Swinkels, J. Electrochem. Soc. 117, 1267 (1970).
49. P.A. Butzelaar and L.P.J. Hoogeveen, Philips Tech. Rundschau 34, 112 (1974/75).
50. G. Bianchi, G. Faita, and T. Mussini, J. Sci. Instr. 42, 693 (1965).
51. G. Barna and R. Jasinski, Anal. Chem. 46, 1834 (1974).
52. R. Jasinski, C. Barna, and I. Trachtenberg, J. Electrochem. Soc. 121, 1575 (1974).
53. P.B. Hahn, M.A. Wechter, D.C. Johnson, and A.F. Voigt, Anal. Chem 45, 1016 (1973).

References

54. H. Huck, Z. Anal. Chem 270, 266 (1974).
55. H. Huck, J. Electroanal. Chem. Interfac. Electrochem. 53, 121 (1974).
56. H. Jahnke, B. Moro, and H. Siebke, Bunsentagung Wien, 1975.
57. H. Jahnke, Ber. Bunsenges. phys. Chem. 72, 1053 (1968).
58. R. Jasinski, J. Electrochem. Soc. 112, 526 (1965).
59. Rockwell Int. Corp., US Patent No. 3821090, June 28, 1974.
60. Westinghouse Electric Corp., Deutsche Offenlegungsschrift No. 1954663, May 6, 1970.
61. Farbenfabriken Bayer AG, French Patent No. 2012511, March 20, 1970.
62. R. Chand and P.R. Cunningham, Trans. Geosci. Electron. 8, 158 (1970).
63. K. Brill, 3es Journeés Int. d'Etude des Piles à Combustible, Comp. Rend., Brussel, 1969, p. 39.
64. R. Chand and R.V. Marcote, AIChE Meeting 1971, Preprint No. 68c.
65. D.P. Lucero, Anal. Chem. 40, 707 (1968).
66. K. Kordesch and A. Marko, Mikrochim. Acta 36/37, 420 (1951).
67. W.J. Baker, J.F. Combs, T.L. Zinn, A.W. Wotring, and R.F. Wall, Ind. Eng. Chem. 51, 727 (1959).
68. Fa. Beckman, US Patent No. 2913386.
69. Beckman Bulletin No. 7143.
70. Beckman Instructions No. 1637.
71. G. Halpert and R.T. Foley, J. Electroanal. Chem. Interface Electrochem. 6, 426 (1963).
72. D.T. Sawyer, R.S. George, and R. Rhodes, Anal. Chem. 31, 2 (1959).
73. Ch. Fabjan, Electrochim. Acta 20. 863 (1975).
74. Licentia Patent-Verwaltungs-GMBH, Deutsche Offenlegungsschrift No. 2240350, 21.3.1974.

75. Energetics Science Inc., US Patent No. 3776832, Dec. 4, 1973.
76. H.W. Bay, K.F. Blurton, H.C. Lieb, and H.G. Oswin, Amer. Laboratory 4, 57 (1972).
77. K.F. Blurton and J.M. Sedlak, J. Electrochem. Soc. 121, 1315 (1974).
78. A.B. La Conti and H.J.R. Maget, J. Electrochem. Soc. 118, 506 (1971).
79. Dynasciences Corp., US Patent No. 3622487, Nov. 23, 1971.
80. D. Dutta and D. Landolt, J. Electrochem. Soc. 119, 1320 (1972).
81. Dynasciences Corp., US Patent No. 3622488, Nov. 23, 1971.
82. F. Tödt, Arch. Metallkunde 1, 469 (1947).
83. L.C. Clark, R. Wolf, D. Granger, and Z. Taylor, J. Appl. Physiol. 6, 189 (1953).
84. M.D. Lilley, J.P. Storey, and R.W. Raible, J. Electroanal. Chem. Interfac. Electrochem. 23, 425 (1969).
85. R.W. Pittman, Nature 195, 449 (1962).
86. R. Huch, A. Huch, and D.W. Lübbers, Biomed. Tech. 19, 87 (1974).
87. R. Huch, D.W. Lübbers, and A. Huch, in D.F. Bruley and H.I. Bicher (Eds.), Oxygen Transport to Tissue, Plenum Publishing Corp., New York, 1974, p. 1121.
88. R.B. Reeves, D.W. Rennie, and J.R. Pappenheimer, Fed. Proc. 16, 693 (1957).
89. R.V. Marcote, R. Chand, and T.H. Johnston, Proc. ISA Analys. Instr. Symp. Instrument Soc. of America, Pittsburgh, 1973, p. 31.
90. P. Hersch, Nature 169, 792 (1952).
91. P. Hersch, Anal. Chem. 32, 1030 (1960).
92. F.A. Keidel, Ind. Eng. Chem. 52, 490 (1960).
93. F. v. Sturm, personal communication.

References

94. P. Hersch, US Patent No. 3223597, Dec. 14, 1965.
95. P. Hersch, US Patent No. 3291705, Dec. 13, 1966.
96. P. Hersch, US Patent No. 3314863, April 18, 1967.
97. E.L. Eckfeldt and E.W. Shaffer, Anal. Chem. 36, 2008 (1964).
98. Battelle-Institut Frankfurt/Main, Deutsche Offenlegungsschrift No. 2327825, 12.12.1974.
99. E.F. Gibbons, A.H. Meitzler, L.R. Foote, P.I. Zacmanidis, and G.L. Beaudoin, SAE-Paper No. 750224 (1975).
100. E.M. Logothetis, K. Park, A.H. Meitzler, and K.R. Land, Applied Phys. Letters 26, 209 (1975).
101. M.D. Earle, Phys. Rev. 61, 56 (1942).
102. D.S. Tannhauser, Solid State Commun. 1, 223 (1963).
103. R.N. Blumenthal, J. Coburn, J. Baukus, and W.M. Hirthe, J. Phys. Chem. Solids 27, 643 (1966).
104. R.N. Blumenthal, J. Baukus, and W.M. Hirthe, J. Electrochem. Soc. 114, 172 (1967).
105. R.N. Blumenthal, J.C. Kirk, and W.M. Hirthe, J. Phys. Chem. Solids 28, 1077 (1967).
106. B. Fisher and D.S. Tannhauser, J. Chem. Phys. 44, 1663 (1965).
107. N.G. Error and J.B. Wagner, J. Phys. Chem. Solids 29, 1597 (1968).
108. J. Bransky and J.M. Wimmer, J. Phys. Chem. Solids 33, 801 (1972).
109. H. Meier, Z. phys. Chem. 212, 73 (1958).
110. R.J. Cherry and D. Chapman, Nature 215, 956 (1967).
111. B. Rosenberg, T. Misra, and R. Switzer, Nature 217, 423 (1968).
112. T. Misra, B. Rosenberg, and R. Switzer, J. Chem. Phys. 48, 2096 (1968).
113. T. Seiyama and S. Kagawa, Anal. Chem. 38, 1069 (1966).

114. M. Dimbat, P.E. Porter, and F.H. Stross, Anal. Chem. 34, 1502 (1962).
115. H.P. Siebert, Elektronik 21, 155 (1972).
116. R. Zechnall and G. Baumann, MTZ Motortechnische Zeitschrift 34, 7 (1973).
117. R. Zechnall, G. Baumann, and H. Eisele, SAE-Paper No. 730566 (1973).
118. Calculations of BASF for Robert Bosch GmbH (1973).
119. L. Steinke and H. Weyl, Deutsche Offenlegungsschrift No. 2315444, 28.3.1973.
120. W.J. Fleming, D.S. Howarth, and D.S. Eddy, SAE-Paper No. 730575 (1973).
121. J.N. Reddy, Bendix Tech. J. 6, No. 1, 1 (1973).
122. R.D. Carnahan, K.J. Youtsey, and D.H. Spielberg, Deutsche Offenlegungsschrift No. 2233299, 6.7.1972.
123. R.A. Spilski and W.D. Creps, SAE-Paper No. 750371 (1975).
124. G. Härtel, Proc. Sec. Symp. on Low Poll. Power Syst. Development, Düsseldorf (1974).
125. T.Y. Tien, H.L. Stadler, E.F. Gibbons, and P.J. Zacmanidis, Amer. Ceram. Soc. Bulletin 54, 280 (1975).
126. J.A. Rietdiyk, Sec. Autom. Emmission Conf., Ann Arbor (1973).
127. A.L. Cederquist, E.F. Gibbons, and A.H. Meitzler, SAE-Paper No. 760202 (1976).

Electrochemical Aspects of the Photographic Processes

W. JAENICKE

Institute of Physical Chemistry, University of Erlangen-Nürnberg, Erlangen, Germany

1	Introduction: General Survey of Photography with Silver Halides	93
2	Photographic Efficiency from the Aspect of Materials	95
3	Ionic Processes in Silver Halides	97
	3.1 Equilibria in Macro- and Microcrystals	97
	3.2 Ion Transport Phenomena	106
4	Electronic Properties of Photographically Active Substances	108
	4.1 Electronic Equilibria in Silver Halides and their Dependence on the Redox Potential	108
	4.2 Traps for Electrons and Holes; Chemical Sensitization	113
	4.3 Mobility and Lifetime of Electronic Charge Carriers	117
5	The Latent Image as a Sequence of Ionic and Electronic Rections	122
	5.1 Indirect Methods of Examination and Kinetic Approaches	122
	5.2 Mechanism of Latent Image Formation	124

Dedicated to Professor C. Wagner on the occasion of his 75th birthday.

	5.3 Stability and Size of Latent Image	128
6	Optical Sensitization	134
	6.1 Primary and Secondary Processes of Optical Sensitization	134
	6.2 The Question of Energy and Electron Transfer during Sensitization	136
	6.3 The System Dye, Substrate	141
	6.4 Mechanism of Electron Transfer	142
	6.5 The Determination of the Highest Occupied and the Lowest Vacant Levels of Sensitizers, E_{HO} and E_{LV}	146
	6.6 Experiments on the Mutual Position of the Energy Levels of Dye and Substrate and its Variability	150
	6.7 Supersensitization	156
7	Amplification and Bleaching of the Latent Image	160
	7.1 Thermodynamics of the Development Process; Principle of Chemical and Physical Development	160
	7.2 Developing Substances	161
	7.3 Electron and Hole Injection by Redox Systems	163
	7.4 Electrode Theory of Development; General Aspects	166
	7.5 The Conditions and the Mechanism of Filamentary Growth of Silver	172
	7.6 Kinetics of Photographic Development	175
	7.6.1 Physical Development	175
	7.6.2 Chemical Development	178
	7.7 Some Special Electrocatalytic Effects in Development	185
	7.7.1 Lith Development	185
	7.7.2 Superadditivity	186
	7.7.3 Catalytic Intensification in Color Development	188
	7.7.4 The Silver Dye Bleach Process	189
8	Electrochemical Aspects of Fixation	192
9	List of Symbols	193
10	References	197

1 INTRODUCTION: GENERAL SURVEY OF PHOTOGRAPHY WITH SILVER HALIDES

To date, the most common method of recording light signals is the conventional silver halide photography (1-3). Only by using silver halides can a primary, light-induced effect be amplified by consumption of chemical energy up to a factor of about 10^9.

Both the primary processes and the amplification involve the separation of charges and chemical reactions of charged particles. Therefore, they can be described in terms of electrochemistry. This has been shown since the publication of the classical paper of Gurney and Mott in 1938 (4), the theory of which has been improved and modified by an enormous number of papers, but in principle confirmed.

There are four steps to be discussed in the photographic process, all of which are of an electrochemical nature: the primary process, the optical sensitization, the development, and the fixation.

1. During the primary process in the light-sensitive crystal a "latent image" is formed by a sequence of electronic and ionic reactions. At first excitons are formed that dissociate into electron-hole pairs:

$$\text{lattice} \xrightarrow{h\nu} e' + h^\cdot \qquad (1)$$

Recombination is prevented by the trapping of both charge carriers.* The trapped electrons e'_t then react with mobile ions, for example, with interstitials $Ag_i^\cdot$:

$$e'_t + Ag_i^\cdot \longrightarrow Ag_i^x \qquad (2)$$

---

* For the discussion in the solid state in general the nomenclature of Kröger (235) has been used: V = vacancy; superscripts $\cdot$ × ' = net charge of the defect +1, 0, -1; subscript = lattice site, occupied by the defect; subscripts $_i$ and $_t$ = interstitial and trapped.

These atomic centers are enlarged and simultaneously stabilized by a repetition of Eqs. 1 and 2, until a speck of silver is formed. In Fig. 1a, a simple picture of this process is given with the aid of an energy-level diagram.

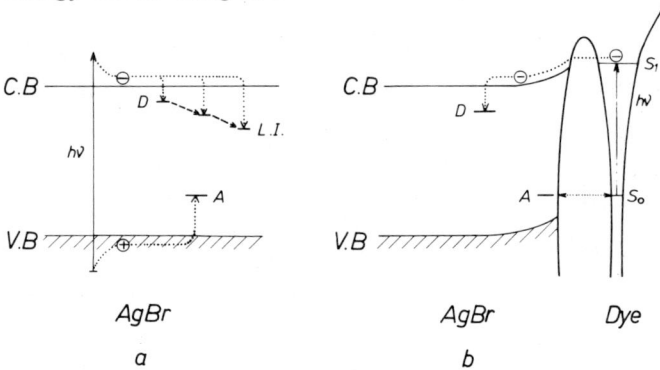

Fig. 1. (a) Scheme of latent image formation in silver halides. (b) Scheme of optical sensitization with electron transfer. C.B., conduction band; V.B., valence band; A, acceptor level; D, donator level; L.I., latent image; S_0, S_1, ground and excited (singulet) level of the dye. Dotted line, electronic step; broken line, ionic step.

There are several problems of electrochemical interest concerning the primary process:

a. The surface and the subsurface properties of the crystals.
b. The importance of chemical pretreatment of the crystals in enlarging the sensitivity.
c. The mechanism of latent image growth.
d. The role of photoholes.

2. By optical sensitization electronic carriers in the crystals are generated via the transfer of energy or electrons from an

excited dye, adsorbed on the crystals. An energy-level diagram of the electron-transfer mechanism is shown in Fig. 1b. Related phenomena of electrochemical interest are the mechanisms of desensitization and supersensitization.

3. During the photographic development electrons are transferred from a redox system to the latent image speck, where they react with silver ions coming either from the silver halide crystal ("chemical development") or from the solution ("physical development"). In this manner the speck is enlarged autocatalytically. Color photography is possible with developers whose oxidation products are able to form dyes with added coupling substances.

Although the primary processes, the light absorption by the crystal of the dye, have quantum efficiencies $\Phi = 0.1$ to 1, the efficiency of the total process depends mainly on the supply of developer or of silver ions; therefore $\Phi_{total} \gg 1$. For chemical development the size of the silver halide grains (10^7 to 10^{10} silver ions) is the determining factor.

4. In the fixation process developed or partially developed crystals are dissolved in complexing agents. This process too has some electrochemical aspects, especially if development and fixation are combined ("monobath" process).

2 PHOTOGRAPHIC EFFICIENCY FROM THE ASPECT OF MATERIALS

As can be seen from Section 1, several conditions must be met by materials suitable for photography: They should have no electronic dark conductivity; the recombination rate of electron-hole pairs, generated by light, must be small; with one of the carriers it must be possible to induce a subsequent reaction, suitable for the discrimination of exposed and unexposed material. Those substances can be used in two ways:

1. Formation of a metallic phase M_m of m atoms by light, the photoelectrons reacting in accordance with the overall reaction:

$$m \cdot (e' + M^\cdot) \longrightarrow M_m \qquad (3)$$

2. Destruction of a metallic phase by light, using photoholes, in accordance with the reaction

$$M_m + m \cdot h^\cdot \longrightarrow m \cdot M^\cdot \qquad (4)$$

The first process gives negative pictures and is possible if photoholes do not interfere. It can proceed by the Gurney-Mott mechanism, requiring consecutive ionic and electronic processes. As well as with silver halides, analogous processes were performed with Hg(II)- (5), and Pb(II)- halides (in absence of oxygen) (6), Pd(II)- and Fe(II)- oxalates (where the photoholes react with $C_2O_4^{2-}$, giving 2 CO_2) (7).

In the second process positive pictures are formed directly. It can be realized by metal layers, evaporated upon photosensitive crystals like As_2S_3, Pb(II)- or Tl(I)- halides (8). In this case no ionic processes are required, but the photoelectrons must be trapped to prevent recombination. By special pretreatment ("chemical sensitization") both mechanisms can be obtained at will in the same material. Untreated substances are mostly photographically insensitive, since the two photoprocesses interfere with each other (9).

If maximum sensitivity is desired, the first mentioned mechanism with silver halides (especially with mixed crystals of AgBr + several percent of AgI) is unsurpassed (10). For this reason the rest of this chapter deals mainly with the silver halides.

3 IONIC PROCESSES IN SILVER HALIDES

3.1 Equilibria in Macro- and Microcrystals

Although silver halides have a marked covalent character, they can be treated for most practical purposes as ionic compounds. They are characterized by an almost ordered anion lattice and a Frenkel disorder of the cations (Fig. 2).

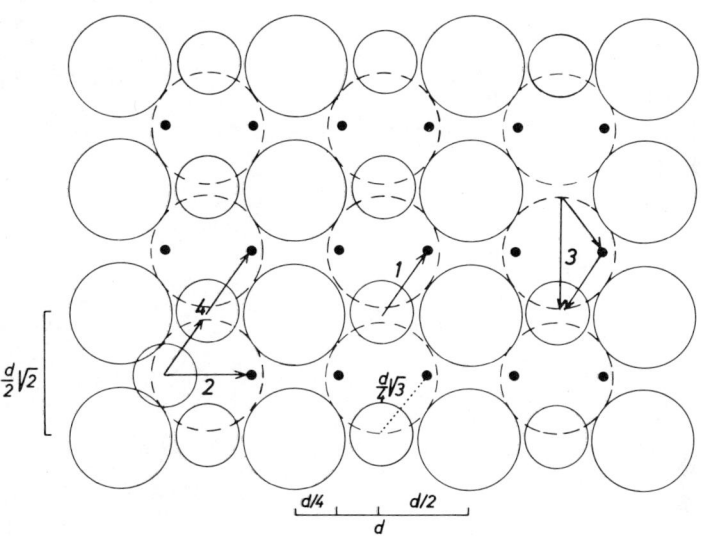

Fig. 2. View on a 110 plane of AgBr (11). Large circles, Br_{Br}^{x}; small circles, Ag_{Ag}^{x}; dots, interstitial states. 1, generation of a Frenkel defect; 2, interstitial diffusion; 3, vacancy diffusion; 4, interstitialcy diffusion.

In a finite crystal in a vacuum, interstitials $Ag_i^{\cdot}$ result from evaporation of silver ions from positive kink sites (K_+) into the bulk of the crystal, silver ion vacancies V_{Ag}' from deposition of lattice silver ions upon negative kink sites (K_-):

$$K_+ + I \rightleftharpoons Ag_i^{\cdot} + K_-, \quad \text{free enthalpy} = \Delta G_i \qquad (5)$$

$$K_- + L \rightleftharpoons V'_{Ag} + K_+, \quad \text{free enthalpy} = \Delta G_v \qquad (6)$$

where I is the interstitial position and L is the lattice position of silver.

From Eq. 5 or 6 one obtains (12) for K_+ and K_- the charges $+\frac{1}{2}$ and $-\frac{1}{2}$. This is obvious since at kink sites only one-half of the charge is compensated by excess charges in the surroundings. If s_j is the surface concentration, c_j is the bulk concentration, Z, Z_i ($Z_i = 2Z$), are the concentrations of lattice and interstitial positions and ϕ_s, ϕ_b are the surface and bulk potential, then the equilibrium condition of the interstitials is obtained from Eq. 5 by

$$\mu_i^o + RT \ln c_i + F\phi_b + \mu_-^o + RT \ln s_- - F\phi_s/2$$
$$= \mu_+^o + RT \ln s_+ + F\phi_s/2 + \mu_I^o + RT \ln Z_i \qquad (7)$$

$$c_i = 2Z \frac{s_+}{s_-} \cdot \exp \frac{-\Delta G_i^o}{RT} \cdot \exp \frac{F(\phi_s - \phi_b)}{RT} \qquad (8)$$

In the same way it follows from Eq. 6 that

$$c_v = Z \frac{s_-}{s_+} \cdot \exp \frac{-\Delta G_v^o}{RT} \cdot \exp \frac{-F(\phi_s - \phi_b)}{RT} \qquad (9)$$

where

$$\Delta G_i^o = \mu_i^o - \mu_I^o + \mu_-^o - \mu_+^o; \quad \Delta G_v^o = \mu_v^o - \mu_L^o - \mu_+^o - \mu_-^o \qquad (10)$$

From Eqs. 8 and 9 we obtain the Frenkel equilibrium:

$$c_i c_v = 2Z^2 \exp\left(\frac{-\Delta G_i^o + \Delta G_v^o}{RT}\right) = 2Z^2 \exp\left(\frac{-\Delta G_{11}}{RT}\right) = K_{11} \qquad (11)$$

independently of the potential.

For the potential difference we obtain from Eqs. 8 and 9

$$2 F (\phi_s - \phi_b) = \Delta G_i^o - \Delta G_v^o + RT \ln \frac{c_i}{2 c_v} \qquad (12)$$

Therefore, if $\Delta G_i^o \neq \Delta G_v^o$, a potential difference between surface and bulk is found (13,14), which results in a surface charge, that is compensated by a space charge. For doped crystals we have $c_i \neq c_v$; therefore, $\phi_s - \phi_b$ can be changed by doping the crystals. $\phi(x)$ and $c(x)$ are calculated by means of the Poisson equation.

Space charges and electrical fields within the crystal are assumed to be of great importance for the photographic process. For this reason they have been investigated in single crystals and evaporated layers in different ways: by surface conductivities (15), by measurements with a dynamic capacitor technique (16), and by the compensation of inner fields with external fields and measurement of the distribution of the latent image as a function of field strength (17,18). With photographic emulsions other methods must be used, for example, dielectric losses as a function of the frequency (19,20) or relaxation measurements to elucidate the conductivity distribution within the grains (21,22).

Some results with large and small crystals of AgCl and AgBr are compiled in Table 1. Of special interest is the fact that the free enthalpies of single defect formation depend on the crystal orientation (16,25,28). The deviations in the results are large; in AgCl even the sign of the surface potential is questionable. It has to be remembered that emulsion crystals are rather impure, and that they are embedded in gelatin. After drying, gelatin is able to exert large pressures on the crystal, and this can influence the disorder; gelatin can also form compounds with silver ions (30).

The thickness of the space-charge layer within the crystal, the Debye length δ, depends on the total concentration of charge carriers:

TABLE I

Bulk and Surface Effects, Resulting from the Cation Disorder in AgBr and AgCl Macro- and Microcrystals. (Surface conditions for different crystal planes). Free Enthalpy of the Frenkel Disorder (Eq. 11: $\Delta G_{11} = \Delta G_i + \Delta G_v$. Surface Potential $\phi_s - \phi_b$. Activation energies U_i, U_v of the Ionic Migration. Ionic Mobilities u_i, u_v (300K). Ionic conductivity κ_{ion} (300K) (i = interstitial, v = vacancy). (References are given in parentheses).

		Silver chloride		Silver bromide	
		Macrocrystals	Microcrystals (d = 0.4 μm)	Macrocrystals	Microcrystals (d = 1.0 μm)
ΔG_{11}/eV		$1.44 - 8.1 \cdot 10^{-4} T$ (24)		$1.06 - 7.6 \cdot 10^{-4} T$ (24)	
ΔG_v/eV	<111>	$0.8 - 1.7 \cdot 10^{-3} T$ (16)		$0.9 + 7 \cdot 10^{-5} T$ (16)	
	<110>	$-0.2 + 8.6 \cdot 10^{-4} T$ (16)		$0.3 + 8.6 \cdot 10^{-4} T$ (16)	
ΔG_i/eV	<111>	$0.6 + 9 \cdot 10^{-4} T$ (16)		$0.15 - 8.3 \cdot 10^{-4} T$ (16)	
	<100>	$1.6 - 1.6 \cdot 10^{-3} T$ (16)		$0.8 - 1.6 \cdot 10^{-3} T$ (16)	
ΔG_{11}/eV (300K)		1.21		0.83	
ΔG_v/eV (300K)	<111>	0.28 (16)	0.69 (25)	0.9 (16)	0.65 (28)
	<100>	0.06 (16)		0.56 (16)	
	–	0.28 (24)	0.79 (20)		0.79 (20,38)

ΔG_i/eV (300K)	<111>	0.93 (16)	0.52 (25)	0.46 (20)	-0.07 (16)	0.18 (28)	0.27 (20)	0.2 (38)
	<100>	1.1 (16)			0.27 (16)			
	-	0.93 (24)						
$(\phi_s - \phi_b)/V$ (300K)	<111>	0.16 (16)^a		-0.17 (20)	-0.52 (16)^a		-0.26 (20)	
	<100>	0.22 (16)			0.14 (16)^a			
	<200>				-0.16 (18)			
	-	0.32 (26)			-0.15 (26)			
U_v/eV		0.37 (27)			0.34 (23)			
U_i/eV		0.16 (23)			0.15 (23)			
U_v/cm^{2}V^{-1}sec^{-1} (300K)		$1.16 \cdot 10^{-6}$ (27)			$4.5 \cdot 10^{-6}$ (23)			
U_i/cm^{2}V^{-1}sec^{-1} (300K)		$4 \cdot 10^{-4}$ (27)			$5.3 \cdot 10^{-4}$ (23)			
κ_{ion}/Ω^{-1}cm^{-1} (300K)		$2 \cdot 10^{-9}$ (20)		$9 \cdot 10^{-9}$ (20)	$2 \cdot 10^{-8}$ (20)		$1 \cdot 10^{-6}$ (20)	

^{a} Calculated for pure crystals (for doped crystals see Eq. 12).

$$\delta^2 \approx \varepsilon_o \varepsilon \frac{RT}{F^2 \cdot \sum c_j z_j^2} \tag{13}$$

where ε_o is the permittivity of vacuum and ε is the dielectric constant.

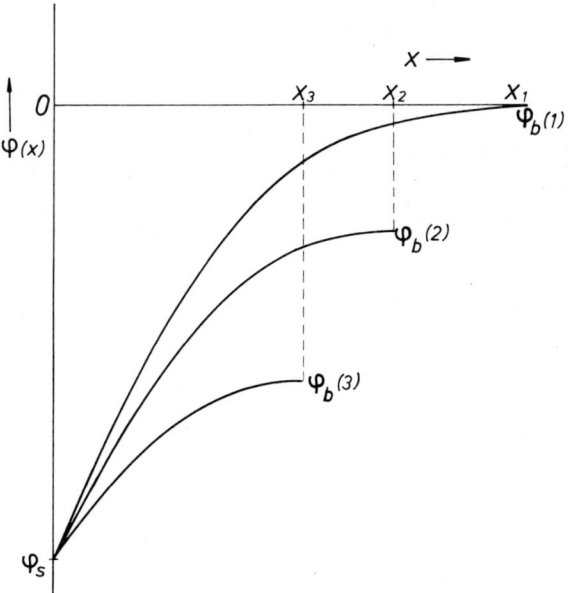

Fig. 3. Space-charge potential for different crystal thickness x_i (31). ϕ_s, surface potential; ϕ_b, potential in the center of the crystal.

If the diameter of emulsion crystals is comparable with δ, then $\phi_s - \phi_b$ decreases (Fig. 3) (12,31). In small crystals ϕ can approximately be assumed to be constant:

$$\phi_s - \phi_b \to 0$$

Since the total concentration of kink sites, s_o, is a function of temperature [e.g., at room temperature in AgBr, $s_o \approx 10^{12}$ cm^{-2}

(32)], exhaustion of kinks must be considered for small crystals. If their surface concentration s_j is taken as proportional to the volume concentration c_j, we obtain from Eq. 8 with $\phi_s = \phi_b$:

$$c_i \frac{c_-}{c_+} \approx 2Z \exp\left(\frac{-\Delta G_i^o}{RT}\right) = K_{14} \tag{14}$$

With the condition of electroneutrality

$$c_i + \frac{c_+}{2} = c_v + \frac{c_-}{2} \tag{15}$$

a relation between c_i and c_o ($c_o = c_- + c_+$) can be derived from Eqs. 11 and 14:

$$c_i - \frac{K_{11}}{c_i} - \frac{c_o}{2} + \frac{c_i c_o}{K_{14} + c_i} = 0 \tag{16}$$

In Fig. 4 values of c_-, c_+, c_i, and c_v as a function of c_o, calculated from Eq. 16 are given (12). The scale of the top axis denotes the size of grains with 1% of the surface sites being kinks. In the range of photographic interest, there is an average of less than one vacancy per crystal (dashed line), whereas the concentration of interstitials and negative kinks is inversely proportional to the grain diameter (19).

In addition to the point defects in the interior of the crystal, especially in mixed crystals Ag(Br,I), dislocations are always present; these enlarge the number of growth sites and allow the deposition of inner latent images. On the other hand, doping with divalent ions, which is of great influence on the disorder of large crystals, does not play a significant role in microcrystals.

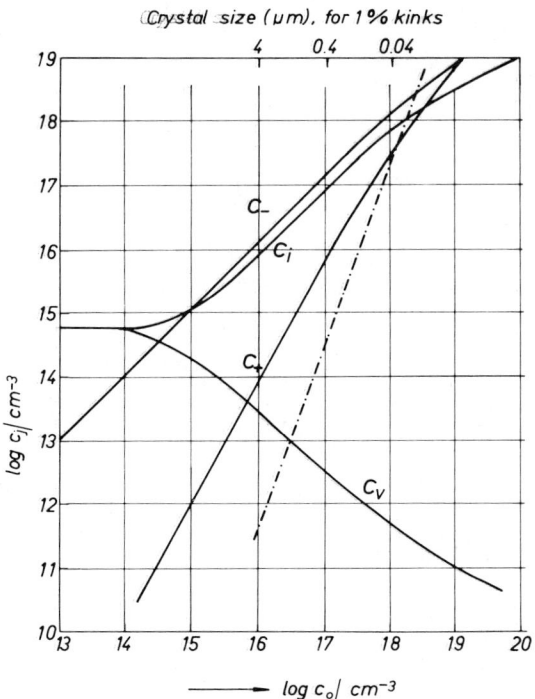

Fig. 4. Concentration of point defects and positive and negative kinks as a function of the total concentration c_o of kinks. Broken line, one vacancy per crystal for 1% kinks and different crystal sizes (top axis) (12).

So far our considerations only apply to the subsurface region of crystals in a vacuum. Photographic emulsions, however, are prepared in solutions of distinct p_{Ag} in the presence of gelatin and such additives as dyes and antifoggants. In contact with other phases, the free enthalpies of complex formation must be considered. Since the solvation energy of cations exceeds that of anions, in contact with pure solvent the crystal surface remains negatively charged; this is compensated by a positive space charge in the fluid phase. The equilibrium, however, depends on the p_{Ag} of the solution and is established by the adsorption of ions. Therefore, the crystal gets a net charge.

In combination with the effects mentioned above within the crystal, a negative surface charge between positive space charges both in the crystal and in the solution may be possible. It is not clear whether all equilibria, solution-surface and surface-bulk of the crystal, are established (31,33).

Experiments have been carried out with monocrystals in which p_{Ag} in the solution did not influence the electric field within the crystal (17,18). It seems that the number of adsorption sites that establishes the equilibrium crystal-solution is much smaller than the number of kink sites responsible for the charge distribution within the crystal. In this case the excess ions at the surface may be at distances from each other of the same magnitude as δ. A "discrete charge" double-layer structure, therefore, must be considered especially in the boundary crystal-solution (34), but possibly within the crystal too.

Other considerations, however, indicate a strong influence of the composition of the solution upon the crystal disorder. As a limiting case it was assumed that there is no special property of the surface (13,35). Then two space charges within phases of different dielectric constants are connected.

Both a.c. and d.c. experiments with silver halide membranes between solutions of variable p_{Ag} showed a minimum conductivity at a distinct p_{Ag} (isoelectric point) (36,37). The minimum was found to be dependent on doping (36,220). Also, from dielectric loss experiments it has been shown that the conductivity of emulsion grains increased with adsorption of positive dyes and decreased with negative dyes (38).

A critical review of the experiments, however, leads to the

conclusion that the observed conductivity changes are insufficient to change the sign of the surface potential.

3.2 Ion Transport Phenomena

Of special importance for the photographic primary processes are the rates v_j of exchange of places of point imperfections. They are connected with the diffusion coefficient by the equation

$$D_j = \frac{RT}{z_j F} u_j \approx \frac{1}{6} v_j d^2 \qquad (17)$$

where u_j is the electric mobility, z_j is the charge number, and d is the jump distance.

At room temperature the probability of the exchange of place of cation vacancies is about two orders of magnitude smaller than that of interstitials (see Table 1), while any movement of anions can be neglected (39).

Energetically favored is a transport mechanism by which an interstitial motion into a lattice position is coupled with a collinear motion of the lattice ion into an interstitial position (interstitialcy motion) (40); see Fig. 2, step 4.

To obtain an insight into the light-induced ionic processes within silver halides it is important to know the relaxation time of ionic processes. From Eqs. 5 and 6 the Frenkel equilibrium is

$$I + L \underset{k_{-18}}{\overset{k_{18}}{\rightleftarrows}} Ag_i^{\cdot} + V_{Ag}' \qquad (18)$$

It follows that (see Eqs. 5, 6 and 11)

$$-\frac{dc_i}{dt} \approx k_{-18} \cdot c_i c_v - k_{18} Z Z_i = k_{-18} (c_i c_v - \bar{c}_i \bar{c}_v) \qquad (19)$$

since the equilibrium concentrations $\bar{c}_i$ and $\bar{c}_v \ll Z_i$, Z.
With $x = c_i - \bar{c}_i = c_v - \bar{c}_v$ and $x \ll \bar{c}_i$ or $x \ll \bar{c}_v$ we have

$$-\frac{dx}{dt} \approx k_{-18} \, (\bar{c}_i + \bar{c}_v) \cdot x \qquad (20)$$

By integration the relaxation time τ is found:

$$\tau^{-1} = k_{-18} \, (\bar{c}_i + \bar{c}_v) \qquad (21)$$

The rate constant k_{-18} can be calculated, the reaction being diffusion controlled (41), as

$$k_{-18} = 4\pi (D_i + D_v) \cdot N_L \cdot d \cdot g \qquad (22)$$

where d approximates the jump distance, g is the factor that makes allowance for the Coulomb energy $U(r)$ between the defects. For the present case of attraction,

$$g \approx \frac{U(r)_{r=d}/kT}{\exp(U(r)_{r=d}/kT) - 1} = \frac{F^2}{4\pi \, \varepsilon \varepsilon_o N_L \, dRT} \qquad (23)$$

The sum of the diffusion coefficients is given by the ionic conductivity κ_{ion}:

$$\kappa_{ion} = F \left(\bar{c}_i u_i + \bar{c}_v u_v \right) ; \text{ for pure crystals: } \kappa_{ion} = F \cdot \bar{c}_i \left(u_i + u_v \right) \qquad (24)$$

From Eqs. 17, 21, 22, 23, and 24, the relaxation time is

$$\tau^{-1} = \frac{\kappa_{ion}}{\varepsilon_o \varepsilon} \left(1 + \frac{\bar{c}_v u_i + \bar{c}_i u_v}{\bar{c}_i u_i + \bar{c}_v u_v} \right); \text{ for pure crystals: } \tau^{-1} = \frac{2 \, \kappa_{ion}}{\varepsilon_o \varepsilon} \qquad (25)$$

At room temperature we have with the values of κ_{ion} in Table 1 for single crystals: $\tau(AgBr) \approx 3 \cdot 10^{-5}$ sec; $\tau(AgCl) \approx 2.5 \cdot 10^{-4}$ sec; for emulsion crystals: $\tau(AgBr) \approx 6 \cdot 10^{-7}$ sec. This is in agreement with measured values (10^{-6} sec) (21). By direct measurements at elevated temperatures with microwave pulses, however, larger values were obtained for AgCl: $\tau = \tau_o \exp(U/kT)$; $\tau_o = 9.33 \cdot 10^{-15}$ sec; $U = 0.83$ eV (42).

4 ELECTRONIC PROPERTIES OF PHOTOGRAPHICALLY ACTIVE SUBSTANCES

4.1 Electronic Equilibria in Silver Halides and their Dependence on the Redox Potential

Photographically active substances should not have electronic dark conductivity, but their band gap must be small enough to be excited by visible light. As is shown in Fig. 5, the silver halides occupy a middle position between covalent compounds such as sili-

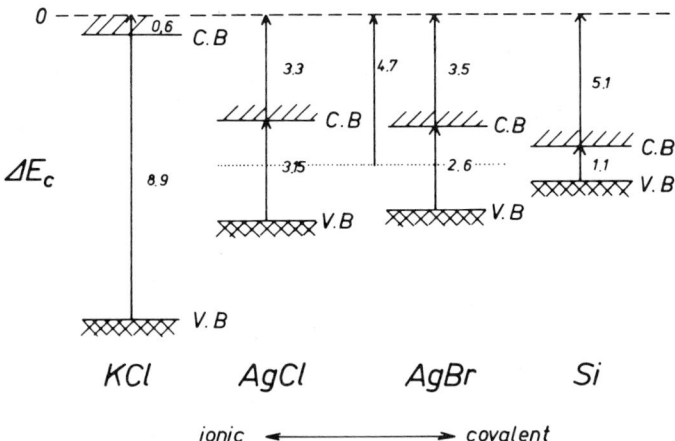

Fig. 5. Simplified band models of KCl, AgCl, AgBr, and Si. C.B., conduction band; V.B., valence band; ΔE_C, band gap; dotted line, Fermi energy of metallic silver (167); numbers, energy in eV.

con and ionic compounds such as the alkali halides. In contrast to the latter, photoexcitation of electrons in silver halides is always combined with photocurrents. The rather small band gap is caused by an almost exact coincidence of the energies of the d-shell of silver and the p-shell of the halide. In this way a large splitting of the filled states occurs and the additional energy for optical excitation into the 5 s band of silver becomes small (10). The dependence of band gap midpoint energy E_T (Eq. 57) and chemical potential of the electrons (Fermi energy E_F) is given by

$$E_F = -E_T + \frac{kT}{2} \ln \frac{n}{p} \qquad (26)$$

where n,p are the concentrations of electrons and holes. In pure, stoichiometric crystals n = p.

In oxidizing or reducing media, however, the compounds become nonstoichiometric. For AgBr as an example, the following equilibria are valid (44) (for the sake of simplicity the lattice positions are omitted in the equations):

$$Ag(s) + \frac{1}{2} Br_2(g) \rightleftharpoons AgBr(s) \qquad (27)$$

$$Ag(s) \rightleftharpoons Ag_i^x \rightleftharpoons Ag_i^{\cdot} + e' \qquad (28)$$

$$\frac{1}{2} Br(g) \rightleftharpoons V_{Ag}^x \rightleftharpoons V_{Ag}' + h^{\cdot} \qquad (29)$$

From Eqs. 27 to 29 and the Frenkel equilibrium (Eq. 4 and 5 or 11) the electronic equilibrium is obtained:

$$Br_{Br}^x \rightleftharpoons e' + h^{\cdot} \qquad (30)$$

Equations 28 and 29 show that traps are formed by the excess component in the surroundings that can dissociate and influence the equilibrium Eq. 30 (equilibrium constant = K_{30}).

According to Eq. 27, however, the bromine activity in Eq. 29 is related to that of silver in Eq. 28. At room temperature the equilibrium constant of Eq. 27 is given by

$$\log K_{27} = \log (P_{Br_2}/\text{atm})^{1/2} \cdot a_{Ag} = -16.8 \qquad (31)$$

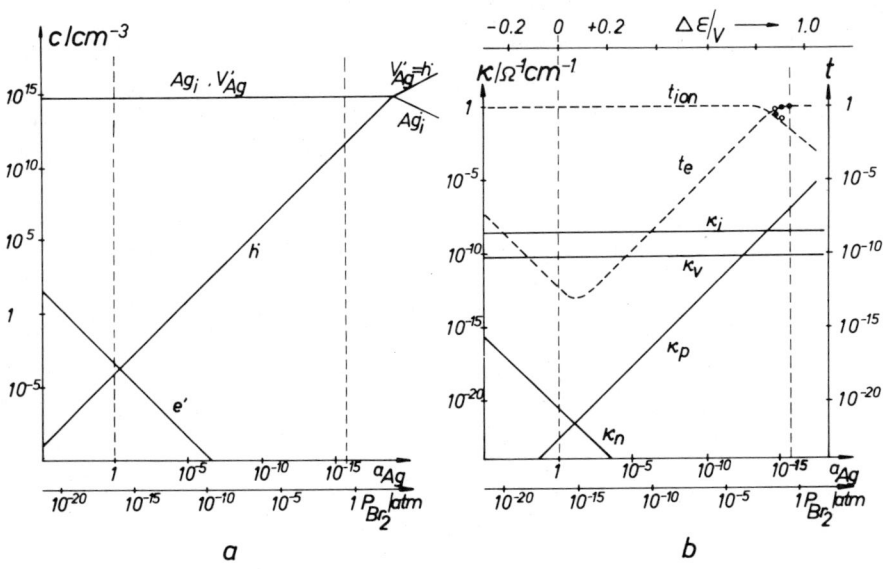

Fig. 6. (a) Concentrations of point defects, free electrons, and holes in AgBr as a function of silver and bromine activity at 25°C. (b) Conductivities κ_j of point defects, electrons and holes in AgBr as a function of silver and bromine activity at 25°C, calculated for macrocrystals (43). Dashed vertical lines, contact with metallic silver and with saturated bromine vapor. Top scale, redox potential against a silver metal electrode. t_{ion}, t_e, transference numbers of point defects and electronic carriers; solid circles, measured values of t_e; open circles, measured values of t_{ion}.

If the condition of electroneutrality in the bulk of the crystal is applied, all equilibrium concentrations can be calculated as functions of metal or anion activity (43,44), as done in Fig. 6. In contact with bromine (Eq. 29) the crystal is a p-type conductor ($p \gg n$); in contact with silver (Eq. 28) we have $n > p$, especially if $a_{Ag} > 1$, that is, in contact with small silver particles (latent images). This is of importance during development and amplification (see Chapter 7). The concentrations of the minority carriers can be obtained at elevated temperatures from conductivity measurements with the aid of ion-blocking electrodes (244).

The results given in Fig. 6 can also be expressed in terms of the Fermi energy. From Eq. 26 it follows with Eqs. 29 and 30 that

$$E_F = -E_T + \frac{kT}{2} \ln K_{30} - kT \ln p = \text{const} - kT \ln p_{X_2}^{1/2} \qquad (32)$$

(Fig. 7). The influence of doping can be seen (238) if p is expressed by the equilibrium constant K_{29} of Eq. 29

$$p = K_{29} \cdot \frac{p_{Br_2}^{1/2}}{c_v} \qquad (33)$$

and if c_v is expressed by K_{11} (Eq. 11) and the condition of electroneutrality (c_M is the concentration of divalent foreign metal)

$$c_M + c_i + p = c_v \qquad (34)$$

If p can be neglected in Eq. 34 we have for

$$c_M = 0 : p_o = \frac{K_{29}}{K_{11}^{1/2}} \cdot p_{Br_2}^{1/2} \qquad (35)$$

$$c_M \gg K_{11}^{1/2} : p = \frac{K_{29}}{c_M} \cdot p_{Br_2}^{1/2} \qquad (36)$$

Therefore if P_{Br_2} = const,

$$\frac{p_o}{p} = \frac{n}{n_o} = \frac{c_M}{K_{11}^{1/2}} \qquad (37)$$

In Fig. 7 the change of the Fermi energy after doping with Cd^{2+} is shown as a dashed line.

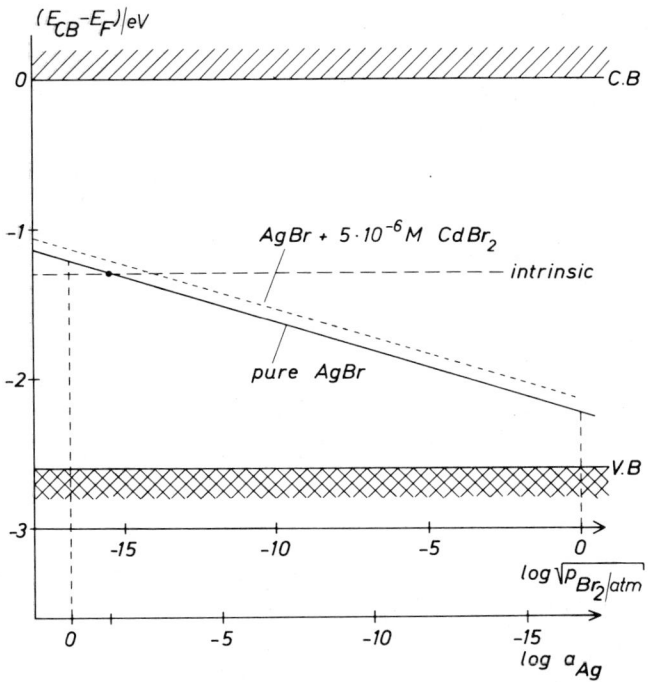

Fig. 7. Fermi energy change of pure and doped AgBr as a function of silver activity or bromine pressure at $25^{\circ}C$.

The influence of the surroundings upon the properties of photosensitive substances is often observed. For example, when oxygen is adsorbed, electron traps are formed at the surface and free holes in the interior (46). According to another explanation, oxygen can inject holes into deep traps at the surface (83) (see Fig. 11). The recombination rate of electron-hole pairs generated by illumination therefore is enhanced (see Section 4.2). Thus (especially with lead halides) the photographic response is greatly enhanced in a vacuum (6,46).

4.2 Traps for Electrons and Holes; Chemical Sensitization

As is seen in Eqs. 28 and 29 the electronic carriers are partially trapped. If the depth of the traps is ΔE, the lifetime τ_t of carriers in the traps is given approximately by (47,48)

$$\tau_t \text{ (sec)} \approx 10^{-12} \cdot \exp(\Delta E/kT) \qquad (38)$$

If at room temperature, $\Delta E = 0.1$ eV, $\tau_t \approx 0.1$ sec; if $\tau_t = 1$ yr (lifetime of the latent image), $\Delta E = 1.2$ eV.

There are many kinds of traps: electron and hole traps, uncharged and charged traps (with chemical and Coulomb interaction), mobile and immobile traps, and traps in the bulk or at the surface. Most work has been done with the silver halides. Experimental methods to characterize different kinds of traps are spectra at low temperatures, glow curves, amplification by light or development, and measurements of temperature dependence of electronic mobilities (see Section 4.3).

Trapping is influenced by boundary layers: If the surface is negatively charged, electrons are repelled and cannot be trapped close to the surface (49).

Since the decoration experiments of Hedges and Mitchell (50) it has been assumed that in the interior of the crystals dislocations act as electron traps. Jogs with a positive charge should be preferentially effective (48). An indication of this is that higher amounts of photolytic silver are deposited at dislocation lines if the crystals are bathed in $AgNO_3$ than if they are bathed in KBr. The number of dislocations increases with iodide concentration in the Ag(Br,I) crystals (239).

The effective excess charge at a single dislocation caused by the asymmetry of the lattice is much smaller than that of a point defect. Therefore, the energy of trapping is only of the order of 0.05 eV (51). On the basis of the Coulomb interaction the deposition of silver at such places is difficult to understand, especially if the large dielectric constant of silver halides (ε = 10 to 13) is considered. Perhaps foreign substances, enriched at such inner dislocations or at similar surface sites, are necessary (52). Quantum-mechanical calculations (99) give ΔE = 0.158 V for a silver atom in the vicinity of a dislocation.

Hole traps are assumed to be mostly point defects. In the bulk of pure silver halides self trapped holes (53) according to the reaction
$$Ag_{Ag}^{\times} + h^{\cdot} \rightarrow Ag_{Ag}^{\cdot} \tag{39}$$
are possible; V centers $[V'_{Ag} h^{\cdot}]$ are also possible (54), but they are rather instable (55) and present only in low concentrations in emulsion crystals. A temporary existence, however, can be assumed. This is essential since V centers are mobile hole traps, able to diffuse by coupled motion of $h^{\cdot}$ and V'_{Ag}.

Other important traps are impurities present also in purified crystals, such as $I_{Br}^{\times}$ in AgBr or $Br_{Cl}^{\times}$ in AgCl, which act as uncharged traps ($\Delta E \approx$ 0.3 eV) (48), or $Fe_{Ag}^{\cdot}$ ($\Delta E \approx$ 0.4 ± 0.1 eV) (43) and especially $Cu_{Ag}^{\times}$ ($\Delta E \approx$ 0.7 eV) (56,57,58), which probably can be temporarily stabilized by the reactions
$$Cu_{Ag}^{\times} + h^{\cdot} \rightarrow Cu_{Ag}^{\cdot} \tag{40}$$

$$Cu_{Ag}^{\cdot} + Ag_{Ag}^{\times} \rightarrow [Cu_{Ag} V_{Ag}]^{\times} + Ag_{i}^{\cdot} \tag{41}$$

The copper concentration in emulsion crystals is rather high (> 10^{17} cm^{-3}, that is, 100 atoms in a crystal of 0.1 µ m diameter).

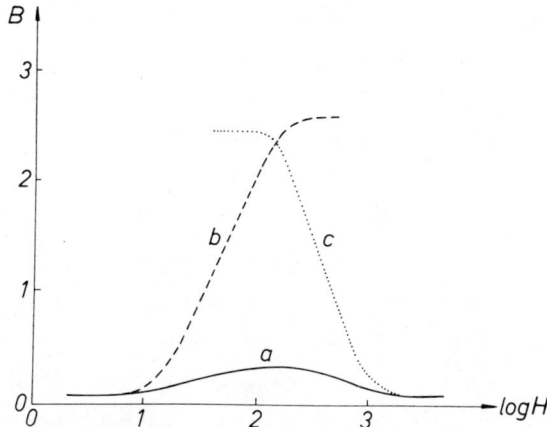

Fig. 8. Characteristic curves of layers of AgBr, evaporated at 600 to 700°C, crystal size 0.3 μm (243). B, optical density (B = -log T); H, exposure; (a) unsensitized; (b) chemically sensitized with Au-Ir (85); (c) fogged by evaporation with a developable layer of silver.

Crystals with only the traps mentioned above are insensitive to light, as is shown by the photoresponse of pure evaporated AgBr layers in Fig. 8, curve a (59). The most important traps must be artificially introduced into the crystals by "chemical sensitization". This is done either by slight reduction of the grain surface or by treatment with sulfur-containing substances in the presence of, for example, Au ions. In the first case very small particles of silver are formed on the surface (62,63,64), in the second case probably mixed (Au, Ag)-sulfides (49,60), complexes of atomic Ag or Au with single charged S ions (61), or Au specks (60).

During the illumination the silver centers act as hole traps (62) (or possibly as halogen acceptors (59)), as can be seen in Fig. 8, curve c: silver evaporated on AgBr layers is oxidized during exposure, so that positive pictures are obtained after development. The electrons must be trapped in the interior perhaps because the surface becomes negative in contact with silver. By gold-sulfur sensitization, however, deep electron traps are formed at imperfections of the crystal (63, 247) (the nucleation of silver can also be catalyzed by gold (59, 60)). This is shown by Fig. 8, curve b: Illumination of gold sensitized evaporated AgBr layers give negative pictures after development, since photolytic silver is formed at electron traps.

It is difficult to explain why, on illumination, silver particles can either be enlarged (forming latent images) or destroyed (acting as hole traps). Perhaps only silver at sites with negative excess charge is able to trap holes (64). In this case a quick diffusion of silver ions is necessary after trapping.

New quantum-mechanical calculations of silver aggregates on different silver halide structures, carried out with the extended Hückel method or Pople's CNDO-method (65,66), have shown under what conditions the electrons or holes can be localized most effectively. The results indicate preferred hole trapping at a parallelepiped (low ionization potential) (see Table 2).

TABLE II

Calculated Energies of Quantum-mechanical Models of Silver on AgBr (66)

Model	Binding energy U/eV	Ionization potential I/eV	Electron affinity A/eV
AgBr parallelepiped	6.70	9.54	3.14
Positive kink site	6.56	11.44	3.77

4.3 Mobility and Lifetime of Electronic Charge Carriers

The transport of electronic charge carriers generated by light is determined by their diffusion coefficients D and lifetimes τ. Corresponding to Eq. 17, D can be obtained from the mobilities u. The microscopic mobility u_H, measured by the Hall effect (67), is greater than the drift mobility u_d, which is influenced by temporary trapping. u_d is measured by the displacement x of photoelectrons or holes during synchronous short light and field pulses of duration τ and field strength $d\phi/dx$: $u_d = (x/\tau)/(d\phi/dx)$. During the pulses the slow ions do not move, the reducing electrons give rise to colloidal or developable silver depositions up to a distance x from the place of absorption, and the holes oxidize evaporated metal layers up to a distance x. A schematic representation is given in Fig. 9.

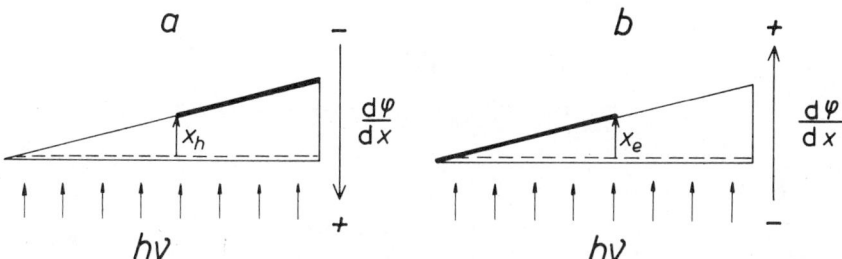

Fig. 9. Principle of the drift mobility measurement with light and field pulses in wedge-shaped crystals. (a) Hole mobility; (b) electron mobility. Thick line, silver after development. $d\phi/dx$, electric field; x, drift distance during the pulse duration τ. Dashed line, penetration depth of UV radation.

For electrons in large crystals, u_d and u_H are approximately equal since only shallow traps are present. In microcrystals, however, small drift mobilities are observed owing to multiple trapping on interstitials (69,70) (Table 3). Similar results are always obtained when high concentrations of deep traps are present, which are filled and emptied during the experiment, for example, if holes are injected into AgBr crystals in contact with bromine vapor, or if a boundary of holes moves through a silver halide crystal (57). This can be done if a galvanic cell with a silver halide diaphragm between a donor and an acceptor solution is polarized, for example, the cell

$$\text{Pt} \mid \text{Br}_2, \text{Br}^- \mid \text{AgBr (single crystal)} \mid \text{Br}^- \mid \text{Ag}$$

When a constant current is switched on, a front of holes moves through the crystal, which separates a zone of higher and one of lower conductivity. The migration rate dx/dt is proportional to the measurable change of potential difference $d\Delta\phi/dt$. With the crystal thickness b, the potential differences $\Delta\phi_o$ and $\Delta\phi_h$ for the states without holes (o) and filled with holes (h), we have (57)

$$u_t = \frac{b}{\Delta\phi_h} \cdot \left(\frac{dx}{dt}\right)_{t=o} = \frac{b^2}{\Delta\phi_h(\Delta\phi_o - \Delta\phi_h)} \cdot \left(\frac{d\Delta\phi}{dt}\right)_{t=o} \quad (42)$$

A low mobility u_t, caused by deep traps, was observed in accordance with the relation

$$\frac{u_t}{u_H} = \frac{p}{p_\bullet} \quad (43)$$

where $p_\bullet$ is the total concentration of holes, and p is the concentration of free holes. If there is an equilibrium of trapped holes with a total concentration of traps $N_\bullet$ ($N_\bullet \gg p_\bullet$) and a dissociation constant K_{29} (see Eq. 29), the ratio Eq. 43 can be expressed by

$$\frac{u_t}{u_H} \approx \frac{1}{(N_\bullet/K_{29})+1} \quad (44)$$

where $K_{29} = K_o \cdot \exp(-\Delta E/kT)$. ΔE is found to be 0.7 to 0.78 eV (57,71). At high temperatures the traps are dissociated ($K_{29} \gg N_\bullet$) and the Hall mobility is measured, as it is seen in Fig. 10.

TABLE III

Properties of Electronic Carriers in Silver Bromide Macro- and Microcrystals at Room Temperature (48,69,70)

	Electrons		Holes	
	Macrocrystals	Emulsion crystals	Macrocrystals	Emulsion crystals
Mobilities $u/cm^2V^{-1}sec^{-1}$				
Hall: u_H	66		1.7	
drift u_d	60	0.2	0.5 to 1.7	10^{-3}
moving boundary u_t			10^{-3} to 10^{-4}	
Diffusion coefficient D/cm^2sec^{-1}				
$D_H = u_H kT/e_o$	1.7	$(5 \cdot 10^{-3})$	$4.4 \cdot 10^{-2}$	$(2 \cdot 10^{-5})$
$D_t = u_t kT/e_o$			$3 \cdot 10^{-7}$ to $3 \cdot 10^{-6}$	
Mean free path w/nm	7		0.2	
Lifetime τ/sec	10^{-8} to 10^{-5}	$3 \cdot 10^{-6}$	10^{-8} to 10^{-5}	$1.5 \cdot 10^{-5}$
	10^{-7} (73)	1.2 to $2.5 \cdot 10^{-6a}$	$2 \cdot 10^{-6}$ (73)	
Diffusion distance $x/\mu m$ $x = \sqrt{D_H \tau}$ (73)	3.7	1	2.6	0.2
Quantum yield[b] (73)	1 ± 0.13		0.7 ± 0.2	

[a] Depending on p_{Ag} (236). [b] Of free charge carriers, measured at 254 to 436 nm.

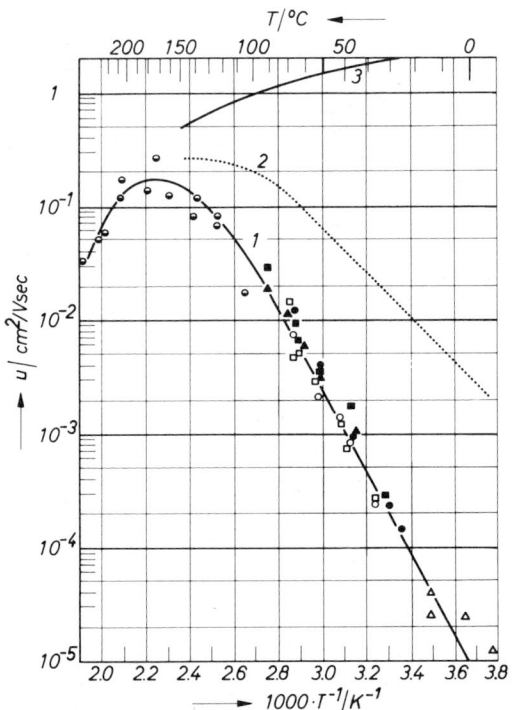

Fig. 10. Hole mobility in AgBr as a function of temperature (57). Curve 1, mobility u_t (Eq. 42), △○□ ▲●■ , measured with moving boundary at different crystals; ◐, measured with diffusion. Curve 2, calculated values from Eq. 45. Curve 3, Hall mobility (67).

These experiments prove the existence of deep hole traps in AgBr (probably Cu^+ or Fe^{2+}), which are necessary for the latent image.

Another slow transport mechanism is the coupled diffusion of uncharged complexes such as $[Ag_i^{\cdot} e']$ or $[V'_{Ag} h^{\cdot}]$. From the transport equations we have for this case (57,72)

$$u_t = u_H \cdot t_{ion} = u_H(1 - t_e) \qquad (45)$$

where t_{ion} is the total transference number of the disordered ionic species, and t_e is the transference number of the respective electronic charge carrier. The diffusion of silver through the phase boundary Ag/AgBr was found to obey Eq. 45 (72). The experiments with diffusion of bromine, however, cannot be explained by Eq. 45, as is shown from the dashed curve in Fig. 10 (which gives a trap energy of merely 0.43 eV) (57).

The lifetime τ of light-generated charge carriers can also be measured by combined pulses of electric field and of strongly absorbed light, for example, if square potential pulses are applied to the crystal through blocking electrodes, and the decay curves of photocharges or the photomoment M versus field strength $d\phi/dx$ are recorded (73) (M = $q\bar{x}$ where q is charges per unit area and $\bar{x}$ is average separation of electron-hole pairs). In these experiments slow decays with the field relaxation time of $Ag_i^{\bullet}$ ($\approx$100 µsec) are also observed; they result from the neutralization of trapped electronic charges by the association of ions (21,73). A microwave technique, combined with light pulses, was also used for emulsion crystals (236).

With τ (see Table 3), and the trap concentration N, we can calculate the cross section σ of the traps that are responsible for the lifetime: from the decay of carriers dn along the distance dx,

$$-dn = n \cdot (\sigma/S) \cdot SNdx \qquad (46)$$

where S is the area and the mean thermal velocity v = dx/dt, it follows that

$$n = n_o \cdot \exp(-\sigma Nvt) \qquad (47)$$

Therefore

$$\tau = \frac{1}{\sigma Nv} \qquad (48)$$

σ is strongly dependent on the material of the trap and the pretreatment of the crystal (74,243). If N is unknown, for charged point defects a mean value of $\sigma \approx 10^{-13}$ cm^2 and for uncharged defects $\sigma \approx 10^{-15}$ cm^2 can be used to estimate N from τ. For σ_n and σ_p see (243).

5 THE LATENT IMAGE AS A SEQUENCE OF IONIC AND ELECTRONIC REACTIONS

5.1 Indirect Methods of Examination and Kinetic Approaches

The formation of the latent image can be treated as nucleation and growth of a new phase, similar to a cathodic metal deposition. The sequence of (fast) electronic and (slow) ionic processes is illustrated by experiments on large emulsion crystals, which are treated with a combination of light pulses and electric fields. If light pulses are used shortly before ($\leq 10^{-5}$ sec) or during field pulses, the ions remain unaffected. The photoelectrons, however, are transported during their lifetime and the latent image is formed only at the anodic site of the crystal (21). If, however, a stationary field is applied, interstitials move in the direction of the cathode. During a short light pulse, therefore, latent images are preferentially found at the cathodic site of the crystal (75).

Until now, a great deal of knowledge about the latent image has been obtained by the examination of the reciprocity diagram, that is, the optical density (absorbance) B = log (1/T) after development as a function of intensity J and exposure H. In particular, the reciprocity law failure at different temperatures, or the effects of multiple exposures of different intensity, duration, or wavelength (intermittency and Herschel effect) can be explained by means of coupled electronic and ionic processes. Metastable, not developable preimage specks can be formed if at least two quanta are absorbed by a grain within 5 sec (76).

From a statistical theory of the characteristic curve B(log H), the minimum number of quanta required to make a crystal developable can be calculated: about four quanta are sufficient to initiate the development of one-fifth of the most sensitive crystals of a high-speed emulsion (77). To develop all crystals of a given size, a 10 to 100 times greater exposure is necessary (77).

There have been many attempts to treat the formation of the latent image as a sequence of chemical reactions, either by solving a system of differential equations with the aid of an analog computer (78) or by using a Monte Carlo method (79,83,247). In the latter approach the irradiation and therefore the electron and hole concentrations were assumed to be stationary and the reservoir of silver ions and traps to be large. The recombination and all growth processes of the nucleus were taken as second-order reactions, all decompositions as first-order reactions (Fig. 11). For simplification the trapping processes up to the formation of single silver and bromine atoms (framed in Fig. 11) were assumed to be in equilibrium. Therefore, definite constant probabilities can be chosen for the actual concentrations of n_f, n_t, p_f, p_t, Ag, and Br (see legend to Fig. 11).

For the nucleation and growth processes the number of the interstitials was taken to be large. Therefore, the ionic steps were taken to be of pseudo-first order. Trapping of an electron at an aggregate of m atoms leads, without delay, to an aggregate of m + 1 atoms, whereas the rate of electron trapping was assumed not to depend on m. During nucleation and growth, no back reactions were considered; therefore the size of the nucleus depends only on the random initial conditions of the grain and on the exposure time. With a Monte Carlo simulation, finally, a characteristic curve was obtained that was compared with experimental data.

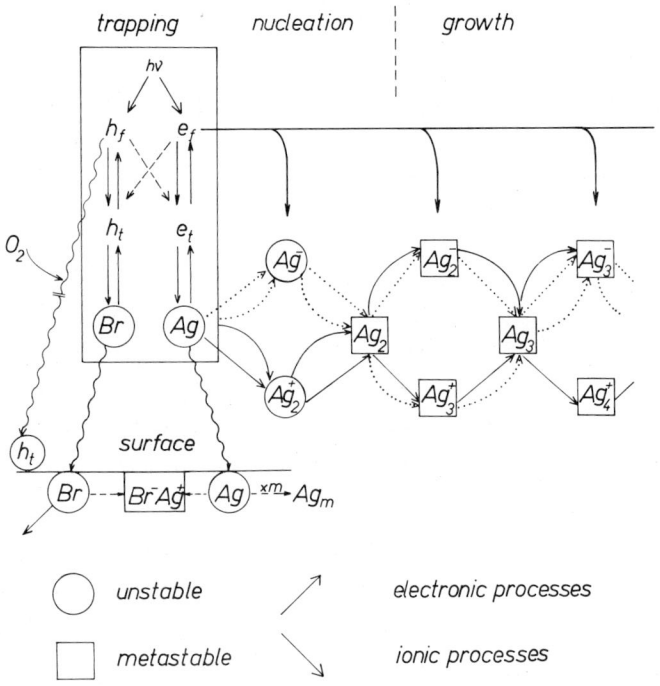

Fig. 11. Scheme of different views of latent image formation. Framed, stationary equilibria after Hamilton (79); horizontal, Gurney-Mott mechanism; dotted straight arrow, Gurney and Mott (4); solid straight arrow, Trautweiler (97); dotted curved arrow, Baetzold, plain surface (100); solid curved arrow, Baetzold, kink site (66). Left, vertical, Malinowski's mechanism (9). Index t, trapped; f, free. Dashed lines, recombination processes. Extreme left side, effect of oxygen after Hamilton (83).

5.2 Mechanism of Latent Image Formation

The kinetic concepts mentioned are adequate to describe a formalism of latent image formation, but they give little information as to the characteristic peculiarities of the process. For a better understanding, the results of the preceding sections with respect to the special ionic and electronic properties of the silver halides must be used.

Bromine is highly soluble, but silver is only slightly soluble in silver bromide (Fig. 6). There are deep hole traps of integral charge at point defects in the interior of the crystals, and shallow electron traps of a fractional part of a charge at dislocations or surface structural defects. Also, the reaction rate of charged traps with ions is high, because of the great mobility and the high disorder of ions.

In this way the photoholes are bound in deep traps, which can be totally discharged by association of a vacancy. Therefore, a recombination with photoelectrons, as well as an agglomeration of holes in the bulk of the crystal become improbable. The uncharged vacancy-hole complex can slowly diffuse through the crystal and reach the surface, where it is decomposed, giving rise to bromine.

The trapped electrons, however, cannot be discharged totally by diffusing interstitials. Therefore, silver atoms can be agglomerated by subsequent trapping of electrons and interstitials, but the Coulomb energy of such an agglomerate is low. It is further diminished by the rather high dielectric constant of the crystal. Therefore, the condensation of the photosilver in few specks according to the Gurney-Mott mechanism becomes questionable. One can argue that, as a result of the lower dielectric constant at the crystal surface, the cross section of electron traps (Eq. 48) is also increased (3). Then no permanent trapping occurs within the volume; the photoelectrons hop randomly from one shallow trap to another, until they reach the surface (8). But, as shown in Fig. 8, there is no surface latent image in chemically unsensitized crystals (17,26,59).

There are two possibilities of overcoming this difficulty: space-charge effects or growth of the nucleus by uncharged particles. It has been observed from measurements of the dielectric constant as a function of frequency (80) that the negative sign of the surface charge on silver halide crystals is inverted if they are covered with silver sulfide. This effect is already evident if the diameter of the covered area is less than 10 nm, which is such a small part of the total grains surface that it cannot be detected by a measurement of the surface potential. The sensitivity specks act then like small windows in the surface charge that allow the photoelectrons to pass, but repel the photoholes by Coulomb forces (81). In this way the latent image can grow according to the Gurney-Mott mechanism. At the same time the diffusion of the photoholes is favored by the surface charge of the sulfide-free part of the crystal surface. If this mechanism obtains, the transport of holes does not need hole-vacancy complexes. At the surface the holes can be destroyed by the centers that result from reduction sensitization (see Fig. 8).

Another possible mechanism that has been discussed is that trapped electrons, together with silver ions, initially form unstable silver atoms, which diffuse with multiple association until they condense at a sensitivity speck, which acts as a crystallization center (9) (see Fig. 11 on the left). In accordance with this view, the latent image does not grow by alternating electron and ion trapping, but by condensation of atoms. The latter mechanism is also suggested by experiments on development of evaporated silver nuclei (169).

The two discussed mechanisms are analogous to the Volmer-Horiuti and the Volmer-Tafel mechanism of hydrogen evolution. In both cases, however, single specks are only formed at moderate intensities of light. At high intensities a spontaneous condensation of small nuclei at many places within the crystal takes place and the sensitivity is reduced (79,82).

Impressive model experiments on the processes during latent image formation have been made with thin sheet crystals of AgBr (thickness up to 100 μm), first used by Mitchell (50,84), which after illumination with penetrating or with strongly absorbed light were etched in the form of a wedge and developed in order to make visible the distribution of photolytic or latent silver within the crystal (17,56,85). In the absence of chemical sensitization, no surface image was found. If, however, the crystals were immersed in a developer and illuminated with high-intensity light, some silver was reduced. Therefore, the poor photoresponse is partially due to a secondary reaction of the products silver and bromine at the crystal surface. If the crystal is exposed to penetrating radiation, visible photolytic silver is formed at both surfaces, whereas latent images are formed throughout the crystal, also laterally beyond the illuminated area (85). It follows that temporarily trapped holes migrate to the surface, probably in form of uncharged complexes $[h^\cdot V'_{Ag}]$. Therefore, the subsurface silver centers are not attacked and grow to photolytic particles. The diffusion of holes from the interior of the crystal, however, is so slow that they partially recombine with silver. Therefore, the surviving specks are only of latent image size.

If the crystal is doped with divalent cations (Cd^{2+}, Pb^{2+}), the concentration of interstitials is reduced. In this case no photolytic silver is formed in the illuminated area, but some latent image can be detected beyond this area. This can be explained by the different mobility of photoholes and photoelectrons: mobile electrons diffuse and form latent images at sites where photoholes are absent. The slow holes remain within the illuminated area and annihilate the remaining electrons. At the same time the illuminated surface on the crystals is etched. This is the result of a migration of $[h^{\cdot}V'_{Ag}]$ complexes to the surface where the bromine can escape, leaving square etch pits.

If the crystals are doped with Cu^+, photolytic silver is formed throughout the exposed crystal, since Cu^+ is a deep hole trap, which associates V'_{Ag} (see Eq. 41), whereas the liberated $Ag_i^{\cdot}$ reacts with the photoelectrons.

5.3 Stability and Size of Latent Image

Latent images formed in silver halides are possible in the interior and on the surface of the crystals, but in lead halides they are possible only on the surface (86).

Surface silver on pure silver halide is rather unstable; if evaporated on silver bromide it diffuses into the bulk of the crystal (9,87). The stability of the surface latent images can also be reduced by a recombination process of the photo products, Ag and Br, which is enhanced by the high surface mobility of the atoms (9). Surface images can be attacked in particular by oxygen. Therefore, if illuminated in a high vacuum, no fading of the latent image is observed even in chemically unsensitized layers (46, 88). The latent image is stabilized if it is bound to sensitivity specks like Au, Ag_2S, or Pb (9) or to gelatin (9,89).

Experiments on the stability of latent images generally are based on a comparison of the redox potentials ε_r of the latent image and ε_∞ of a silver electrode in the same solution:

$$RT \ln a \text{ (nucleus)} = -F(\varepsilon_r - \varepsilon_\infty) = -F\Delta\varepsilon_r \qquad (49)$$

$\Delta\varepsilon_r$ can be found if exposed layers are bathed in solutions of defined redox potential $\Delta\varepsilon$, measured against the potential ε_∞ of a silver electrode. The image is bleached if $\Delta\varepsilon - \Delta\varepsilon_r > 0$ and it is intensified if $\Delta\varepsilon - \Delta\varepsilon_r < 0$; this can be established by developing the bathed layers. Such experiments have often been repeated since the work of Reinders (90) but with contradictory results. In earlier papers (90,91) $\Delta\varepsilon_r = -50$ to -100 mV was found. Later (92,93) it was shown that the method of Reinders was subject to errors, because of complex formation of silver with the redox system used; furthermore, undersaturation of the solutions with regard to silver bromide was overlooked. If these errors are avoided, and if the measured reduction rate as a function of the redox potential is extrapolated to zero, the equilibrium potential of a metallic silver electrode is attained, that is, $\Delta\varepsilon_r = 0$.

New experiments, however, have shown another source of error (94). The gelatin, present in all experiments, is a redox buffer that lowers the redox potential at the grain for some time up to -0.3 V. During this time, image specks, which would be unstable in gelatin-free solutions, can reach a stable size. The effect therefore largely depends on the method of gelatin hardening. These experiments have shown that the former results were of the correct order of magnitude since the two errors cancelled each other out.

The critical redox potential depends on the photographic material and on the exposure (Fig. 12). It increases with intensity (Fig. 12, curves a and b), since smaller image specks are formed at higher intensities. Specks of gold-sensitized emulsions are more stable than metallic silver (Fig. 12, curve 4) (247). The growth of such specks by deposition of silver is equivalent to an underpotential deposition of Ag on Au, which was confirmed by model experiments (242).

With the aid of the experiments mentioned, the free energy of the latent image can be obtained; this determines the developability of the exposed grain. For the mechanism of latent image

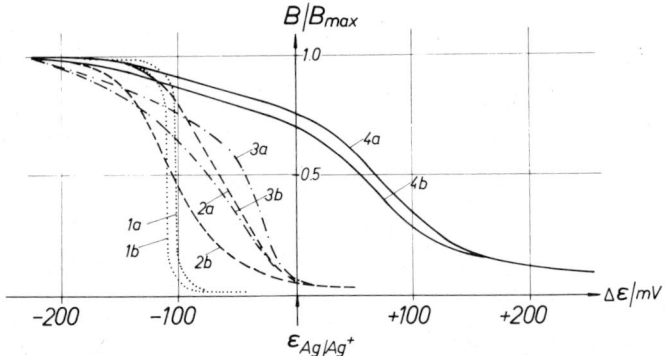

Fig. 12. Stability of development centers as a function of their redox potential ε_r (94). Abzissa $\Delta\varepsilon_r$ (Eq. 49) $\varepsilon_\infty = \varepsilon_{Ag/Ag^+}$. Ordinate, relative optical density B after development. 1 – 4, different emulsions of which 4 = Au-sensitized. (a) Low intensity exposure, (b) high intensity exposure.

formation, however, the number of silver atoms in the speck is of interest. To calculate the size of the latent image from its activity (Eq. 49), Thomson's equation has usually been employed: For a spherical nucleus of the radius r and a surface tension γ in a homogeneous medium one has

$$RT \ln a = \frac{2\gamma V_M}{r} \qquad (50)$$

where V_M is the molar volume of silver. The validity of this equation, however, is questionable since the most active latent images are on the surface of the crystals (see Section 7.3), forming a triple phase system in which strongly adsorbed two-dimensional nuclei are expected on the halide surface, embedded in gelatin or solution.

For two-dimensional nucleation (29) instead of γ the boundary free energies ρ_i must be used with the equilibrium condition

$\sum \rho_i dl_i \leq 0$ (l_i is the length of the edges). A simple estimate is obtained by a circular two-dimensional nucleus of radius r. In this case Eq. 50 must be replaced by

$$RT \ln a = \frac{\rho S_M}{r} \qquad (51)$$

where S_M is the molar area ($S_M \approx N_L d^2/2$; d is the lattice constant). Since ρ can be expressed by

$$\rho = \gamma \cdot d \qquad (52)$$

the result is equal to Eq. 50, but now valid for a circle of radius r. The meaning of γ or ρ may be questionable in systems of only a few atoms, but in recent experiments (95,256) the validity of Eq. 49 has been demonstrated for evaporated silver layers of grain size of 1 to 100 nm. Theoretically, it was shown, and by evaluation of the grain distribution under the electron microscope it was also proved, that during treatment with redox systems of the potential $\Delta \varepsilon'$ containing silver ions, the size class temporarily disappears for which $-F\Delta\varepsilon' = 2\gamma V_M/r'$. In this way a linear relation between $-\Delta\varepsilon'$ and $1/r'$ was found, from which $\gamma = 9.2 \cdot 10^{-5}$ Jcm^{-2}, and with gelatin adsorbed (256) $\gamma' = 4.0 \cdot 10^{-5}$ Jcm^{-2} was obtained. If the latter value is used, one obtains from the experimental $\Delta\varepsilon = -0.1$ to -0.2 V (see Fig. 12) the radius of the latent image speck: 0.4 nm $< r < 0.8$ nm. From this the number m of silver atoms can be calculated. One has from Eq. 50 for spherical nuclei that $20 < m < 120$, and from Eq. 51 for two-dimensional nuclei that $8 < m < 30$. This is only a rough estimate, but it may be sufficient to demonstrate that reasonable sizes of latent images, which are in agreement with the nonthermodynamic results (see below) are obtained especially if two-dimensional nuclei are assumed.

Until now the only attempt directly to measure the size distribution of latent images in terms of the number of silver atoms was made by Matejec and Moisar (96). They shifted a latent image into the interior of the grain by crystallization of new silver halide upon the exposed grains. Then they bleached the image by potential-controlled injection of holes into the crystal (see Section 7.2). After varied bleaching times the grains were dissolved and the remaining specks physically developed. From the decrease of the number of specks as a function of bleaching time the size distribution of the specks was obtained (Fig. 13). From the curves it is confirmed that latent images consist of about

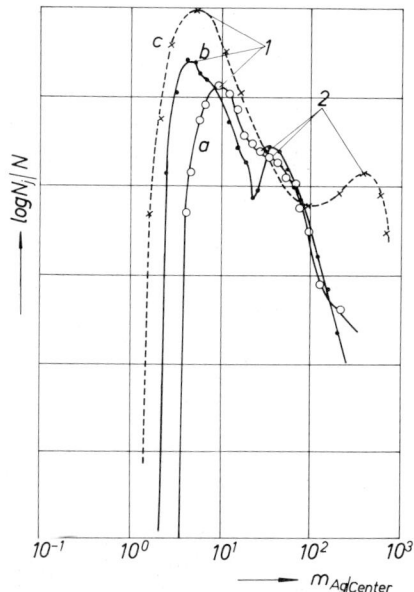

Fig. 13. Size distribution of latent images, calculated from Eq. 87 with κ_e-values of Fig. 6b (96). Abscissa, number of silver atoms per nucleus; ordinate, relative number of nuclei; (a) high-intensity exposure (flash); (b) low intensity, short time (Hg lamp, 30 sec); (c) low intensity, long time (Hg lamp, 300 sec); 1, surface latent image; 2, inner latent image.

4 to 40 silver atoms. Similar differences of the distributions for high- and low-intensity exposures were observed as seen in the redox potential measurements (Fig. 12).

The minimum size of specks for physical development was investigated by a controlled evaporation method (169). With a statistical approach the size distribution of evaporated silver and gold nuclei was obtained, which were enlarged and made visible in the electron microscope either by evaporation of Zn or by physical development with a diffusion transfer method (see Section 7.1). A minimum size for developability of four atoms for Ag and only two atoms for Au was found.

Since there is no direct proof of the sequence in which ions and electrons are trapped during the latent image formation, an attempt was made by quantum-mechanical methods to calculate the energy of silver aggregates as a function of size.

The first calculation with a simple particle-in-a-box model (97) is only applicable to metal clusters in vacuo. After a number of improvements (98), rather involved calculations (65,66,100) with semiempirical methods were used to calculate the subsequence of energies of silver clusters upon a rigid AgBr lattice of up to 17 ion pairs. Because of the delocalization of the valence electron, single Ag atoms on a plain surface are found to be positively charged. They were preferentially situated at an interstitial position. Ag clusters cannot trap an electron unless a silver ion is added first. The energetically favored sequence is

$$Ag \xrightarrow{e'} Ag^- \xrightarrow{Ag^+} Ag_2 \xrightarrow{Ag^+} Ag_3^+ \xrightarrow{e'} Ag_3 \xrightarrow{e'} Ag_3^- \xrightarrow{Ag^+} Ag_4 \xrightarrow{Ag^+} Ag_5^+ \quad \text{etc.} \quad (53a)$$

If a model of a kink site of AgBr is used as a base for the silver, the electron affinity of all even-sized aggregates is changed, so that they can act as electron traps. In this way the thermodynamic pathways for growth of silver aggregates at a surface defect are

$$\text{Ag} \xrightarrow{\text{Ag}_{1/4}^+} \text{Ag}_{2/4}^+ \xrightarrow{e'} \text{Ag}_{2/4} \xrightarrow{e'} \text{Ag}_{2/4}^- \xrightarrow{\text{Ag}_{1/4}^+} \text{Ag}_{3/4} \xrightarrow{e'} \text{Ag}_{3/4}^- \xrightarrow{\text{Ag}_{1/4}^+} \text{Ag}_{4/4} \xrightarrow{e'} \text{Ag}_{4/4}^- \text{ etc.} \quad (53b)$$

Both alternatives, together with Trautweiler's assumption (97) are shown in Fig. 11.

6 OPTICAL SENSITIZATION

6.1 Primary and Secondary Processes of Optical Sensitization (101)

If silver halides are irradiated with light energies greater than the band gap ΔE_C, free electrons in the conduction band and free holes in the valence band are generated. If, however, suitable dyes having excitation energies (absorption maxima) $\Delta E < \Delta E_C$ are adsorbed and excited, either free electrons in the conduction band of the crystal and trapped holes or free holes in the valence band of the crystal and trapped electrons are found (102). The first process mentioned is the primary process of optical sensitization, the latter that of optical desensitization.

The primary processes are followed by secondary reactions (103). Free electrons can generate and free holes can bleach latent images. In the system crystal-dye, however, other electronic processes also are possible, especially if the substrate is also excited. Many sensitizing dyes can trap free electrons and holes and thus act as recombination centers. This irreversible desensitizing ability is, in particular, observed at high concentrations of dyes. If only electrons are trapped (e.g., in triplet states) they can be transferred to adsorbed oxygen. This kind of desensitization can be removed by evacuation. By trapping photoholes on the other hand, the lifetime of photoelectrons can be enlarged and a sensitizing effect results. These disturbing secondary effects always interfere if optical sensitization is observed with photographic methods such as development.

The primary processes were observed exclusively by measuring photocurrents in silver halide membranes (104) and gelatin-free layers (105), or photomoments (73) (see Section 4.3) in sensitized single crystals (102, 106) and photographic emulsions (107). The measurement of photocurrent was made in a galvanic cell with a polarized AgBr membrane between two electrolytes, one of which contained a sensitizing dye. A lock-in technique with light modulation was necessary to discriminate the small electronic current of about 10^{-9} A/cm^2 from the large ionic portion. A quantum yield of only 10^{-3} was observed since the greatest part of the photocarriers was trapped or compensated by reverse ionic currents during the period of illumination. Photocurrents were only observed if the dye-containing electrolyte was negatively polarized against the crystal, showing generation of mobile photoelectrons during the sensitization act. The photocurrent spectrum in the visible range was virtually identical to the absorption spectrum of the dye.

The photomoments (73) in photographic layers were measured in a four-cell capacity bridge. Sensitized photographic layers were used as dielectric in the cells; one of them was illuminated with light flashes coincident with voltage pulses. If illuminated in the range between 310 and 390 nm, where only the silver halide was sensitive, the photoresponse was independent of the adsorbed dye (Fig. 14, curve a), while with photographic measurements a strong desensitization was found with some of the dyes. If illuminated in the absorption region of various dyes (> 520 nm) some photoconductivity of the silver halide was still found with desensitizing dyes (Fig. 14, curve b, to the left of dye number 8).

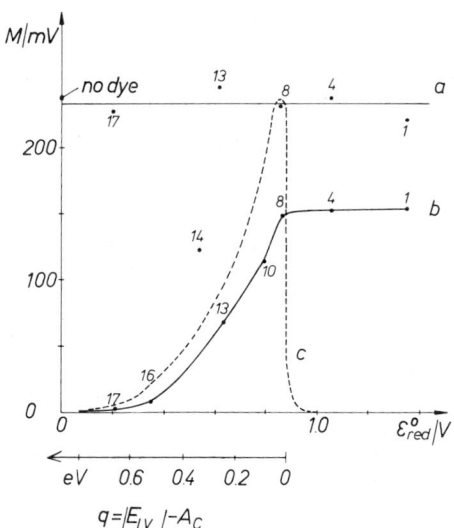

Fig. 14. Curve a, b, photomoment M of electrons during combined light and field pulses in sensitized AgCl-emulsions (p_{Ag} = 8) as a function of the reduction potential of sensitizers No. 1-17 (upper scale of the abscissa) (107). (a) illumination range 310 - 390 nm. (b) illumination > 520 nm. Curve c, densitiy of states in the excited dye (arbitrary scale) as a function of energy q (Eq. 54) (lower scale of abscissa), estimated from curve b.

6.2 The Question of Energy and Electron Transfer during Sensitization

To explain the primary process of optical sensitization, energy and electron transfer can be assumed between dye and substrate.

Energy transfer is a phenomenon of resonance; therefore, an overlap of states within the dye and the crystal is necessary. For distances r greater than several nanometers, a dipole interaction of the centers as described by Förster (108) must be discussed. This interaction is proportional to r^{-6} or, for plane layers, proportional to x^{-4}. For direct contact of the centers, however, (overlap of the electron clouds) an effect (109) is valid which decreases exponentially with r.

Since the excitation energy $\Delta E = h\nu_{max}$ of a sensitizing dye is smaller than the bandgap $\Delta E_C = I_C - A_C$ (I is the ionization energy and A is the electron affinity), the resonance states in the crystal must be located in the tail of the absorption spectrum. This area with small extinction coefficients is strongly influenced by doping or chemical sensitization (see Section 4.2), but it was found that optical sensitization is independent of these treatments (110). If there was any influence of an Au-S treatment, it was due primarily to decreased desensitization rather than spectral efficiency changes (111).

Although impressive experiments with insulating layers between sensitizing dye and crystal demonstrate that energy transfer in accordance with Förster's theory is possible (110,112), and that centers of resonance are probably located about 3 nm below the crystal surface, the theory can scarcely predict any of the observed sensitization effects.

The electrochemical theory of optical sensitization in its simplest form postulates a direct electron transfer from the excited dye to the crystal. In this theory it is not a question of whether two frequencies are equal, but whether two energy levels correspond: Sensitization is observed if the conduction band receives an electron from the excited dye molecule; desensitization is observed if the valence band receives a hole. It is, however, also possible that electrons are injected directly into the surface states without reaching the conduction band (113). In this case silver atoms in the surface can be formed as a secondary reaction and either are able to dissociate electrons into the conduction band (a reaction that cannot be distinguished from thermal assisted transfer), or may condense, in accordance with Malinowski's (9) assumption (see Fig. 11), to form latent images.

According to the electron transfer mechanism the dye is primarily oxidized during sensitization, and primarily reduced during desensitization.

Whether the dye is finally regenerated is an old question. In some early experiments (114) it was found that the number of silver atoms formed by prolonged sensitized photolysis exceeded the number of adsorbed dye molecules. In the case of energy transfer it is obvious that the dye returns to its initial state, but if electrons are transferred a slow regeneration of the dye is only possible via empty acceptor levels within the band gap (e.g., the levels that are located about 0.7 eV above the valence band; see Section 4.2). A regeneration by preimage centers or sensitivity specks would result in partial desensitization. On the other hand, regeneration can be caused by reactions with substances in the gelatin matrix, which is also in contact with the dye.

It is remarkable that in the electrochemical experiments of sensitization (104,115) no regeneration was found, but destruction of the dye with a simultaneous decrease of the photocurrent. Regeneration occurred, however, if reducing agents, such as hydroquinone, sulfite, or $Fe(CN)_6^{4-}$ were added (see Section 6.7).

In practical photography no regeneration is necessary, because the number of adsorbed quanta is small compared with the number of adsorbed dye molecules.

There are many results that favor the electron transfer mechanism, but to date no experimentum crucis is known. A convincing argument would be the existence of radical intermediates

during the sensitization process. So far this has not been proved
in photographic layers, but it has been shown in the mentioned
electrochemical model experiments (104,115).

A rather convincing experiment to prove the importance of
energy levels rather than energy differences was made with four
fluorescing dyes of different ionization energies I but similar
excitation energies ΔE (116) (Fig. 15a). During irradiation into
the maximum of absorption the fluorescence is maintained if the
dyes are adsorbed on alkali halides (with no electron transfer
possible; see Fig. 5). If adsorbed on AgBr or AgCl, only dye IV
shows fluorescence, while the fluorescence of dyes I-III is
quenched. Only with dyes I and II, however, photolytic silver was

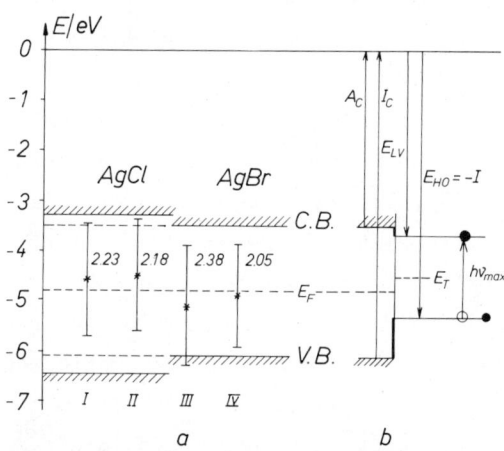

Fig. 15. (a) Energetic position of four fluorescing sensitizers
(I-IV) compared with the valence and the conduction band of AgCl
and AgBr, V.B. and C.B. (116). E_F, bandgap midpoint ($\approx$ Fermi
energy); numbers, excitation energies; $\Delta E = h\nu_{max}$ (eV) of the
dyes. * = E_T (Eq. 57); I, erythrosine; II, thiocarbocyanine; III,
phenosafranine; IV, pinacryptol green. (b) Position of the energy
gap midpoint E_T after Tani (128) of a sensitizing dye compared
with the Fermi energy E_F, including the respective energy values.

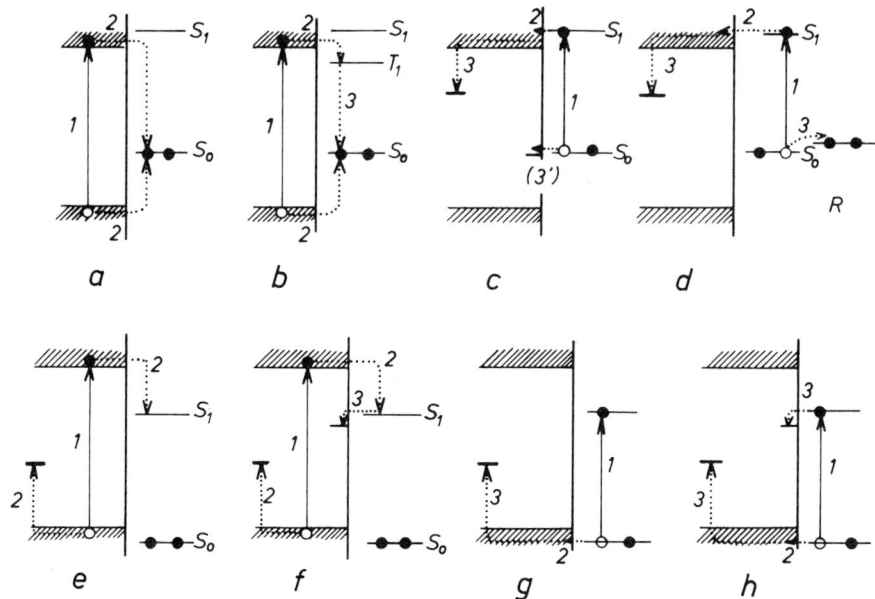

Fig. 16. Some possibilities of electron transfer processes between valence and conduction bands of the substrate (hatched) and adsorbed dyes (S_0, ground state; S_1, excited singulet state) in the band model (schematic, $E_{LV} \approx -A$).
1,2,3, sequence of reactions; solid circle, electron; open circle, hole. a-d, sensitizing dye. a,b, irradiation in the absorption range of the substrate. a, dye as a recombination center; b, like a, but via triplet state T_1; c,d, irradiation in the absorption range of the dye; c, electron transfer and latent image formation, regeneration possibly via surface states; d, like c, with a dye aggregate. Regeneration via supersensitizer R. e-h, desensitizing dye; e,f, irradiation in the absorption range of the substrate; e, dye as an electron trap, bleaching of inner latent image centers; f, like e, but formation of surface centers (Capri blue effect); g,h, irradiation in the absorption range of the dye; g, hole transfer, bleaching of inner latent image center; h, hole transfer like g, surface center formation like f.

found, because their excitation levels are near or above the conduction bands of the crystals (electron injection). In contrast, the ground level of dye III is below the valence band of AgBr. Therefore, hole injection takes place (Fig. 15a). It is remarkable that desensitization (quenching of fluorescence) was also found with AgCl, although the band gap of AgCl is so large that the ground state of dye III becomes higher than the valence band (see Fig. 15a). This is an indication of an energy distribution in the states of the adsorbed dye (see Section 6.4).

If the electron transfer mechanism is accepted, many observed effects can be illustrated with a simple band model, as is shown in Fig. 16 (see legend).

6.3 The System Dye, Substrate

In the ground state, sensitizing substances must neither be able to uptake electrons nor to transfer electrons into trapped holes because both properties would cause desensitization. In particular, the cyanines and related dyes are suitable substances. The general formula of the cyanines is given in Fig. 17. The flat molecules of those dyes are able to form characteristic aggregations

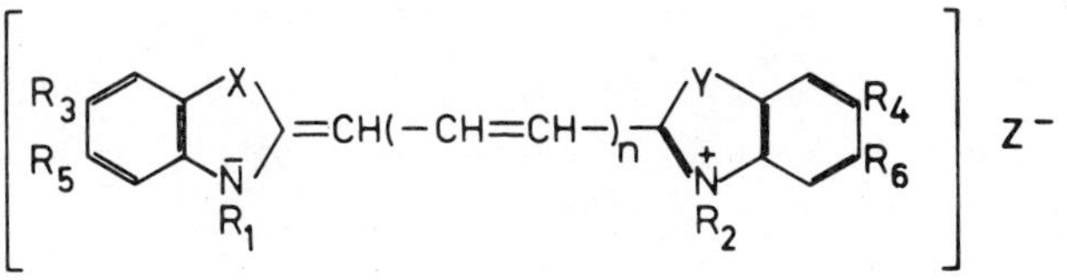

$$\left[\begin{array}{c} R_3 \\ R_5 \end{array} \!\!\!\! \bigcirc\!\!\!\!\!\!{}^X_{\underset{R_1}{\bar{N}}}\!\!\!\! \rangle =CH(-CH=CH-)_n \!\!\!\! {}^Y\!\!\!\! \bigcirc \!\!\!\! {}^{R_4}_{\underset{R_2}{\overset{+}{N}}R_6} \right] Z^-$$

n = 0·····5 ; $R_{1,2} = C_j H_{2j+1}$; $R_{3,4,5,6} = C_j H_{2j+1}$ ($j \geq 0$); Cl, ϕ;

or $R_3 = R_5$ and $R_4 = R_6 = -CH=CH-$

X,Y = $-CH=CH-$; O; S; Se; $N-CH_3$; $C(CH_3)_2$

Fig. 17. General formula of cyanine dyes.

(117) (H and J aggregates) with narrow absorption maxima and high mobility of excitons (see Section 6.7).

Sensitizing dyes are strongly adsorbed at the crystal surface. This is considered to be of great importance for the electron transfer process. The adsorption is dependent on epitaxy (118) and on the interaction energy of special groups like -S- with the substrate (119). In some cases the structure of dye aggregations upon the substrate is known (119, 120).

The interaction between adsorbed dye and substrate is small, since the action spectrum of the adsorbed dye in general is found to be identical with the spectrum in solution, with the exception of a small displacement, which is mostly bathochromic and which rises with increasing refractive index of the substrate.

6.4 Mechanism of Electron Transfer

In order to explain the mechanism of electron transfer, different concepts have been developed: It is supposed either that the sensitizer is a band-model semiconductor (121) or that the dye molecules act as individual sites. In the latter case they can be regarded as impurities with energy levels in the forbidden band of the substrate or as a particular phase that is separated from the substrate by an energy barrier (122) (see Fig. 1b).

The band-model is supported by the observation that most crystallized sensitizing dyes are semiconductors that can be doped by impurities (123), mostly forming p-type systems. Then the interface crystal-sensitizer is described as a p-n junction. Therefore, it has to be assumed that to equalize the Fermi energies, electrons are first transferred in the dark from the substrate to the dye, forming a macropotential that enhances an electron current in the opposite direction during the irradiation of the dye layer (121,105).

Since the usual dye layers are less than monomolecular and since the electronic dark conductivity of the silver halides is extremely poor (see Section 4.1), the semiconductor model probably is inconvenient for photographic sensitization. Moreover, it has been shown experimentally that if there is electron transport in the dark, it is directed from the dye to the halide (124).

A model considering the dye as an impurity center is not compatible with the similarity of the absorption spectra of solvated and adsorbed dyes. Therefore, the energy-barrier model should be preferred with a barrier depth of about the thickness of the π-electron cloud of the dye (≈ 0.3 nm). The transfer into the conduction or valence band is then caused by tunneling, at a rate ($k \approx 10^{11}$ sec^{-1}) that is great compared with the rate of fluorescence or intersystem crossing ($k \approx 10^8$ to 10^{10} sec^{-1}), even if the height of the barrier is equal to the binding energy of the excited electron (ca. 3 eV).

An energetic criterion for the sensitization can be given using the term (125)

$$q = |E_{LV}| - A_C = I - (\Delta E - E_{FC}) - (I_C - \Delta E_C) \qquad (54)$$

(see Fig. 15b) where E_{LV} is the energy of the lowest vacant level of the dye in the adsorbed state and A_C is the electron affinity of the crystal (equal to the level of the conduction band). The energy level E_{LV} can be expressed by the ionization energy I, the excitation energy ΔE (= $h\nu_{max}$), and the relaxation energy E_{FC} of the Franck-Condon transition into the equilibrium configuration. All energies must be calculated for the adsorbed state. A_C can be expressed by the ionization energy I_C and the band gap ΔE_C. $q \leq 0$ is equivalent to a sensitizing transfer without thermal assistance.

This simple expression, however, must be improved, since the energies are distributed over a whole range of values. To obtain the overall efficiency of sensitization, a procedure such as has been used for electron transfer processes in solution (172) must be applied.

The ionization energy of the sensitizer is statistically distributed since the first product of the photoionization is a vibrationally excited ion with a relaxation energy of statistical character (125,126) and also since the ionization energy is influenced by the randomly distributed electric charges of the substrate in the vicinity of the dye molecules. The adjustment of bond distances and angles for dye molecules at different sites must also be considered (111,127). A distribution of the form

$$\frac{N(E)dE}{N_\bullet} = \frac{b}{2} \cdot \exp(-b|E_o - E|) \, dE \tag{55}$$

was derived theoretically for the relative number of molecules, having ionization energies between E and E + dE (126) (b = 8 to 12/eV). From this, one obtains an approximation for the overall efficiency W of the sensitization

$$W \sim \int_0^{I_o} N(E) \, dE \tag{56}$$

I_o is the value of E where q (Eq. 54) equals zero.

In other approaches (127,128) thermal-assisted transfer is also considered, at least qualitatively: From an empirical dependence between reduction potential ε^o_{red} and maximum of absorbance (see Section 6.5) Tani (128) has concluded that sensitization or desensitization prevail if for the energy gap midpoint E_T of the dye the following inequalities are fulfilled (see Fig. 15):

$$E_T = \frac{E_{HO} + E_{LV}}{2} \quad \begin{matrix} > E_F \; : \; \text{sensitization} \\ < E_F \; : \; \text{desensitization} \end{matrix} \tag{57}$$

where E_{HO} and E_{LV} are the energies of the highest occupied and the lowest vacant level and E_F is the Fermi energy of the substrate. According to Eq. 57 it is not necessary that E_{LV} be smaller than A_C as was required in Eq. 54. In photographic layers the Fermi energy of the silver halide is given approximately by

$$E_F \approx -\frac{I_C + A_C}{2} \qquad (58)$$

The sensitizing properties of a dye, however, can be influenced by a variation of the Fermi energy (129,130). The sensitizing ability, for example, of phenosafranine (dye III of Fig. 15a) is significantly improved by doping of the silver halide with Cd^{2+} or by removal of hole injecting oxygen (130) (see Section 6.6).

The results obtained by Tani (128) lead to the concept that for some dyes sensitization and desensitization are possible simultaneously, for example, injection of electrons into the conduction band and of holes into the valence band (with the result of a partial quenching).

The injection rates j_n and j_p can be assumed to depend on the respective barrier heights $|E_{LV}|-A_C$ (= q) for electrons and $I_C - |E_{HO}| = I_C - I$ for holes (131) (see Fig. 15b). Simple expressions for this statements are

$$j_n = \nu_n \cdot \exp\frac{-(|E_{LV}| - A_C)}{kT} = \nu_n \exp\frac{-q}{kT} \qquad (59a)$$

$$j_p = \nu_p \cdot \exp\frac{-(I_C - |E_{HO}|)}{kT} \qquad (59b)$$

Then the condition for prevailing electron or hole injection is

$$\frac{j_n}{j_p} = \frac{\nu_n}{\nu_p} \cdot \exp\frac{(A_C - |E_{LV}| + I_C - |E_{HO}|)}{kT} \gtrless 1 \qquad (60)$$

If the frequency terms ν_n and ν_p do not differ very much ($\nu_n \approx \nu_p$), one has the condition as expressed in Eq. 57,

$$I_C + A_C - (|E_{HO}| + |E_{LV}|) \gtrless 0 \text{ or } -E_F + E_T \gtrless 0 \qquad (61)$$

A similar result can be derived if the energy distribution of the Marcus theory of redox reactions is used; see (129).

An experimental method of determining the energy distribution curve can be obtained from the measurements of the photomoment M, given in Fig. 14 (107). The abscissa is linearly dependent on the energy E_{LV} of the dyes 1-17 (see Section 6.5). Curve b is a measure of the sensitized photoconductivity as a function of the relative position of $|E_{LV}|$ and A_C of AgCl. The limiting value (dye 1-8) is reached when all dye levels lie above the conduction band. If the photoconductivity is zero (dye 16-17), all levels lie below the conduction band (the hole conductivity is negligible). In the middle section the levels lie around the conduction band. Therefore, the slope of curve b (dashed curve c) gives an approximate picture of the densitiy of states as a function of the energy.

6.5 The Determination of the Highest Occupied and the Lowest Vacant Levels of Sensitizers, E_{HO} and E_{LV}

According to the electron transfer mechanism it is necessary for the examination of a dye as a sensitizer to know E_{HO} or E_{LV} in the adsorbed state (see Eq. 54). Direct measurements of $I = -E_{HO}$ were made by means of the photoelectric emission threshold of at most monomolecular dye layers (132,133), but such measurements are difficult and sensitive to impurities.

Quantum-mechanical calculations [semiempirical Hückel MO (111, 127,134) or Hartree-Fock SCF calculations (135)] result in values of $I' = -E'_{HO}$ in the gaseous state. To calculate $I = -E_{HO}$ in the system substrate-dye-gelatin the following process can be used [D = dye, (s) = solid, (g) = gaseous]:

$$D(s) \xrightarrow[+L_V]{} D(g) \xrightarrow[+I']{} D^+(g) + e'(g) \xrightarrow[-L_V'-U(r)]{} D^+(s) + e'(s) \quad (62)$$

Therefore, one has

$$I = I' + L_V - L_V' - U(r) \quad (63)$$

The heats of evaporation per molecule, L_V and $L_{V'}$, are taken as approximately equal; for the calculation of the electrostatic energy $U(r)$ the Born equation is used:

$$U(r) = 1 - \frac{1}{D_\infty}\left(\frac{z^2 e_o^2}{2 \cdot 4\pi\varepsilon_o \bar{r}}\right) \quad (64)$$

where D_∞ is the (optical) dielectric constant of the system (calculated as a suitable mean value), e_o the unit charge, and $\bar{r}$ a measure of the radius of the molecule. From this it follows (134) that $I - I' \approx 2$ eV.

More convenient and often sufficient is the estimate of E_{HO} and E_{LV} with the aid of the measured standard oxidation and reduction potentials of the dyes, ε_{ox}^o and ε_{red}^o (128,136-139). To check the reversibility of the charge-transfer reaction, cyclic voltammetry with solid electrodes is recommended (138). The assumption is that values of ε_{ox}^o and ε_{red}^o can be related directly to E_{HO} and E_{LV}. If, however, a free enthalpy of a reaction is related to an electronic energy of a molecule, the entropy contributions to the free enthalpy must be small or at least sufficiently constant. Also, protolytic equilibria must be considered or eliminated.

The relation between ε_{red}^o and E_{LV} is a special problem. Even if ε_{red}^o is connected with the electron affinity A, it cannot be assumed for molecular dispersed systems that A is equal to $-E_{LV}$. There are experiments showing that the electron affinity level lies 0.1 to 0.3 eV above the energy level of the optical electron

(140), which is plausible from the point of view of electrostatics. From other experiments the reverse sequence was concluded (103), which would be expected if the electron is transferred during a Franck-Condon transition. Considering the experimental and theoretic uncertainties, it may be sufficient to equalize $|E_{LV}|$ and A (107,116,134).

The electrochemical processes used are

$$D^{\overline{\cdot}} \rightleftharpoons D + e', \qquad \varepsilon^o_{red} \qquad (65)$$

$$D \rightleftharpoons D^{\overset{+}{\cdot}} + e', \qquad \varepsilon^o_{ox} \qquad (66)$$

To evaluate ionization potential and electron affinity in the solvated states (f) (I'' and A''), the following steps are necessary:

$$D^{\overline{\cdot}}(f) \underset{-L}{\longrightarrow} D^{\overline{\cdot}}(g) \underset{+A''}{\longrightarrow} D(g) + e'(g) \underset{+L - \psi}{\longrightarrow} D(f) + e'(\text{electrode}) \quad (67)$$

and

$$D(f) \underset{-L}{\longrightarrow} D(g) \underset{+I''}{\longrightarrow} D^{\overset{+}{\cdot}}(g) + e'(g) \underset{+L^+ - \psi}{\longrightarrow} D^{\overset{+}{\cdot}}(f) + e'(\text{electrode}) \quad (68)$$

L, L^+, and L^- are the molecular solvation enthalpies and ψ the work function of the electrode (e.g., for carbon electrodes ψ_C = 4.39 eV). From Eqs. 67 and 68 one has for the standard potentials

$$e_o \varepsilon^o_{red} = A'' + [L - L^-] - \psi + e_o \cdot \varepsilon^\bullet \approx A'' + \Delta L + \text{const} \quad (69)$$

$$e_o \varepsilon^o_{ox} = I'' - [L - L^+] - \psi + e_o \cdot \varepsilon^\bullet \approx I'' - \Delta L + \text{const} \quad (70)$$

$\varepsilon^\bullet$ is the absolute potential of the reference electrode.

The treatment of the solvation energy is a special problem. An estimate is often made by means of Eq. 64 [for corrections see (137)]. It can be expected that the $\Delta L(f)$ of chemically similar substances does not vary to any great extent. A test of the validity of the treatment can be obtained by subtracting

Eq. 69 from Eq. 70,

$$e_o (\varepsilon^o_{ox} - \varepsilon^o_{red}) = I'' - A'' - 2 \Delta L = \Delta E - 2 \Delta L \qquad (71)$$

ΔE is approximately equal to the excitation energy (absorption maximum) of the dye in solution. The experimental relation between $\Delta \varepsilon^o$ and ΔE for a number of sensitizing dyes is given in Fig. 18 (136). If $1 < \Delta E < 3$ eV the results (in eV) for cyanine dyes can be represented by (136)

$$e_o (\varepsilon^o_{ox} - \varepsilon^o_{red}) = 1.33 \Delta E - 1.1 \qquad (72)$$

The slope of the straight line is 1.33, although ≈ 1 would be expected from Eq. 71 [or <1, since $|\Delta L|$ should be smaller with increasing chain length (decreasing ΔE)]. The electrochemical

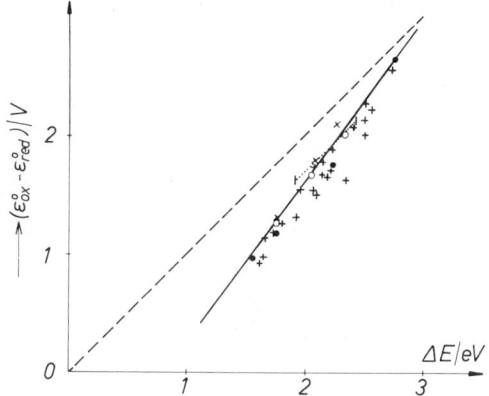

Fig. 18. The difference of oxidation and reduction potentials $\varepsilon^o_{ox} - \varepsilon^o_{red}$ of a number of dyes as a function of the excitation energy ΔE (in solution). Broken line, Eq. 71 (with $\Delta L = 0$); solid line, curve after Dähne (136) (Eq. 72); dotted line, Eq. 74 in the ΔE range of dyes used by Berriman and Gilman (141). Open circles, 2.2' quinocyanines (n = 0,...,2) (n, see Fig. 17); closed circles, thiacyanines (n = 0,...,3); crosses, dyes, measured by Large (138).

method, therefore, is useful only for relative measurements and as a first approximation.

6.6 Experiments on the Mutual Position of the Energy Levels of Dye and Substrate and its Variability

Not only is the difference of the standard potentials linearly dependent on ΔE but also each potential separately, especially if dyes of equal structure but different chain length are compared. In Fig. 19, ε^o_{red} and ε^o_{ox} are plotted for numerous sensitizing and desensitizing dyes as a function of $\Delta E = h\nu_{max}$. Sensitizing dyes are assembled within the hatched area, and desensitizing dyes are found in the area below (128). The straight lines $\varepsilon^o_{red}(\Delta E)$ can be expressed by

$$- 2 e_o \varepsilon^o_{red} = \Delta E + \text{const} \tag{73}$$

where for sensitizing dyes const = ± 0.2 eV, while for desensitizers, const > 0.4 eV (all potentials are given against NHE).

From the empirical equation 73 the already mentioned relation Eq. 57 can be derived, if in Eq. 69 $\Delta L(f)$ and also $|E_{LV}| - A'$ is taken as constant. However, the straight lines $\varepsilon^o_{ox}(\Delta E)$ and $\varepsilon^o_{red}(\Delta E)$ in Fig. 19 are not symmetric, although they would be expected to be so from Eqs. 69 and 70 (compare the difference between Eqs. 72 and 71).

In order to use relations such as Eq. 57, 69, or 70 for the examination of the substances as sensitizers, the values of the constants must be numerically known. Generally, this is not possible for the system substrate-dye-gelatin. Therefore, it is convenient to use solvated systems to get relative values and to fix one point in the system empirically with a test substance.

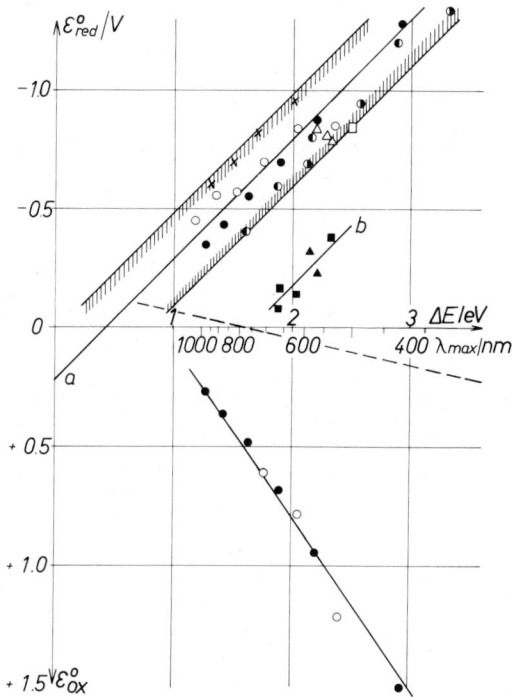

Fig. 19. Reduction and oxidation potentials ε^o_{red} and ε^o_{ox} (versus NHE) of sensitizing and desensitizing dyes, adsorbed on AgBr as a function of the excitation energy ΔE (128). Hatched area, region of sensitization. Curve a, Eq. 73 for sensitizers (mean value); curve b, Eq. 73 for desensitizers; dashed line, values of $(\varepsilon^o_{ox} - \varepsilon^o_{red})/2$ as a function of ΔE. Croses, 4.4´ quinocyanines (n = 0,...,3) (n, see Fig. 17); open circles, 2.2´ quinocyanines (n = 0,...,5). ●, thiacyanines (n = 0,...,5); ◐, oxacyanines (n = 0,...,3); ◉, selenacyanines (n = 0,...,3); △, xanthene dyes; □, basic dyes (sensitizing); ■, basic dyes (desensitizing); ▲, styryl dyes.

Using dyes of different ε_{ox}^{o} and ε_{red}^{o}, Berriman and Gilman (107,141) have tried to fix the range of energy levels, in which

1. Electrons are injected into the conduction band (latent images are formed);
2. Holes are injected into the valence band (internal fog is bleached);
3. Electrons are trapped from the conduction band (the blue region is desensitized);
4. Holes are injected into the surface silver (surface fog is bleached).

The layers were irradiated with a spectrum of variable intensity (wedge spectogram method) and developed. In other experiments the quenching of luminescence as a function of temperature and p_{Ag} (see below) was used (142).

For the examination of hole injection, special photographic emulsions with internal fog were prepared, which in the case of hole injection form positive images after development. Since there are dyes that are able to inject holes and electrons simultaneously (see Eq. 59), parallel experiments were carried out in which, apart from the dye, a colorless reversible electron acceptor was adsorbed, which is able to trap electrons from the excited dye before they are injected into the conduction band (1,1´-di-n-butyl 4,4´-bipyridinium dibromide, a substance that is also suitable as redox catalyst during development (143); see Section 7.6.2).

On the left of Fig. 20 the dyes used in (141) are arranged in order of increasing ε_{red}^{o}; on the right are the dyes used in (107) in order of decreasing ε_{ox}^{o}.

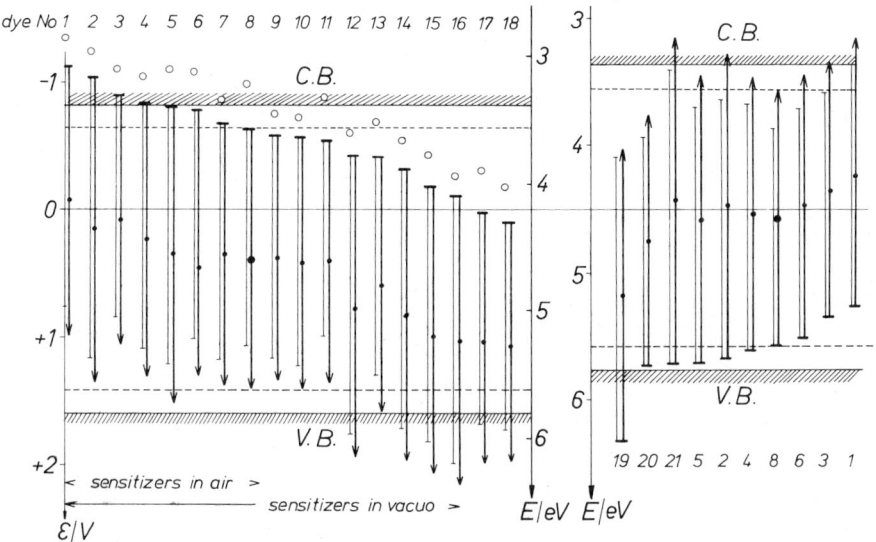

Fig. 20. Potential values (left ordinate) of various sensitizers (No 1-21) and their relation to the energy band of Ag(Br,I). Right, arrangement according to decreasing ε^o_{ox} (141); left, arrangement according to increasing ε^o_{red} (107); small vertical lines, $\varepsilon^o_{ox} - \varepsilon^o_{red}$; thick vertical lines, $\Delta E = h\nu_{max}$. Ordinates E, values of the band energies in relation to NHE, according to the two modes of calibration. Dashed lines, beginning of electron and hole transfer. Small solid circles, E_T values; large solid circle, E_T value of dye 8 (for calibration); open circles, E_{LV}, if counted from ε^o_{ox} (calibration on the left, adjusted to that on the right).

It was found that in Ag(Br,I) sensitization begins at ε^o_{red} < - 0.64 V, and hole injection at ε^o_{ox} > 1.07 V. Several dyes, the potentials of which are beyond these limits, are able to inject holes and electrons, although their excitation energy is smaller than the band gap. For calibration, dye 8 is used which just has the potential values mentioned above. According to

Tani's criterion (Eq. 57) for this substance ΔE (2.01 eV) should be situated symmetrically within the bandgap ΔE_C (2.39 eV). Since the ionization energy of Ag(Br,I) is known (I_C = 5.76 eV), E_{HO} and E_{LV} of the dye are fixed (E_{HO} = $-$ 5.57 eV).

The problem of calibration is the inequality of $e_o(\varepsilon^o_{ox} - \varepsilon^o_{red})$ and ΔE (compare Section 6.5). For the dye used

$$e_o(\varepsilon^o_{ox} - \varepsilon^o_{red}) = \Delta E - 0.3 \text{ (eV)} \tag{74}$$

If generalized, this relation is incompatible with Eq. 72, but within the small range of excitation energies used, the experimental function can also be expressed by Eq. 74 (Fig. 18, dotted line).

If the absolute position of ε^o_{ox} is assumed to be equal to $|E_{HO}|$, the relative position of the crystal bands is given on the right in Fig. 20. If, on the other hand, the absolute position of ε^o_{red} is assumed to be equal to $|E_{LV}|$, the calibration shown on the left in Fig. 20 is obtained. For the first case the calibration of Eq. 70 is (values against NHE)

$$E_{HO}/\text{eV} = - e_o \, \varepsilon^o_{ox} - 4.50 \tag{75}$$

The constant is in agreement with the calculations of Lohmann (144) for the standard hydrogen electrode.

The sequence according to ε^o_{ox} or ε^o_{red} in Fig. 20 was made in order to call special attention to the potentials responsible for hole injection (right) or electron injection (left). If, however, all levels are related to E_{HO} and ε^o_{ox} (Eq. 75) (circles on the left in Fig. 20) the sequence is scarcely altered. Considering the uncertainties of the definition of the band edges and of the potential measurements, and also the poor theoretic background of the evaluation of levels, one method of calibration as given in Eq. 75 may be sufficient.

The same threshold of potentials (ε^o_{red} = $-$ 0.6 to $-$ 0.7 V) was found for the sensitization of all silver halides, but it is questionable whether the same energetical position of the conduction band can be derived from this observation, as long as no measurements of the hole injection into AgCl and AgI are available. Other experiments (111,127) are contradictory and show that for optical sensitization of silver chloride a more negative ε_{red}^{o} is demanded (compare Figs. 5 and 15).

The experiments on hole injection of excited dyes (141) are evaluable only if hole traps at the crystal surface do not interfere. If, however, silver bromide layers covered with some evaporated silver are dye-sensitized (145), it was found that the reaction of holes with the surface traps is faster, the more negative the ground level of the dye is. With such substances no mobile holes were found during irradiation. If sensitizers are used, with a ground level so high that the injection level of holes is about 0.7 eV above the valence band, the holes do not react but remain mobile (with u_t; see Eq. 44) and can be transported by an electric field (see Section 4.3).

Finally, the sensitization is also dependent on the conditions in the outer phase, especially on the redox potential (presence of oxygen), the p_H (168), and the p_{Ag} (146,147).

Oxygen is an effective electron acceptor and therefore a strong desensitizer (103). An influence of oxygen upon the Fermi energy of silver halide can also be used to explain this effect (see Section 4.1). If the irradiation is effected in a vacuum, the sensitization is extended to rather low values of E_{LV} (small ε_{red}^{o}) (107) (see Fig. 20).

An effective method for altering the range of dyes that can sensitize is the adjustment of the p_{Ag} in the surroundings of the silver halide (146,147). This effect may be one of the most direct arguments for the electron transfer mechanism. Since sensitization corresponds to a reduction of Ag^+ and desensitization to an oxidation of Ag, it is clear that, with a more positive potential of the system Ag/Ag^+ (decreasing p_{Ag}), the reduction is promoted. Therefore, desensitizers are able to sensitize if the p_{Ag} of the surroundings is lowered. With a given silver activity the threshold potential of sensitization should be lowered by 0.06 V/p_{Ag}. The experimental value is 0.1 V/p_{Ag}. So far the deviation cannot be explained.

6.7 Supersensitization

If a second substance together with a sensitizing dye is adsorbed on the substrate, in favorable cases a greater photographic sensitivity for a given wavelength can be reached than with the dye alone. The added substance often has no absorption within the range of spectral sensitization, but it also can be a sensitizer. The effect is analogous to the superadditivity in photographic development (see Section 7.6.2) and is called supersensitization (148). It is found in particular with sensitizers in a state of aggregation (e.g. formation of J aggregates, see Section 6.3).

The observed rise in sensitivity can be caused by a change of the absorption spectrum in the presence of the supersensitizer, or by an increase in efficiency of the sensitization. Only the latter effect of the added substance is of interest for the mechanism of supersensitization. It can be found by measurements of the quantum efficiencies Φ of the nonsensitized and the sensitized layer. The relative quantum efficiency Φ_r is given by the ratio of the numbers of quanta of blue light (400 nm), N_{400}, and of light of the wavelength λ(nm), N_λ, which give the same specified developed density:

$$\Phi_r = \frac{N_{400}}{N_\lambda} = \frac{400 \cdot E_{400} \cdot a_{400}}{\lambda \cdot E_\lambda \cdot a_\lambda} \tag{76}$$

where E is the incident energy, and a is the adsorbed fraction of the incident energy.

True supersensitization is announced by a rise of ϕ_r in the presence of an added substance. It also improves the spectrally sensitized photoconductivity.

There are numerous theories to explain this effect: It was assumed (149), for example, that J aggregates containing desensitizing impurities were dispersed by an incorporated supersensitizer into areas of pure dye acting at high degree of efficiency. Rather similar in its result is a theory in which the supersensitizer acts as an exciton trap (150), impeding the propagation of excitons within the dye aggregate, and in this way facilitating the transfer of electrons into the substrate.

The efficiency of positively charged supersensitizers in combination with negativley charged dyes (or vice versa) led to a "glue" theory (151) in which the adsorption of the dye is enhanced by the supersensitizer, together with a loss of J aggregation. At the same time the separation of the dye from the substrate by the supersensitizer could be effective to inhibit a desensitizing reaction of the hole, which remains in the dye after the electron transfer.

From measurements of electron affinities of the dye, A, and ionization energies of supersensitizers, I_R, a direct charge transfer between both substances in contact was also discussed (152). This might enhance the sensitization efficiency if

$$I_R - A = h\nu' < h\nu \text{ (dye)}.$$

The three last theories mentioned can be considered as closely connected with an electrochemical view of the supersensitization (153) according to which the excited dye is reduced by the supersensitizer, which, therefore, acts as a hole-trapping agent. This theory is supported by electrochemical experiments (115,129,153) with semiconductor electrodes. It was found that agents as hydroquinone retard or inhibit the deactivation of the excited state of the dye by formation of CT complexes or by direct electron transfer, thus raising the tunnel probability. Similar results were obtained by measurements of fluorescence quenching with dyes in homogeneous solution (154) as well as adsorbed on silver halides (155). Quenching of the fluorescence was found in presence of reducing substances that did not react with the ground state of the dye.

In systematic experiments (156) with photographic layers that contained a given dye combined with supersensitizers of variable oxidation potentials, it was shown that Φ_r increases sharply if ε_{ox}^o is lowered below a certain value, that is, if E_{HO} of the supersensitizer becomes closer to vacuum than E_{HO} of the sensitizer.

From the experiments (Table 4) it was concluded that the (not directly measurable) oxidation potential of the dye aggregate used on the substrate surface must be 1.05 ± 0.05 V. It seems that if the condition ε_{ox}^o (supersensitizer) < ε_{ox}^o (dye) is fulfilled, the value of Φ_r also increases with the decrease of the reduction potential of the supersensitizer (Table 4). This can be explained by the assumption that electron trapping by the supersensitizer is prevented if ε_{red}^o (supersensitizer) < ε_{red}^o (dye).

The hole-trap theory is also compatible with the fact that, in presence of supersensitizers, the injection of free holes from excited desensitizers into the valence band (see Section 6.6) can be inhibited (148). Therefore, the bleaching of internal fog is diminished or prevented.

With some dyes self-supersensitization was observed (&155,157). This was explained by a supersensitizing action of monomers,

TABLE IV

Dependence of the Relative Quantum Efficiency Φ_r of a Sensitizing Dye on the Potentials ε^o_{ox} and ε^o_{red} (versus NHE) of Different Supersensitiziers (156)

		ε^o_{ox}/V	ε^o_{red}/V	Φ_r
Dye Monomer		+ 1.21	− 0.81	0.03
Supersensitizer	1	+ 1.22	− 1.10	0.04
	2	+ 1.16	− 1.04	0.03
	3	+ 1.10	− 0.84	0.07
	4	+ 1.00	− 0.78	0.39
	5	+ 0.86	− 1.32	0.82
	6	+ 0.84	− 0.90	0.54
	7	+ 0.76	− 1.13	0.76

Dye aggregate (calculated)   + 1.05 ± 0.05

Dye: 1,1´-diethyl-2,2´-cyanine

Supersensitizers: e.g., 5 = [benzothiazole]−C−CH=CH−[phenyl]−N Et$_2$

which on the surface of the substrate were in equilibrium with the aggregate. So far the experiments show that this effect is only found if ε^o_{ox} (monomer) < ε^o_{ox} (aggregate). The dye used in Table 4 therefore shows no self-supersensitization.

Summarizing, it can be stated that the enhancement of the spectral sensitizing efficiency obtained by supersensitization is due to an inhibition of recombination effects within the dye.

7 AMPLIFICATION AND BLEACHING OF THE LATENT IMAGE

7.1 Thermodynamics of the Development Process; Principle of Chemical and Physical Development

During the development process, just as in optical sensitization, electrons are transferred from an outer phase to the latent image speck, but now the redox system O/R^- is in its ground state. The electron transfer reaction is catalyzed by the speck:

$$AgX + R^- \xrightarrow{\text{lat. image}} Ag + O + X^- \qquad (77)$$

The affinity A of the process is given by

$$A = zF\Delta\varepsilon = zF(\varepsilon_{AgX} - \varepsilon_R) = zF(\varepsilon^o_{AgX} - \varepsilon^o_R) - RT \ln a_{Ag} + RT \ln \frac{a_R \cdot a_{AgX}}{a_O \cdot a_X} \qquad (78)$$

with $a_{AgX} = 1$. The index R denotes developer. For the activity of the speck, a_{Ag}, see Section 5.3.

Besides the "chemical development", which is given by Eq. 77, the amplification can also be obtained if the exposed crystal is first dissolved by a complexing agent, while the nucleus remains unaffected, especially if it is adsorbed on gelatin. This "post-fixation physical development" demands a donor of silver ions in the solution, for example, a complex with the ligand Y:

$$[AgY_m]^{1-m} + R^- \xrightarrow{\text{lat. image}} Ag + O + mY^- \qquad (79)$$

where

$$A = zF(\varepsilon^o_{Ag} - \varepsilon^o_R) - RT \ln a_{Ag} + RT \ln \frac{a_R \cdot a_{AgY_m}}{a_O \cdot K \cdot a_Y^m} \qquad (80)$$

where K is the formation constant of the complex.

Most developers contain complexing additives like the antioxidant SO_3^{2-}. Therefore, both mechanisms can occur simultaneously in the exposed grains, the complex being formed by the reaction

$$AgX + mY^- \rightarrow [AgY_m]^{1-m} + X^- \qquad (81)$$

In special cases (diffusion-transfer processes, e.g. Copyrapid and Polaroid Process) chemical and physical development is effected at the same time, but at different phases: an exposed layer is chemically developed, whereas the unexposed grains are dissolved by a complexing agent added to the developer. The silver complex diffuses to a second layer, which is in close contact with the first. This layer contains small catalyzing nuclei. The silver complex coming from the unexposed area of the first layer is physically developed on these nuclei. In this way a positive image is obtained in a single operation.

In a special kind of physical development the redox system AgY_m^{1-m}/Ag in Eq. 79 is substituted by another redox system that is able to react electrocatalytically at the surface of the latent image. In this way colored images can be obtained without intermediate silver formation (see Section 7.7.3).

7.2 Developing Substances

To transfer electrons into the latent image speck, numerous inorganic and organic redox systems with a wide range of redox potentials can be used (158). As inorganic developers, Fe(II)-complexes or the V(II)/V(III) system are of interest, especially for regenerable solutions. The structure of organic developers (mostly aromatic systems) can be represented by the general formula (159)

$$\alpha - (-X=\overset{|}{C}-)_n - \alpha'$$

where

α, α' : NR_1R_2 or OH; n: 0, 2, 4, ...; X: $\overset{|}{C}$ or N;

A discussion of the electrochemical oxidation mechanisms of the different developers is beyond the scope of this chapter. In general, protolytic and redox equilibria simultaneously occur following a scheme as shown in Eq. 82:

$$
\begin{array}{ccccc}
R-H_2 & \xrightleftharpoons{-e^-} & \cdot R-H_2^+ & \xrightleftharpoons{-e^-} & O-H_2^{2+} \\
{\scriptstyle -H^+} \updownarrow K_1 & & \updownarrow & & \updownarrow \\
RH^- & \rightleftharpoons & \cdot R-H & \rightleftharpoons & O-H^+ \\
{\scriptstyle -H^+} \updownarrow K_2 & & \updownarrow & & \updownarrow \\
R^{2-} & \rightleftharpoons & \cdot R^- & \rightleftharpoons & O
\end{array}
\qquad (82)
$$

The redox potentials therefore are dependent on the pH, and the standard potentials ε_R^o are connected with the transfer of two electrons. The transfer proceeds in two steps at the potentials ε_1^o, ε_2^o with radical intermediates and a symproportionation reaction with the equilibrium constant K_s, where

$$\varepsilon_R^o = \frac{\varepsilon_1^o + \varepsilon_2^o}{2}, \qquad RT \ln K_s = \varepsilon_2^o - \varepsilon_1^o \qquad (83)$$

Those radical intermediates have also been observed in photographic layers during development (248,249). They are important for some electrocatalytical mechanisms in the development process (see Section 7.6).

To characterize the different developing substances to date the standard potentials ε_R^o (as a measure of E_{HO}; see Section 6) are usually used [for a calculation of E_{HO} of an adsorbed developer molecule see (160)].

Some simple correlations were found, for example, a linear dependence between $-\log j$ (j is the rate of development) and the half-wave potential of the developer, which is valid as long as the rates are not diffusion controlled. This was found with derivatives of p-phenylenediamine (161), p-aminophenol (162), and hydroquinone (163). Several properties of development reactions, however, can be better correlated with ε_I^o than with ε_R^o (see Section 7.6.2).

7.3 Electron and Hole Injection by Redox Systems

It has been observed that during development of surface images or fog, inner latent images are intensified simultaneously (170, 171). This was detected in macrocrystals by means of a graded etch technique between a first and a second development (170) and also in emulsions if, after a short development, the surface silver was oxidized (bleached) with a diluted acid solution of potassium dichromate and silver nitrate and then the remaining inner latent images developed with a developer containing a complexing agent. The intensification was only observed with special developers, mainly substances with one-electron transfer reaction (1-phenyl-3-pyrazolidone, Fe(II)-EDTA); it was inhibited by adsorbed desensitizers (hole acceptors) (171). In contrast, oxidizing substances are able to inject holes and to bleach inner latent images (96). Those injection processes are to be expected since the electronic disorder of the silver halides depends on the redox potential of the surroundings (Section 4.1).

Therefore, it would seem obvious to treat the system silver-halide developer in terms of semiconductor electrodes (172). Depending on the position and densitiy of the respective energy states, electrons are transferred with constant energy, either into the conduction band or in the form of holes out of the valence band.

The Fermi levels of both phases in contact are balanced by an exchange of electronic and (in contrast to the usual semiconductors) ionic movements, resulting in space or surface charges in both phases. According to this view, the latent image is a state that can accept electrons and that grows step by step by adsorption of silver ions, forming positively charged Ag_m aggregates with a new empty orbital of lower energy, capable of trapping another electron from the redox system, exactly as in the primary process (97) (see Fig. 21). These injection processes are especially effective if the energy bands are bent off in the neighborhood of the image specks in direction of increased n-conductivity (81,173,174).

An attempt has been made to explain the induction period that is always observed at the begin of the development process by

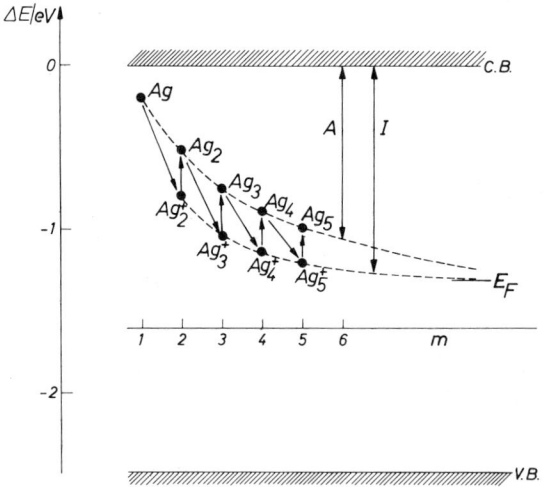

Fig. 21. Estimated ionization energies I and electron affinities A of small silver aggregates in AgBr (m = 1,...,5), related to the conduction band (97,65); E_F, Fermi energy of metallic silver (a_{Ag} = 1).

assuming an injection mechanism for the first growth of the speck, for example, up to the moment when it has reached the surface of the grain (173,174). This, however, is untenable. The injection rate is far too small to play any role during the development process (183). This can be shown by calculating the maximum injection rate of electrons to a spherical nucleus of the diameter $2r_o$ within the crystal. Since the nucleus is surrounded by a great surplus of foreign ions (interstitials and vacancies), the boundary conditions are the same as in polarography in the presence of a supporting electrolyte (175), the diffusion rate of electrons j_d being

$$j_d = 4\pi r^2 D \cdot n \cdot \left(\frac{1}{\sqrt{\pi D t}} + \frac{1}{r}\right) \tag{84}$$

where n is the bulk concentration of electrons and, at the surface of the nucleus, n = 0. Because r is small (< 1 nm), a stationary state is reached within a very short time t' ($r_o < \sqrt{\pi D t'}$). The first term of the sum in Eq. 84 can therefore be neglected. With $D = \kappa_e RT/F^2$ (κ_e is the electronic conductivity of the crystal),

$$j = \frac{4\pi r \kappa_e RT}{F^2} \tag{85}$$

κ_e is given by the potential $\Delta\varepsilon$ of the developer with respect to a silver metal electrode (see Fig. 6b):

$$\kappa_e = \kappa_o \exp \frac{\Delta\varepsilon F}{RT} \tag{86}$$

with $\kappa_o = 5 \cdot 10^{-21} \, \Omega^{-1} \, cm^{-1}$ (300 K). With Eq. 86 one has by integration of Eq. 85 (V_M is the molar volume of silver)

$$\left(\frac{r}{r_o}\right)^2 = 1 + \frac{2 V_M RT \kappa_o \exp(\Delta\varepsilon F/RT)}{F^2 r_o^2} \cdot t \tag{87}$$

A similar formula has been used by Matejec and Moisar (96) to calculate the size of latent images (see Section 5.3).

To avoid spontaneous nucleation of silver at the unexposed crystal, $\Delta\varepsilon$ should not exceed 0.3 V (203). With this value it can be calculated from Eq. 87 that a nucleus having an initial diameter of 0.6 nm grows by $4 \cdot 10^{-3}$ nm within 10 min. (Since in microcrystals $[Ag_i^{\cdot}] \gg [V'_{Ag}]$, according to Eq. 28 the concentration of electrons is even further reduced.) Even if surface charges forming a positive surface potential of 0.3 V are assumed, the size of the nucleus after 10 min development does not exceed 6 nm.

Therefore, injection is not appreciable during the development and the semiconductor view is inadequate. Only surface images can be developed directly.

7.4 Electrode Theory of Development; General Aspects

Since only surface specks can be enlarged at a sufficient rate, the latent image can be regarded as a surface state (97) of an energy that decreases with increasing size (Fig. 21).

The activity of development centers depend not only on the size but also on the material. It was found that physical development is possible with colloidal dispersions of numerous electron-conducting substances (176-178). When related to the surface the reaction rate increased according to the sequence (178)

$$NiS, PbS, ZnS < Ag_2S < C < Au$$

The special effectiveness of Au is in agreement with the stability measurements of latent images (see Fig. 12).

The deposition of silver begins at discrete points of the surface of the particles but after certain times they are covered with silver. Then the reaction rate becomes proportional to the particle surface. The results mentioned are of special importance for the diffusion transfer process (see Section 7.1).

Today it is generally accepted that photographic development is an electrocatalysis (179). During physical development (Eq. 79) the latent image speck is a twofold electrode with the mixed potential ε_m (180). During chemical development (Eq. 77) (and also during the initial state of physical development at a non-silver nucleus), the latent image acts as a local voltaic couple (181,182) (Figs. 22a and 22b), in which silver is reduced at the cathodic area (phase boundary AgX|Ag) whereas the developer is oxidized at the anodic area (phase boundary Ag|solution).

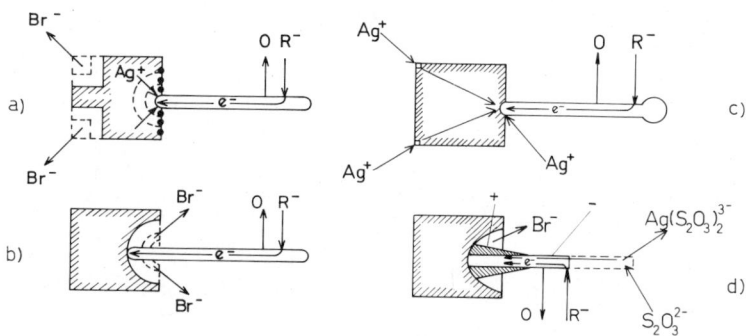

Fig. 22. Scheme of chemical development of filamentary silver. R^-, developer; O, developer oxidation product; (a) ion transport through the silver halide, solid circles are adsorbants, inhibiting the grain dissolution; (b) decomposition of the halide in the vicinity of the filament. At the base of the filament physical development is also possible in part; (c) ideal development (sphere at the top: first development, see text); (d) monobath development, electrochemical dissolution of silver halide; minus, anodic site; plus, cathodic site.

In this case two almost equal potentials are established at both electrodes ($\varepsilon_c \approx \varepsilon_a \approx \varepsilon_m$).

If both mechanisms (Eqs. 77 and 79) are possible at the same time their relative proportions depend on the complexation rate of the silver halide and the reduction rate of the silver complex, compared with the reduction rate of silver ions in AgX and the diffusion rate of the anions X^-. This proportion is also a function of the respective reaction area.

With regard to the mixed potential and the composition of the developer, therefore, four limiting cases can be discriminated (184):

1. During "chemical development" the concentration of halide at the surface of AgX is comparable with the bulk concentration of developer. In this case the potential is low ($\varepsilon_m \approx + 0.2$ V), and the halide disappears at the same rate as silver is formed: the crystal is decomposed, and the silver is deposited in the form of fine filaments.
2. If the potential ε_m is kept at about the same value (0.2 V) by addition of complexing solvents, the dissolution rate of the halide exceeds the reduction rate of the silver ions. In this case silver is deposited via the solution phase, the reaction area is that of the silver-solution interface, and "physical development during solution" is found. The developed silver is in the form of short thick filaments.
3. If the halide is dissolved first, and then the nuclei are developed in the presence of silver ions at positive potentials ($\varepsilon_m \approx + 0.6$ V), the conditions of true physical development are fulfilled, and the deposited silver is spherical.
4. Similar conditions of potential ($\varepsilon_m \approx + 0.6$ V) can be ob-

tained without previous dissolution of the grains if the
developer contains silver ions, for example, in form of a
complex but without addition of complexing agents. At these
potentials the dissolution of the grains is prevented but
silver filaments of any length grow out of the surface image
specks, without any change of the silver halide grain.

During this so-called "ideal development" (184) the silver is deposited at the inner phase boundary AgX|Ag. This was demonstrated if the nuclei were first partly developed by a developer, which formed compact silver. When, afterwards, the specks were enlarged by ideal development, electron micrographs showed the compact silver to be at the top of the filaments, which therefore must be thrown out at the base. It must be concluded that, during the ideal development, the solvated silver ions are not reduced at the silver surface but are desolvated at the crystal surface and become incorporated in the crystal, where they replace the lattice particles consumed during the growth of the filaments (Fig. 22c).

It can be assumed that the site of the silver deposition, also in case (1), is at the base of the filament. If this is true, compact structures are formed if silver is deposited from the solution phase, and filamentary structures are formed if it is reduced at the contact between image speck and crystal. Filaments, however, are not produced at the beginning of the development. In experiments at a low rate of development it was shown by electron micrographs that the nuclei at the grain surface are initially enlarged and that spherical silver is formed. The growth of filaments begins when the nucleus has reached a definite diameter that depends on the experimental conditions (185).

In order to understand the operation of the local voltaic couple in chemical development one must know the place where the silver halide is decomposed. From the electron micrographs of Klein and Matejec (186) both possibilities, given in Figs. 22a and 22b, have to be discussed. Either interstitials move through the crystal and the remaining halide ions diffuse into the solution from energetically favored places, which can be far away from the development center (Fig. 22a), or the nucleus grows at or within an etch groove. In this mechanism no ion transport through the crystal is necessary (Fig. 22b). Both mechanisms have also been observed during electrical reduction of AgBr, embedded in KBr solution by means of a Pt-point cathode touching the crystal (188).

Etch pitch formation as in Fig. 22a can be verified by model experiments if a crystal sheet, mounted between inert electrolyte and developer solutions with an electrolyte bridge of wetted paper (186,187), is exposed at the site of the developer. It has been found, however, that the material removed from the site remote from the developer is only a small fraction of that removed by direct contact with the reducing solution (187).

The mechanism obeyed is in part dependent on the rate of development: At high rates near the base of the filaments a complex dissolution of silver halide by the remaining halide anions is favored and physical development at the base is possible. This has been shown during electrochemical reduction (193) and development (189). At the base of the filament, large current densities are to be expected at high rates of development. With a filament diameter of 20 nm and a development time of 10 min for a grain of 300 nm diameter, a current density of 50 mA/cm^2 is calculated, which causes a high cathodic overpotential. If the dissolution rate is unequal along the crystal surface, a tangential convection of the solution is to be expected. The filaments therefore move along the etch groove and eat up the crystal. This has been observed during electrochemical reduction of AgBr (188) and confirmed by cinematographic studies of the development (190). It is enhanced because the reaction is exothermic and the temperature rises at the point of reaction (191).

If etch grooves are formed during the development, gelatin is desorbed and the retarding influence of gelatin (see Section 5.3) is removed. Perhaps, in this way, new filaments are spontaneously formed so that the number of filaments exceeds the number of surface image specks (192). However, the same result is obtained if inner latent images become developable during the etching process.

Several filaments moving in an etch groove can coalesce at the base. In this way a branching of the filaments can be explained that was sometimes observed but could not be understood if the filaments were to grow at the base from a fixed point.

If, however, the rate of development is low, the crystal is disintegrated merely at the thermodynamic favored places of the crystal. In this case adsorption phenomena (e.g., of gelatin) must also be considered (183). Therefore, the mechanism of Fig. 22a can be favored. The mechanism mentioned first, is, however, preferred since a dependence of the rate of development on the conductivity of the crystals has never been found. Even Hg_2Cl_2, which has a very low cation mobility, is reduced at similar rates as silver halides. For AgBr, however, it was calculated that under normal conditions the supply of silver ions is not rate determining, even if all silver ions are deposited at the base of the filament (257).

It may be mentioned that doubts have been cast on the validity of the electrode theory for the beginning of the development process in a number of controversial papers (240). On the basis of electron-micrographs it was stated that exposed grains are dissolved by the developer on the place of the latent image, forming a filament of dissolved material, which later is decomposed to form silver. The mechanism of this reaction, however, is not clear, and the possibility of ideal development (see Fig. 22c) is at odds with it.

7.5 The Conditions and the Mechanism of Filamentary Growth of Silver

With respect to the structure of developed silver, one must consider how filaments are able to grow at the base and how they can be stable.

The filaments may grow like whiskers from one or two combined screw dislocations, but their diameter must be limited by a covering of inhibiting substances. An adsorption of inhibitors has been shown by experiments (194) in which silver was physically deposited on chemically developed filaments. At first, the deposition occurs only at distinct points of the surface and becomes increasingly inhibited, the more the first chemical development is prolonged. This results from an increasing adsorption of inhibiting substances. Physical development is also inhibited by a short treatment of the developed silver with oxidizing substances in the presence of bromide ions. The "ideal development" mentioned is also only possible if there is a strong inhibition of silver ion discharge at the surface of the developed silver.

The formation of filaments can, therefore, be caused by a competition between a deposition of silver and of inhibiting substances (173) similar to that in electrocrystallization (195). There is a current of silver ions and of inhibitors towards the surface. If the radius of the curvature of the growth step between two adsorbed molecules is small enough to make the two-dimensional nucleus of silver unstable, growth ceases. Therefore, a filament is formed that continues to grow only at the base where the silver concentration is high. The diameter of the filament is a function of adsorption covering; it is much smaller in gelatin than in polyacrylate (196).

In another theory it is assumed that the nucleus is growing at a dislocation of the silver halide. The discharged silver ions are transported as mobile ad-atoms over a spherical interface by surface diffusion. The critical diameter of the sphere is reached when the diffusion length is equal to the diameter of the sphere. When this diameter is reached, a cylindric growth begins at a rate of 0.1 to 0.2 μm/sec (245). According to this theory there is a condition

$$kr^2 = \text{const} \qquad (88)$$

between the rate constant k of silver deposition and the diameter of 2r of the filament that has been confirmed experimentally (197).

The stability of the filamentary structure may also be caused by adsorption phenomena. In the absence of adsorbants the filaments are unstable, since the surface energy is high. In model experiments with cathodic reduction of anodic silver halide layers (especially silver iodide), a highly dispersed black filamentary structure was initially found (198), which became rearranged in the solution until compact white silver was formed. The rate of levelling increased with the concentration of complexing agents in the solution (I^-, CN^-, SCN^-, $S_2O_3^{2-}$). This might be explained by an intermediate complexation of the surface: The surface atoms are partially ionized and solvated by complexing agents. The complexes diffuse along the solution side of the surface together with the electrons that remain in the surface of the metal and form the negative part of the surface dipole.

From the dependence of the levelling rate on the potential, an ionic character of 0.3 of the surface atoms was calculated. The levelling was also enhanced at positive potentials (199). Therefore, during the reduction of AgI and AgBr, but not of AgCl, filaments were observed. This is in agreement with photographic experience.

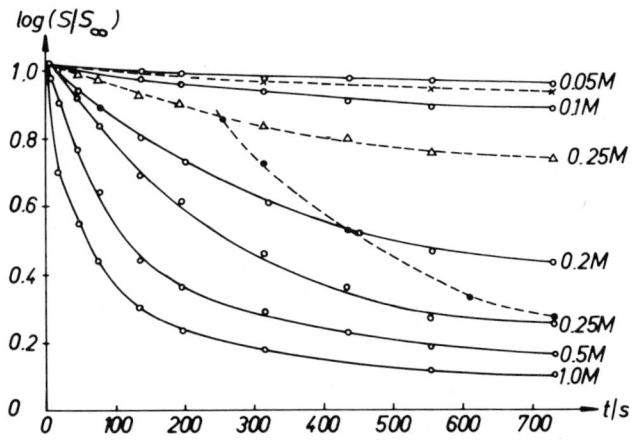

Fig. 23. Levelling of silver filaments, formed by cathodic reduction of silver iodide, as a function of time (198). Ordinate, logarithm of relative silver surface S/S_∞ (S_∞, polished silver). Electrolyte, iodide solutions, saturated with AgI, 25°C. o——o 0.05 to 1.0 M; Δ---Δ 0.25 M with negative potential (- 0.2 V); o---o 0.25 M, potential switched off after 260 sec; x---x 0.25 M without potential, with gelatin.

The stability of filaments is greatly enhanced in the presence of gelatin (Fig. 23) (198). In photographic layers antifoggants, dyes, and developer oxidation products are probably also important as stabilizing substances (200-202).

7.6 Kinetics of Photographic Development

Experiments on the kinetics of development were usually performed in photographic layers. The results, however, are of little value for the kinetics of single grains, since the grains are exposed at different light intensities and developed at different times, depending on their situation within the layer. A simpler interpretation is possible with experiments in fluid colloidal suspensions of silver halide (204) or silver (205). With such systems the change in concentration of the reacting substances can be measured by polarography, potentiometry, or extinction.

The kinetics of growth of single image specks can be deduced from electron micrographs of grains at different stages of development (185,206,207) and from experiments on short-circuited galvanic cells (181,182). Since, during development, the activity of the nucleus decreases (see Section 5.3) and the reaction area increases, the reaction must be autocatalytical (181).

The initial rate further depends on adsorption effects: As a result of adsorption of anions or anionic dyes, the induction period of development is the more extended the greater the negative charge of developer molecules (208). This effect can be cancelled if cationic substances (quaternary salts) or positively charged dyes are adsorbed. The electron transfer can be promoted by adsorption of sensitizers (209); it can be inhibited by adsorption of desensitizers or substances such as benzotriazole or phenylmercaptotriazole (210).

7.6.1 PHYSICAL DEVELOPMENT

A quantitative treatment of physical development (Eq. 79) is simple, because during the growth of the nuclei the reaction area and the transport properties of the cathodic and the anodic reaction are changed in the same way.

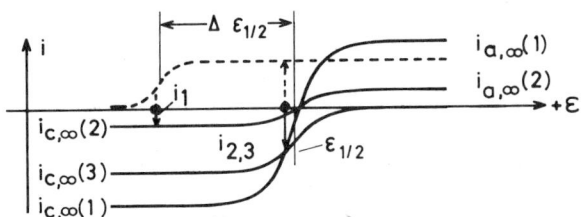

Fig. 24. Superposition of current potential curves during physical development. Broken line, anodic reaction (Eq. 89); solid lines, cathodic reactions (Eq. 90); solid circle, mixed potential ε_m; i, current density; subscript a,c, anodic, cathodic; subscript ∞, limiting current.

The current J of each reaction is given as a function of the potential by the equations of reversible or irreversible polarographic waves (211). In Fig. 24 the superposition is shown for the same anodic reaction

$$R^- \rightarrow 0 + e^- \tag{89}$$

with three different cathodic reactions,

$$[AgY_m]^{1-m} + e^- \rightarrow Ag + m \cdot Y^- \tag{90}$$

that differ in the concentration of the complex and the excess of the ligand Y but are equal in their half-wave potential.

If the difference of the half-wave potentials of both reactions is sufficient, a limiting current of one of the components ($[AgY_m]^{1-m}$ or R^-) is usually expected. For this special condition the law of growth of the nuclei can easily be derived. The rate j_r of the surface reaction is assumed to be proportional to the area $4\pi r^2$ of the spherical nucleus and to the concentration C_j of either the complex or the developer in the liquid at the surface of the crystal (ν_j is the stoichiometric coefficient;

n is the amount of substance):

$$j_r = \frac{\dot{n}}{\nu_j} = \frac{k4\pi r^2 c_j}{\nu_j} \quad ; \quad \dot{n} = \frac{dn}{dt} \quad (91)$$

The transport rate j_d is given by Eq. 84. If a steady state of diffusion is attained ($t \leq 10^{-3}$ sec), Eq. 84 becomes

$$j_d = \frac{4\pi Dr(c_j - C_j)}{\nu_j} \quad (92)$$

where c_j is the bulk concentration. Since $j_r = j_d$, C_j can be eliminated from Eqs. 91 and 92. The integration gives

$$n^{2/3} + n^{1/3} (2D/k) \left(\frac{4\pi}{3V_M}\right)^{1/3} = \left(\frac{4\pi}{3}\right)^{2/3} \frac{V_M^{1/3} 2Dc_j t}{\nu_j} + \text{const} \quad (93)$$

where V_M is the molar volume. The second term on the left (reaction term) is decisive for the start of development; the first term (diffusion term) can be rate determining for the later stages. Both laws, $n^{2/3} \sim c_j t$ (212) and $n^{1/3} \sim c_j t$ (178,213) were found experimentally.

For the general case (outside the limiting current) an expression for the kinetics can be given if the reaction is diffusion controlled, that is, if the equilibrium equation, Eq. 79, applies at the surface of the growing nucleus (183). Then, at the surface (index S is the silver complex) one has

$$\frac{C_O C_Y^m}{C_S C_R} = K = \exp \frac{(\varepsilon_R^O - \varepsilon_{AgY_m}^O) F}{RT} \quad (94)$$

If all C_j are expressed by c_j, using the diffusion rate Eq. 92, Eq. 94 can be integrated. The result is

$$n^{2/3} = n_o^{2/3} + \left(\frac{4\pi}{3}\right)^{2/3} V_M^{1/3} \cdot f(c_j, D_j, K) \cdot t \quad (95)$$

In the special case in which Y is added in excess ($j/4\pi\ rmD_Y \ll c_Y$), and (as usual) $c_0 = 0$, the function $f(c_j, D_j, K)$ is

$$f = A - \sqrt{A^2 - B}$$

$$A = c_S D_S + c_R D_R + \frac{c_Y^m D_Y D_S}{D_0 K} \tag{96}$$

$$B = 4\ c_R D_R c_S D_S$$

The rather complicated dependence of the concentrations is remarkable.

For rate-determining electron transfer reactions a first approximation is given in Eqs. 110 and 113 (Section 6.2), which can also be applied to physical development.

7.6.2 CHEMICAL DEVELOPMENT In chemical development one cannot use the simple condition of either cathodic or anodic limiting current as in physical development since there is no limiting current of the cathodic process, the silver halide reduction. Also, there is no simple relationship between the two reaction areas, their dependence on time, and the conditions of diffusion, but it can be assumed, as in physical development, that the concentrations of the reacting species enter into the kinetic equations in a similar way to that in Eq. 95.

Under these conditions, model experiments with galvanic cells of the form

(cathode) Ag|AgBr|developer|Ag (anode)

are useful (181). On the basis of the dependence of the short-circuit current on the developer composition, the same relation

$$j = k \cdot c_R^a \cdot c_{OH^-}^b \quad (0 \le a, b \le 1) \tag{97}$$

between the rate of development j and the concentrations (R is the developer) was found as applies to photographic layers (214).

This result, the decrease in reaction rate from AgCl to AgI, the dependence of the addition of halide to the developer, and the dependence of an activation energy of development on the developer composition, could all be understood if the current-voltage curves of a diffusion-controlled developer oxidation (reversible polarographic wave) and of a kinetically controlled silver halide reduction were superimposed at the mixed potential ε_m in the short-circuited cell (182).

Recently, Brown and Tong (246) tried to describe chemical development in a similar way by means of two superimposed diffusion-controlled current potential curves, using a model of a hemispherical silver speck of radius r on a spherical silver halide grain of radius r' (r << r'). By simplifying the model of Jaenicke and Sutter (182), a constant mixed potential ε_m on the outer surface of the total system was assumed, whereas in the former work only the anodic and the cathodic surfaces of the speck were considered. Additionally, an irreversible reaction of the oxidized developer has been considered, for example, sulfonation or deamination with a pseudo-first-order rate constant k_d.

Under the conditions mentioned the reaction rate can be derived, taking into account a developer oxidation scheme as in Eq. 82 and a fast symproportionation of the radicals (Eq. 83). For this case (250) the reversible polarographic wave for stationary diffusion up to a hemisphere of radius r (according to Eq. 84) is given by

$$j = \frac{\pi r \cdot D_R c_R (2 + A\sqrt{K_s})}{1 + A\sqrt{K_s} + A^2} \quad (98)$$

with

$$A = ([H^+]^2 + K_1[H^+] + K_1 K_2)^{1/2} \cdot \exp(\varepsilon_R^o - \varepsilon_m)\frac{F}{RT} \quad (99)$$

If the mixed potential is not obtained from the measured cathodic current-potential curve as in (182) but is calculated from a stationary diffusion current of halide ions from the silver crystal of radius r' (246), one has

$$\varepsilon_m = \varepsilon^o_{AgX} - \frac{RT}{F} \ln\ c_X + (\frac{j}{4\pi r' D_X}) \tag{100}$$

where c_X is the halide present in the developer solution. One now obtains

$$A = ([H^+]^2 + K_1[H^+] + K_1 K_2)^{1/2} (c_X + \frac{j}{4\pi r' D_X}) \exp\ (\varepsilon^o_R - \varepsilon^o_{AgX})\frac{F}{RT} \tag{101}$$

If, additionally, the decomposition of the oxidized developer is taken into account (246), A in Eq. 98 must be replaced by A/Q where

$$Q^2 = 1 + \left(\frac{k_d}{D_R}\right)^{1/2} \cdot r \tag{102}$$

Some special cases of Eq. 98 are of interest:

1. $K_s = 0$ (no symproportionation reaction):

$$j = \frac{2\pi r \cdot D_R c_R}{1 + (A/Q)^2} \tag{103}$$

This case was discussed in (246) (a factor of 2π in the numerator is used instead of 4π in the original paper). Three limiting cases can be discussed:

a. $(A/Q)^2 \ll 1$; diffusion-controlled reaction, j being proportional to c_R, independent of $\varepsilon^o_{AgX} - \varepsilon^o_R$.

b. $(A/Q)^2 \gg 1$; $c_X \gg j/4\pi r' D_X$ (large amount of halide in the developer):

$$j \sim c_R \cdot c_X^{-2} \cdot \exp\ (\varepsilon^o_{AgX} - \varepsilon^o_R)\frac{2F}{RT}$$

In this case one obtains $\log j = \text{const} - 2F\ \varepsilon^o_R\ /\ RT$ as was observed in (161-163).

c. $(A/Q)^2 \gg 1$; $c_X \ll j/4\pi r \cdot D_X$ (small addition of halide or no halide at all);

$$j \sim c_R^{1/3} \cdot \exp\ (\varepsilon^o_R - \varepsilon^o_{AgX})\frac{2F}{3RT}$$

The expressions for j in the three limiting cases (a), (b), and (c) should be compared with Eq. 97. One obtains with b dependent on the pH:
a. $a = 1; b = 0$.
b. $a = 1; 0 < b < 2$.
c. $a = 0.3; 0 < b < 0.6$.

In the cases (b) and (c) the rate increases if the developer oxidation product is very unstable, that is, when $(k_d D_R)^{1/2} \cdot r$ is comparable to 1 (see Eq. 102).

2. $K_s \approx 4$ (as for hydroquinone):

$$j = \frac{2\pi r \cdot D_R c_R}{1 + (A/Q)} \qquad (104)$$

In this case we have for the three limiting cases just discussed:
a. $a = 1; b = 0$.
b. $a = 1; 0 < b < 2$.
c. $a = 0.5; 0 < b < 1$.

3. For large values of K_s we have, when $A\sqrt{K_s} > A^2 > 2$,

$$j = \pi r \cdot D_r c_R$$

This limiting current is observed if the reaction can only proceed up to the radical.

Irreversible developer oxidation also has to be considered but does not change the dependence on concentrations.

Photographic experiments show that most practical developer systems fall in an intermediate range of rates where at least two terms in the denominator of Eq. 98 apply (246).

From the model experiments (182) with galvanic cells it was further shown that the exponents a and b (Eq. 97) are dependent on the ratio of cathodic and anodic areas (and consequently on the stage of development).

So far quantitative expressions for the total mechanism of single grain development have not been stated on the basis of the theory of local voltaic couples. An interesting attempt was made by Matejec and Meyer (215) but it can be shown (183) that their boundary condition is only valid under rather improbable suppositions.

A rather simple model for the total reaction of the development of a single grain was used by Pontius, Willis, and Newmiller to interpret their electron micrographs of the growth of development centers (185,206,207). At the beginning of the reaction, as in physical development, they found spherical nuclei that grew proportionally to their surface area in accordance with Eq. 93 in the form

$$n^{1/3} = n_o^{1/3} + f(c) \cdot t \tag{105}$$

(see Fig. 25). When the spheres had reached a given thickness, a filamentary growth began, in accordance with the equation

$$\ln n = \ln n_1 + f(c) \cdot (t - t_1) \tag{106}$$

This law is obeyed if a cylinder of constant radius is lengthened by a rate proportional to the mantle area, that is, the area of the anodic process. Therefore, the anodic reaction must be rate determining.

The experiments usually covered only the first 5% (in a few cases up to 35%) of the total time of development. There are, however, some microscopic measurements (190) from which it can be concluded that in later stages the growth of the filaments is linear:

$$n = n_2 + f(c) \cdot (t - t_2) \tag{107}$$

This law is obeyed if the base of a cylinder growing in length is the area of the rate determining reaction. In later stages, therefore, the cathodic reaction becomes rate determining.

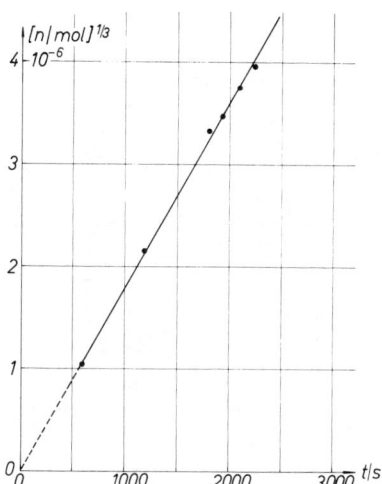

Fig. 25. Dependence of the mass (mol) of development centers on development time, initial stage of development (Eq. 105) (185). Developer, 0.01 M ascorbic acid, 1 g KBr/L, pH = 10.

In the experiments of Pontius and Willis, the rate of development was extremely small. The maximum value of the rate was $1.5 \cdot 10^{-9}$ mol cm^{-2} sec^{-1}. This corresponds to a current density of $1.5 \cdot 10^{-4}$ A cm^{-2}, which is so small that anodic controlled reactions can be superimposed with an unpolarizable cathodic current-potential curve (Fig. 26b), and cathodic controlled reactions with an unpolarizable anodic curve (Fig. 26b). Therefore, under anodic control the mixed potential is (Fig. 26a)

$$\varepsilon_m = \varepsilon_{AgBr} = \varepsilon^o_{AgBr} - \frac{RT}{F} \ln\left(a_{Ag} \cdot a_{Br^-}\right) \qquad (108)$$

The reaction rate becomes

$$j = F \cdot J = S \left[k_f c_R \exp \frac{\alpha z F \varepsilon_m}{RT} - k_b c_O \exp \frac{(\alpha-1) z F \varepsilon_m}{RT} \right] \qquad (109)$$

where S is the surface area of the rate determining process.

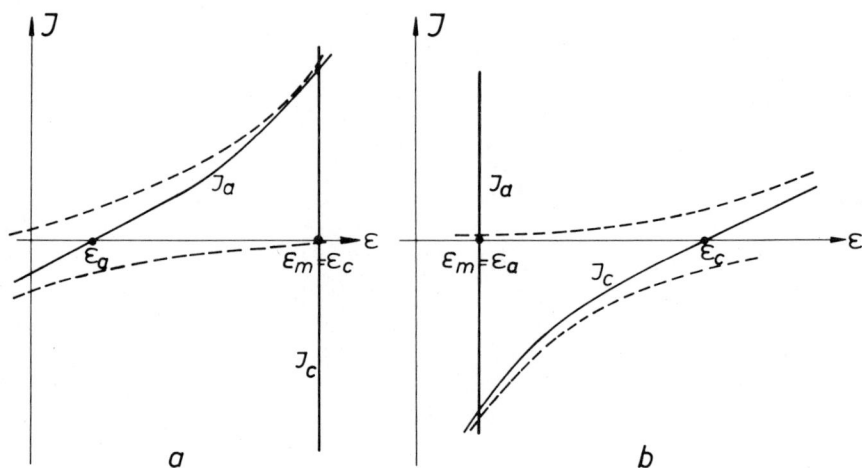

Fig. 26. Superposition of an unpolarizable and a polarizable current-potential curve. (a) anodic control (Eqs. 109 and 110). (b) cathodic control (Eqs. 112 and 113).

If $\varepsilon_{Ag} - \varepsilon_R > 200$ mV, the last term in Eq. 109 can be neglected. If Eq. 108 is substituted in Eq. 109, we have

$$j \approx S\, k_a\, c_R\, a_{Br^-}^{-\alpha z} \cdot a_{Ag}^{-\alpha z}; \quad k_a = k_f \exp \frac{\alpha z F\, \varepsilon_{AgBr}^o}{RT} \qquad (110)$$

In the same way, for cathodic control (Fig. 26b) we have

$$\varepsilon_m = \varepsilon_R^o + \frac{RT}{zF} \ln \frac{c_O}{c_R} \qquad (111)$$

and

$$j = S \left[k_f'\, a_{Ag}\, \exp \frac{\alpha F \varepsilon_m}{RT} - k_b'\, a_{Ag^+}\, \exp \frac{(\alpha-1) F \varepsilon_m}{RT} \right] \qquad (112)$$

Under the same conditions as in Eq. 110 the first term can be neglected:

$$j \approx S\, k_c \left(\frac{c_R}{c_O}\right)^{(1-\alpha)/z} \cdot a_{Ag^+}; \quad k_c = k_b' \exp \frac{(\alpha-1) F \varepsilon_R^o}{RT} \qquad (113)$$

Depending on the shape of the nucleus (i.e., the values of S), by integration of Eq. 110 or Eq. 113 the observed rate laws, Eq. 105, 106, or 107 are obtained.

If the experimental straight line $n^{1/3}(t)$ of the initial growth of development centers in Fig. 25 is extrapolated, the origin of the coordinates is reached. It follows that there is no change in the mechanism for the observed period of the reaction. The so called induction period of development is merely a consequence of the autocatalytic increase of the reaction area.

The experimental conditions, however, were unusual; furthermore, the thickness of the filaments was greater (70 to 100 nm) than normally found (15 to 20 nm). In general, the initial activity a_{Ag} in Eq. 110 cannot be taken as unity, but rather Eq. 50 should be used. Considering this effect, a more exact expression for the initial rate can be derived (216), the result of which is a short induction period.

So far it has been assumed that there is no inhibition of the anion transfer into the solution. This condition is usually fulfilled, but it is also possible to inhibit this reaction, for example, by adsorbed substances such as benzotriazole (241) (see Fig. 22a).

7.7 Some Special Electrocatalytic Effects in Development

7.7.1 LITH DEVELOPMENT For the production of screened half-tone images a developer is necessary that gives low fog and very high contrast γ; γ is the slope of the characteristic curve (cf. Fig. 8): $\gamma = (\partial B / \partial \log H)_t$. In the so-called lith development (217) hydroquinone generally is used as developing substance. The antioxidant sulfite is added in form of sodium formaldehydebisulfite, which permits an equilibrium with a very low stationary sulfite concentration. To explain the lith effect, Suga (218) has suggested an electrochemical mechanism: Because of the low sulfite concentration, the oxidation product quinone is rather stable in the developer solution. It was assumed that the grains with greater latent images are first developed, giving rise to quinone. In the presence of bromide ions, quinone is able to oxidize smaller latent images on adjacent grains, which in this way become undevelopable.

7.7.2 SUPERADDITIVITY If the rate of development is measured as a function of the mole fraction of binary developer mixtures, many systems show pronounced maxima (Fig. 27). This "superadditivity" (219) is found especially in mixtures with hydroquinone and in the presence of an antioxidant such as sulfite, which is contained in most developers.

The superadditivity can be explained by an electron transfer from a reduced molecule A (e.g., hydroquinone) to the speck via an adsorbed molecule B, which seems to be stationary in a kind of semiquinone state. Therefore, the superadditivity depends on the redox potential ε_1^o of the first electron transfer of the catalyst B (143) (see Eq. 83). The superadditivity increases with the stability of uncharged or positively charged semiquinone states of the catalyst B (221), given by the symproportionation constant K_s (Eq. 83). If the rate of development of equal mixtures containing one developer A and various redox indicators B, which work as auxiliary developers, is plotted against the redox potentials of B, a sharp maximum is found within the region of potentials, given by the half-wave potential of A and the silver halide potential (143,222).

Model experiments with linear sweep voltammetry using stationary electrodes (223) or rotating disks (224) of silver, gold, or

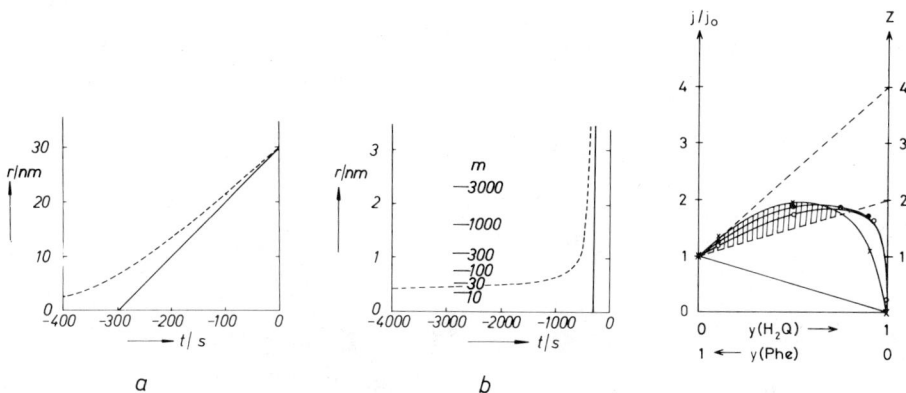

Fig. 27. Superadditivity as a function of the composition of a hydroquinone-phenidone (= 1-phenyl-3-pyrazolidone) developer (223,234). Ordinate, relative rate of development, related to pure phenidone. Abszissa, mole fraction. ●——●, physical development with KBr; o——o, without KBr; x——x, chemical development with KBr. Straight line, expected for additivity, broken line, calculated for reversible two- or four-electron transfer (z = 2, z = 4). Hatched area, minimum fraction of development by hydroquinone monosulfonic acid.

platinum have proved that the regeneration of B by A is a heterogeneous reaction of adsorbed catalyst with the developer (223).

The superadditivity can be formulated electrochemically as an increase in the exchange current of the electrode when the catalyst is added. Slow electron transfer was observed in particular with hydroquinone in the presence of sulfite ions (223,237). In these solutions the oxidation product, quinone, reacts at a high rate, producing hydroquinone monosulfonic acid, which, however, is a poor developing substance. When the auxiliar developer was added, the hydroquinone and the hydroquinone monosulfonic acid were oxidized reversibly (Fig. 28). Therefore, in superadditive mixtures, containing hydroquinone and sulfite, the number z of electrons transferred per molecule of hydroquinone increases from 2 to 4.

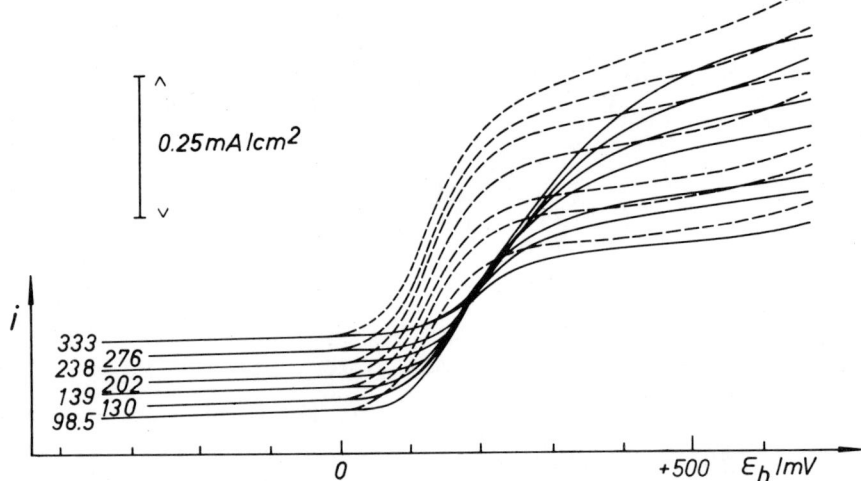

Fig. 28. Anodic current-potential curves of $5.5 \cdot 10^{-3}$ M hydroquinone monosulfonic acid, measured with a rotating disk of Au at 20°C, pH = 11.3 (224); dashed curves, addition of $6.7 \cdot 10^{-5}$ M phenidone. Parameter, rotational speed (sec^{-1}). Potential versus NHE.

Suitable auxiliary developing substances are 1-phenyl-3-pyrazolidone, p-phenylene diamines, p-aminophenols, heterocyclic azines (222), and electron-trapping substances such as desensitizers (143) (see Section 6.6).

7.7.3 CATALYTIC INTENSIFICATION IN COLOR DEVELOPMENT   In color development one can take advantage of another electrocatalytic mechanism, using oxidizing substances which are able to react spontaneously at the surface of latent images. In this way developers can be oxidized, and dyes can be formed by subsequent coupling processes without any participation of silver ions.

Following this idea hydrogen peroxide or other peroxy-compounds were used (251,252) which are known to react electrocatalytically at silver surfaces (253). Better results were obtained with some Co(III)-complexes (254,255) especially with $[Co(NH_3)_6]^{3+}$ - ions.

These complexes are unable to oxidize developers like substituted p-phenylene diamines in a homogeneous reaction but in contact with silver or nuclei of some other metals their reduction is greatly enhanced, which probably is caused by catalytic exchange of ligands, forming complexes of lower stability.

The process may be described by the following reaction scheme:

$$[Co(NH_3)_6]^{3+} + Cl^- + lat.image \rightarrow Ag..[..Cl\ Co(NH_3)_5]^{2+}_{ad} + NH_3 \quad (115)$$

$$Ag..[..Cl\ Co(NH_3)_5]^{2+}_{ad} + H_2O \rightarrow AgCl + [Co(NH_3)_5 H_2O]^{2+} \quad (116)$$

$$AgCl\ (at\ latent\ image) + R^- \xrightarrow{lat.image} Ag + Cl^- + O \quad (117)$$

$$O + coupling\ substance \rightarrow dye \quad (118)$$

This scheme is consistent with the observation that small latent images can be destroyed (bleached) by the complex (Eq. 116) while at large latent images the regeneration reaction (Eq. 117) takes place.

The process can be used to intensify color images (254) and, if fixing agents are added, to carry out one-bath color processes (255).

7.7.4 THE SILVER DYE BLEACH PROCESS In the silver dye bleach process insoluble azo dyes are incorporated into silver halide emulsions. After exposure and normal development, the developed silver is used to reduce the azo dyes to colorless soluble amino compounds that can be removed by washing. In this way positive color images are formed directly if the azo dye is colored complementary to the optical sensitization of the layer (225).

The steps of the process can be carried out successively if the redox potential is adjusted by means of the pH value (226). The reduction of the azo dye is given by the reaction

$$R-N=N-R' + 4\ H^+ + 4\ e' \rightarrow 2\ R-NH_2 \qquad (119)$$

Therefore, $\partial\varepsilon/\partial pH = -0.059$ V, as in the developer oxidation. Since at high pH $\varepsilon_{Ag} > \varepsilon_{developer} > \varepsilon_{dye}$, at pH = 10 only the exposed silver bromide is reduced and silver is formed. ε_{Ag} is independent of the pH. Therefore, the sequence of the potentials can be inverted at low pH. In particular $\varepsilon_{Ag} < \varepsilon_{dye}$ if a low ε_{Ag} is enforced by addition of complexing substances such as thiourea or iodide.

Silver and dye, however, are insoluble. The interreaction of these two substances is rendered possible by a soluble redox catalyst, for example, quinoxaline = B which reacts in accordance with the following (simplified) scheme (Fig. 29).

BH_2 is an electroinactive tautomer formed irreversibly from HBH in a first-order reaction with the rate constant k. The dye is reduced only by the active dihydro compound, not by the radical (227).

To obtain the right color at the right place, the following reaction rates must be controlled: transport of the catalyst to the silver, electron transfer from the silver, transport of the reduced catalyst to the dye, and reductive cleavage of the dye. The total process is controlled by the simultaneous irreversible tautomerization that permits a small reaction volume around the silver. If tautomerization did not take place, dye remote from

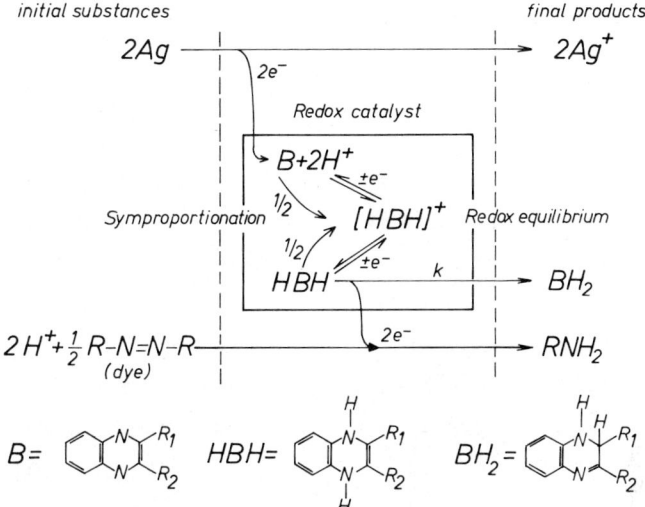

Fig. 29 Reaction scheme of the silver dye bleach process.

the silver might be bleached and undesired colors would result; if it is too fast, only a low efficiency of the silver is obtained for the dye reduction.

The mechanism of the process has now been clarified and the diffusion and reaction rates measured by various electrochemical methods (228).

For a new process (229) that is still in the experimental stage, the claim is made that it avoids the first development and uses the redox catalyst directly as developing substance. In the exposed area the silver bromide is reduced, and in the unexposed area the azo dye is reduced. When the silver is oxidized in a bleaching bath, a color image remains.

8 ELECTROCHEMICAL ASPECTS OF FIXATION

The complex dissolution of the silver halides, especially of strong disordered surfaces, can be described by means of electrochemical surface reactions, followed by diffusion processes (230). It was assumed that the exchange rate of anions and cations between the crystal surface and the solution is different and dependent on concentration and potential. In the case of equilibrium the current-potential curves of the cationic and anionic process intersect at zero current. If net dissolution takes place, an overpotential is established, which guarantees equal net currents of anions and cations. In the experiments (230) the dissolution rate was changed by the rate of rotation of a silver electrode, covered with an anodic layer of silver halides, which dipped into a complexing solvent. The deviation from the calculated equilibrium potential was measured as a function of the rate of solution.

The simple theory is only valid for disturbed surfaces. At single crystals the separation of ions from the surface should be a two-step-process, the rate determining step being the removal of an ion from a growth site. The stationary number of cathodic and anodic growth sites is regulated by the respective potentials, which depend upon the activation energies of separation (231). A number of experiments on monocrystals of salts were performed and confirmed this theory (232).

Another kind of electrochemical dissolution takes place during monobath development, or when partially developed grains are fixed. As a result of the concentration gradient within the diffusion layer, during a complexation process the potential at the crystal surface is always more positive than within the bulk of the solution (187,232).

Therefore, at the end of the filaments that have grown into the solution, silver is dissolved anodically at the top and deposited cathodically at the base of the filaments. This was proved by model experiments with galvanic cells (233). The total reaction is equivalent to an electrochemical dissolution of AgBr, forming a complex, for example, $\left[Ag(S_2O_3)_2\right]^{3-}$ (Fig. 22d). According to this mechanism the contact between the filament and the crystal is maintained during the dissolution. This also permits chemical development during monobath processes. At the same time a rather compact silver structure is formed.

9 LIST OF SYMBOLS

A	electron affinity
$\mathcal{A}$	affinity (molar scale)
a	activity
a_λ	absorbed fraction of energy
α	transfer coefficient
B	optical density, absorbance log (1/T)
b	crystal thickness
C	concentration of a liquid at a phase boundary
c	bulk concentration (ions and uncharged particles)
D	diffusion coefficient
D_∞	optical dielectric constant
d	jump distance, lattice parameter
δ	Debye length
E	energy (molecular scale)
ΔE	excitation energy
ΔE_C	band gap
ΔE_t	trap energy

e'	electron
e_o	unit charge
ε_o	permittivity of vacuum
ε	dielectric constant (in combination with ε_o)
ε	potential of an electrode
$\varepsilon^{\bullet}$	absolute potential of an electrode
ε^o	standard potential
ε^o_{ox}	oxidation potential
ε^o_{red}	reduction potential
F	Faraday constant
(f)	fluid
G	free enthalpy (molar scale) (subscript number: number of equ.)
(g)	gaseous
γ	contrast $(\partial B/\partial \log H)_t$
γ	surface tension
H	exposure $(\int J\, dt)$
h	Planck's constant
$h^{\bullet}$	hole
I	interstitial position
I	ionisation energy
i	current density
J	current
J	radiant intensity
j	rate of reaction
K	equilibrium constant (subscript number: number of equ.)
K_+, K_-	positive, negative kink site
k	rate constant
κ	conductivity
L	lattice position
L	solvation enthalpy (molecular scale)

List of Symbols

L_V	heat of evaporation (molecular scale)
l	length
λ	wave length
M	light-induced electric moment
m	number of atoms in a speck
μ	chemical potential (molar scale)
N	concentration of particles
N_L	Loschmidt's constant
NHE	standard hydrogen electrode
n	concentration of electrons (Sections 1 to 7.3)
n	amount of substance (Sections 7.4 to 8)
ν	frequency
ν_j	stoichiometric coefficient
P	pressure
p	concentration of holes
Φ	quantum efficiency
Φ_r	relative quantum efficiency
ϕ	potential
$\Delta\phi$	potential difference
Ψ	work function
q	$\|E_{LV}\| - A_C$
R	gas constant
r	radius, molecular distance
ρ	boundary free energy
S	area
s	surface concentration in a crystal
(s)	solid
σ	cross section
T	absolute temperature
T	transmittance (ratio of transmitted to incident radiance J/J_o)

t	time
t	transference number (with subscript)
τ	relaxation time, lifetime
U	energy of activation or dissociation
U(r)	electrostatic energy (molecular scale)
u	electric mobility (with subscript)
V	vacancy
V	volume
v	thermal velocity
W	efficiency
w	mean free path
x	distance
Z	concentration of lattice positions
Z_i	concentration of interstitial positions
z	charge number (ion or reaction)

Subscripts

a	anodic		L	lattice position
b	bulk		j	running number
C	crystal		LV	lowest vacant
c	cathodic		M	molar
D	dye		m	mixed
d	diffusion		n	electron
e	electronic		O	oxidized form
F	Fermi		o	at zero value
f	free		p	hole
H	Hall		R	reduced substance, developer
HO	highest occupied		r	radius r, surface reaction
I	interstitial position		S	silver complex
i	interstitial		s	surface, symproportionation

T	energy gap midpoint (Eq. 57)	Y	ligand
t	trapped	∞	limiting, infinite radius
v	vacancy	•	total
X	halogen		

10 REFERENCES

1. C.E.K. Mees, in T.H. James (Ed.), The Theory of the Photographic Process, 3rd ed., The McMillan Comp., New York (1966).
2. H. Frieser, G. Haase and E. Klein (Eds.), Die Grundlagen der photographischen Prozesse mit Silberhalogeniden, Akademische Verlagsgesellschaft, Frankfurt am Main (1968).
3. J.F. Hamilton, in J.O McCaldin, G. Somorjai (Eds.), Progress in Solid State Chemistry 8, Pergamon Press, Oxford (1973).
4. R.W. Gurney and N.F. Mott, Proc. Roy. Soc. (London) 164A, 151 (1938).
5. H.D. Jonker, C.J. Dippel, H.J. Houtman, G.J.G.F. Jansen, and L.K.H. van Beek, Photogr. Sci. Eng. 13, 1 (1969).
6. R.I. Dawood, A.J. Forty, and M.R. Tubbs, Proc. Roy. Soc. (London) A284, 272 (1965); M.R. Tubbs, J. Photogr. Sci. 17, 162 (1969); W. De Gruyter and J. Schoonman, Photogr. Sci. Eng. 17, 382 (1973).
7. H. Jonker, C.K.H. van Beek, C.J. Dippel, G.J.G.F. Jansen, A. Molenaar, and E.J. Spiertz, J. Photogr. Sci. 19, 96 (1971).
8. J. Malinowski, Photogr. Sci. Eng. 15, 175 (1971); R. Stoychewa-Topalova, and J. Malinowski, J. Photogr. Sci. 22, 262 (1974).
9. J. Malinowski, Photogr. Sci. Eng. 18, 363 (1974).
10. L.M. Sliwkin, Sci. Progr. (Oxford) 60, 151 (1972).
11. J. Teltow, Ann. Physik 6 5, 71 (1949).
12. J.F. Hamilton, Photogr. Sci. Eng. 18, 493 (1974).

13. T.B. Grimley and N.F. Mott, Disc. Faraday Soc. 1, 3 (1947); T.B. Grimley, Proc. Roy. Soc. (London) 201A, 40 (1950).
14. K. Lehovec, J. Chem. Phys. 21, 1123 (1953).
15. F. Trautweiler, L.E. Brady, J.W. Castle, and J.F. Hamilton, in in G.A. Somorjai (Ed.), The Structure and Chemistry of Solid Surfaces, 4th International Materials Symposium, University of California, Berkeley, Calif., Wiley, New York, p. 83 (1969).
16. S. Danyluk and J.M. Blakely, Surface Sci. 40, 37, 359 (1973); J.M. Blakely and S. Danyluk, Photogr. Sci. Eng. 18, 489 (1974).
17. V.I. Saunders, R.W. Tyler, and W. West, Photogr. Sci. Eng. 12, 90 (1968).
18. F. Trautweiler, Photogr. Sci. Eng. 12, 98 (1968).
19. J. van Biesen, J. Appl. Phys. 41, 1910 (1970).
20. S. Takada, Japan J. Appl. Phys. 12, 190 (1973); Photogr. Sci. Eng. 18, 500 (1974).
21. J.F. Hamilton and L.E. Brady, J. Appl. Phys. 30, 1893, 1902 (1959).
22. J.F. Hamilton and L.E. Brady, J. Phys. Chem. 66, 2384 (1962); J. Appl. Phys. 35, 1565 (1964); 37, 2268 (1966).
23. P. Müller, Phys. Status Solidi 12, 775 (1965).
24. K.L. Kliewer, J. Phys. Chem. Solids 27, 705, 719 (1966).
25. R.C. Baetzold, J. Phys. Chem. Solids 35, 89 (1974).
26. I.V. Ardashev and D.M. Samoylovich, Photogr. Sci. Eng. 17, 348, 351 (1973).
27. I. Ebert and J. Teltow, Ann. Physik 15, 268 (1955).
28. R.C. Baetzold and J.F. Hamilton, Surface Sci. 33, 461 (1972).
29. See K.J. Vetter, Electrochemical Kinetics, Academic Press, New York, London, 1967, p. 316.
30. Yu. P. Smirnov, Zh. Nauch. Priklad. Fotogr. Kinematogr. 19, 224 (1974).

31. Y. Tan and H. Hoyen, Surface Sci. 36, 242 (1973); H. Hoyen, Photogr. Sci. Eng. 17, 188 (1973).
32. R.B. Poeppel and J.M. Blakely, Surface Sci. 15, 507 (1969); see S. Danyluk and J.M. Blakely, Surface Sci. 40, 37, 359 (1973).
33. E.P. Honig, Nature 225, 537 (1970).
34. M. Mirnik, Croat. Chem. Acta 42, 161 (1970).
35. R. Matejec and R. Mayer, Z. Physik. Chem. N.F. 55, 94 (1967); R. Matejec, R. Mayer, and G. Kaufhold, Z. Physik. Chem. N.F. 59, 251 (1968).
36. R. Matejec, Z. Electrochem. 66, 326 (1962); Photogr. Sci. Eng. 7, 123 (1963); Photogr. Korr. 104, 71 (1968).
37. J.E. Hall and L.J. Bruner, J. Chem. Phys. 50, 1596 (1969).
38. T. Tani and S. Takada, Photogr. Sci. Eng. 18, 620 (1974).
39. A.P. Batra and L. Slifkin, J. Phys. Chem. Solids 30, 1315 (1969).
40. M.D. Weber and R.J. Friauf, J. Phys. Chem. Solids 30, 407 (1969).
41. P. Debye, Trans. Electrochem. Soc. 82, 265 (1942).
42. W. Rind and W. Martienssen, Photogr. Sci. Eng. 17, 58 (1973).
43. F.A. Kröger, J. Phys. Chem. Solids 26, 901 (1965).
44. C. Wagner, Z. Elektrochem. Ber. Bunsenges. Physik. Chem. 63, 1027 (1959).
45. P. Müller, S. Spenke, and J. Teltow, Phys. Status Solidi 41, 81 (1970).
46. W.C. Lewis and T.H. James, Photogr. Sci. Eng. 13, 54 (1969); T.H. James, Photogr. Sci. Eng. 14, 84 (1970); 18, 100 (1974).
47. R.E. Maerker, J. Opt. Soc. Am. 44, 625 (1954).
48. J. Berry, J. Photogr. Sci. 21, 202 (1973).
49. E. Fatuzzo and S. Coppo, J. Photogr. Sci. 12, 138 (1968).
50. J.M. Hedges and W.J. Mitchell, Phil. Mag. 44, 223 (1953).

51. R.C. Brandt and F.C. Brown, Phys. Rev. 181, 1241 (1969).
52. P. Müller, Phys. Status Solidi 12, 392 (1965).
53. M. Höhne and M. Stasiv, Phys. Status Solidi 28, 57 (1968).
54. J. Malinowski, J. Photogr. Sci. 16, 57 (1968); Photogr. Sci. Eng. 14, 112 (1970).
55. J. Malinowski, Contemp. Phys. 8, 285 (1967).
56. W. West and V.I. Saunders, J. Phys. Chem. 63, 45 (1959); D.C. Burnham and F. Moser, Phys. Rev. 136 A, 744 (1964).
57. E. Eisenmann and W. Jaenicke, Z. Physik. Chem. N.F. 49, 1 (1966); Photogr. Sci. Eng. 11 121 (1967).
58. J. Malinowski, J. Photogr. Sci. 16, 57 (1968).
59. J. Malinowski, Photogr. Sci. Eng. 18, 363 (1974); J. Eneva and J. Malinowski, J. Photogr. Sci. 22, 273 (1974).
60. J.F. Hamilton, Photogr. Sci. Eng. 16, 126 (1972).
61. K.W. Chibisov, J. Photogr. Sci. 9, 26 (1961).
62. H.E. Spencer, Photogr. Sci. Eng. 11, 352 (1967); E. Moisar, Photogr. Korr. 106, 49 (1970).
63. H.E. Spencer, L.E. Brady, and J.F. Hamilton, J. Opt. Soc. Amer. 54, 492 (1964).
64. H.E. Spencer, J. Photogr. Sci. 20, 143 (1972).
65. R.C. Baetzold, J. Chem.Phys. 35, 4355, 4363 (1971); J. Solid State Chem. 6, 352 (1972).
66. R.C. Baetzold, Photogr. Sci. Eng. 19, 11 (1975).
67. R.C. Hanson, J. Phys. Chem. 66, 2376 (1962).
68. R.K. Ahrenkiel, Phys. Rev. 180, 859 (1969).
69. J.S. Wei and F.C. Brown, Photogr. Sci. Eng. 17, 197 (1973).
70. J.P. Galvin, Photogr. Sci. Eng. 16, 69 (1972).
71. M. Georgiev and J. Malinowski, J. Phys. Chem. Solids 28, 931 (1967).
72. E. Schöne, O. Stasiv, and J. Teltow, J. Phys. Chem. 197, 145 (1951).

73. V.I. Saunders, R.W. Tyler, and W. West, Photogr. Sci. Eng. 16, 87 (1972); Rev. Sci. Instr. 41, 1466 (1970); 42, 1546 (1971).
74. W. Platikanowa and J. Malinowski, Phys. Status Solidi 14, 205 (1966).
75. E. Klein and R. Matejec, Z. Elektrochem. 63, 883 (1959).
76. W.F. Berg, Photogr. Sci. Eng. 11, 242 (1967).
77. A. Marriage, J. Photogr. Sci. 9, 93 (1961); E. Klein, J.Photogr. Sci. 10, 26 (1962); A.E. Ames, Photogr. Sci. Eng. 17, 154 (1973).
78. See R. Meyer, J. Photogr. Sci. 10, 14 (1962); R. Meyer, H. Langner, and G. Kaufhold, in H. Frieser, G. Haase and E. Klein (Eds.), Die Grundlagen der photographischen Prozesse in Silberhalogeniden, Akademische Verlagsgesellschaft Frankfurt am Main, 1968, p. 1417.
79. B.E. Bayer, J.F. Hamilton, J. Opt. Soc. Amer. 55, 439, 528 (1965); 56, 1088 (1966); J.F. Hamilton, Photogr. Sci. Eng. 12, 143 (1968).
80. E. Fatuzzo and S. Coppo, J. Appl. Phys. 43, 1457, 1467 (1972).
81. E. Fatuzzo and S. Coppo, J. Photogr. Sci. 20, 43 (1972).
82. R. Matejec and E. Moisar, Photogr. Sci. Eng. 12, 133 (1968).
83. J.F. Hamilton, Photogr. Sci. Eng. 18, 371 (1974).
84. P.V. McD Clark and J.W. Mitchell, J. Photogr. Sci. 4, 1 (1956).
85. V.I. Saunders, R.W. Tyler, and W. West, J. Chem Phys. 37, 1126 (1962).
86. J.F. Reber, J.G. Fernandez-Garcia, R. Steiger, and Ch.G. Boissomas, Photogr. Sci. Eng. 18, 630 (1974).
87. W.F. Berg, Photogr. Korr. 107, 194 (1971).
88. T.H. James, J. Photogr. Sci. 20, 182 (1972).

89. E.E. Loening, in J.W. Mitchell (Ed.), Fundamental Mechanisms of Photographic Sensitivity; Proceedings of a Symposium held at the University of Bristol, 1950, Butterworths, London, 1951; H.E. Keller, Photogr. Korr. 103, 69, 86, 104, 117 (1967).
90. W. Reinders, J. Phys. Chem. 38, 738 (1934).
91. P.J. Hillson, J. Photogr. Sci. 6, 97 (1958).
92. R. Matejec and E. Moisar, Photogr. Korr. 100, 39 (1964).
93. I. Schmidt, Z. Wiss. Photogr., Photophys. Photochem. 60, 1 (1967); G.F. van Veelen, Photogr. Korr. 105, 179 (1969).
94. E.A. Frei, Photogr. Korr. 105, 5, 21, 37 (1969); E.A. Frei and W.F. Berg, Photogr. Sci. Eng. 13, 81 (1969).
95. J. Konstantinov, A. Panov, and J. Malinowski, J. Photogr. Sci. 21, 250 (1973).
96. R. Matejec and E. Moisar, Z. Elektrochem. Ber. Bunsenges. Physik. Chem. 69, 566 (1965).
97. F. Trautweiler, Photogr. Sci. Eng. 12, 138 (1968).
98. A.N. Latyshev and M.I. Molotskii, Zh. Nauch. Priklad. Fotogr. 14, 264, 437 (1969); M.I. Molotskii, Zh. Nauch. Priklad. Fotogr. Kinematogr. 16, 38 (1971).
99. M.I. Molotskii, A.N. Latyshev, and K.V. Chibisov, J. Photogr. Sci. 20, 201 (1972).
100. R.C. Baetzold, Photogr. Sci. Eng. 17, 78 (1973).
101. For a historical review see W. West, Photogr. Sci. Eng. 18, 35 (1974).
102. W. West and P.B. Gilman, Photogr. Sci. Eng. 13, 221 (1969).
103. T.H. James, Photogr. Sci. Eng. 16, 120 (1972); 18, 100 (1974); J. Photogr. Sci. 20, 182 (1972).
104. H. Gerischer and H. Selzle, Electrochim. Acta 18, 799 (1973).
105. B. Levy, M. Lindsay, and C.R. Dickson, Photogr. Sci. Eng. 17, 115 (1973).
106. V.I. Saunders, R.W. Tyler, and W. West, J. Chem. Phys. 46, 199 (1967).

107. P.B. Gilman, Photogr. Sci. Eng. 18, 475 (1974).
108. Th. Förster, Disc. Faraday Soc. 27, 7 (1959); Ann. Physik VI, 2, 55, (1948).
109. D.L. Dexter, J. Chem. Phys. 21, 836 (1953).
110. R. Steiger, P. Junod, B. Kilchoer, and E. Schumacher, Photogr. Sci. Eng. 17, 107 (1973).
111. D.M. Sturmer, W.S. Gaugh, and B.J. Bruschi, Photogr. Sci. Eng. 18, 56 (1974).
112. L. v. Szentpaly, D. Möbius, and H.J. Kuhn, J. Chem. Phys. 52, 4618 (1970).
113. S.S. Collier and P.B. Gilman, Photogr. Sci. Eng. 16, 413 (1972).
114. Summarized in J. Eggert, W. Meidinger, and H. Arens, Helv. Chim. Acta 31, 1168 (1948).
115. H. Tributsch, Ber. Bunsenges. Physik. Chem. 73, 582 (1969).
116. T. Tani, Photogr. Sci. Eng. 14, 63 (1970).
117. G. Scheibe, Angew. Chem. 52, 633 (1939).
118. W. Vanassche, Photogr. Sci. Eng. 18, 288 (1974); C. Reich, Photogr. Sci. Eng. 18, 335 (1974).
119. D. Mastropaolo, J. Potenza, and G.R. Bird, Photogr. Sci. Eng. 18, 450 (1974).
120. D.L. Smith, Photogr. Sci. Eng. 18, 309 (1974).
121. H. Meier, Spectral Sensitization, Focal Press, London, New York, 1973, p. 179.
122. R.C. Nelson, J. Opt. Soc. Amer. 48, 948 (1958).
123. See H. Meier, Organic Semiconductors. Dark and Photoconductivity of Organic Solids. Monographs in Modern Chemistry, 2, Verlag Chemie, Weinheim 1974.
124. J.W. Trusty and J.M. Ferrier, Photogr. Sci. Eng. 18, 92(1974).
125. R.C. Nelson, Photogr. Sci. Eng. 18, 485 (1974).
126. P.E. Yianoulis and R.C. Nelson, Photogr. Sci. Eng. 18, 94 (1974); J. Photogr. Sci. 22, 17 (1974).

127. D.M. Sturmer, W.S. Gaugh, and B.J. Bruschi, Photogr. Sci. Eng. 18, 49 (1974).
128. T. Tani, Photogr. Sci. Eng. 12, 80 (1968); 14, 72 (1970).
129. H. Gerischer, Photochem. Photobiol. 16, 243 (1972).
130. T. Tani, Photogr. Sci. Eng. 18, 165 (1974).
131. P. Nielsen, Photogr. Sci. Eng. 18, 186 (1974).
132. A. Terenin and I.A. Akimov, Z. Phys. Chem. (Leipzig) 217, 307 (1961).
133. R.C. Nelson and R.G. Selsby, Photogr. Sci. Eng. 14, 342 (1970).
134. T. Tani and S. Kikuchi, Photogr. Sci. Eng. 11, 129 (1967).
135. R.G. Selsby and R.C. Nelson, J. Molec. Spectr. 31, 1 (1970).
136. S. Dähne, Z. Wiss. Photogr. Photophys. Photochem. 59, 13 (1965).
137. A. Stanienda, Z. Naturforsch. 23b, 1285 (1968).
138. R.F. Large, in R.J. Cox (Ed.), Photographic Sensitivity, Symposium on Photographic Sensitivity, Cambridge, 1972, Academic Press, London, New York 1973, p. 241.
139. L. Gouverneur, G. Leroy, and I. Zador, Electrochim. Acta 19, 215 (1974).
140. R.C. Nelson, J. Phys. Chem. 71, 2517 (1967).
141. R.W. Berriman and P.B. Gilman, Photogr. Sci. Eng. 17, 235 (1973).
142. L. Costa, F. Grum, and P.B. Gilman, Photogr. Sci. Eng. 18, 261 (1974); L. Costa and P.B. Gilman, Photogr. Sci. Eng. 19, 207 (1975).
143. J.F. Willems, Photogr. Sci. Eng. 15, 213 (1971).
144. F. Lohmann, Z. Naturforsch. 22a, 843 (1967).
145. S. Getzov, J. Malinowski, Photogr. Sci. Eng. 19, 184 (1975).
146. S.S. Collier and P.B. Gilman, Photogr. Sci. Eng. 16, 415 (1972); Y. Renotte, Bull. Roy. Soc. Sci. Liège 72, 586 (1973); Chem. Abstr. 80, 125721 (1974).

147. I.H. Leubner, Photogr. Sci. Eng. 18, 175 (1974).
148. For a review see P.B. Gilmen, Photogr. Sci. Eng. 18, 418 (1974).
149. G.R. Bird, B. Zuckerman, and A.E. Ames, Photochem. Photobiol. 8, 393 (1968).
150. W. West and B.H. Carrol, J. Chem. Phys. 19, 417 (1951).
151. O. Riester, Photogr. Sci. Eng. 13, 13 (1969); 18, 295 (1974).
152. R.C. Nelson, J. Photogr. Sci. 22, 56 (1974).
153. R. Memming and H. Tributsch, J. Phys. Chem. 75, 562 (1971).
154. H. Leonhard and A. Weller, Ber. Bunsenges. Physik. Chem. 67, 791 (1963).
155. P.B. Gilman and T.D. Koszelak, J. Photogr. Sci. 21, 53 (1973).
156. P.B. Gilman, Photogr. Sci. Eng. 12, 230 (1968).
157. S.S. Collier, Photogr. Sci. Eng. 18, 430 (1974).
158. For a comprehensive review see J. Eggers, in H. Frieser, G. Haase and E. Klein (Eds.), Die Grundlagen der photographischen Prozesse mit Silberhalogeniden, Akademische Verlagsgesellschaft, Frankfurt am Main, 1968, p. 813.
159. W. Pelz, Angew. Chem. 66, 231 (1954).
160. T. Tani, Photogr. Sci. Eng. 18, 235 (1974).
161. R.L. Bent et al., J. Amer. Chem. Soc. 73, 3100 (1951).
162. R.G. Willis, F.E. Ford, and R.B. Pontius, Photogr. Sci. Eng. 14, 141 (1970).
163. R.G. Willis, F.E. Ford, and R.B. Pontius, Photogr. Sci. Eng. 14, 149 (1970).
164. L.K.J. Tong and M.C. Glessman, Photogr. Sci. Eng. 8, 319 (1964); L.K.J. Tong, C.A. Bishop, and M.C. Glessman, Photogr. Sci. Eng. 8, 326 (1964).
165. B. Ilschner, J. Chem. Phys. 28, 1109 (1958).
166. F.A. Kröger, J. Electrochem. Soc. 117, 69 (1970); D.O. Raleigh and H.R. Crowe, J. Electrochem. Soc. 118, 79 (1971).
167. H.E. Farnsworth and R.P. Winch, Phys. Rev. 58, 812 (1940).

168. S. Boyer and L. Pichon, Photogr. Sci. Eng. 18, 555 (1974).
169. J.F. Hamilton and P.C. Logel, Photogr. Sci. Eng. 18, 507 (1974).
170. V.I. Saunders and W. West, Photogr. Sci. Eng. 11, 35 (1967).
171. T.H. James, Photogr. Sci. Eng. 10, 344 (1966).
172. H. Gerischer, in P. Delahay (Ed.), Advances in Electrochemistry and Electrochemical Engineering, Vol. I, Interscience Publ., New York, 1961, p. 139.
173. W. Jaenicke, Photogr. Sci. Eng. 6, 85 (1962).
174. W. Jaenicke, in W. Berg (Ed.), Photographic Science, Symposium Zürich 1961, Focal Press, London, New York, 1963, p. 78.
175. J. Heyrowsky and J. Kuta, Principles of Polarography, Publishing House of the Czechoslovak Academy of Sciences, Prague, 1965, p. 77.
176. G.P. Faerman and E.D. Voeikova, Uspekhi Nauch. Priklad. Fotogr. Kinematogr. 3, 174 (1955); 4, 150 (1956).
177. T.H. James and D.C. Schuman, Photogr. Sci. Eng. 15, 42 (1971).
178. G.I.P. Levenson and P.J. Twist, J. Photogr. Sci. 21, 211 (1973); 22, 169 (1974).
179. N.F. Mott, Photogr. J. 88B, 119 (1948); see Ref. 2.
180. Kh.S. Bagdasaryan, J. Phys. Chem. URSS, 17, 33 (1943); Acta Physicochem. USSR 19, 421 (1944); Uspekhi Chim. 16, 652 (1947).
181. W. Jaenicke, A. Krüger, and K. Hauffe, Z. Physik. Chem. 197, 161 (1951); W. Jaenicke and G. Schott, Z. Elektrochem. 59, 956 (1955).
182. W. Jaenicke and F. Sutter, Z. Elektrochem. Ber. Bunsenges. Physik. Chem. 63, 722 (1957).
183. W. Jaenicke, J. Photogr. Sci. 20, 2 (1972).
184. H.J. Metz, J. Photogr. Sci. 20, 111 (1972).

185. R.B. Pontius, R.G. Willis, and R.J. Newmiller, Photogr. Sci. Eng. 16, 406 (1972).
186. E. Klein and R. Matejec, Z. Elektrochem. Ber. Bunsenges. Physik. Chem. 61, 1127 (1957).
187. G.I.P. Levenson, W. West, and V.I. Saunders, Photogr. Sci. Eng. 6, 135 (1962).
188. E. Eisenmann and W. Jaenicke, in W. Berg (Ed.), Photographic Sci., Focal Press, London, New York, 1963, p. 178.
189. I. Schmidt, see W. Jaenicke, J. Photogr. Sci. 20, 2 (1972).
190. H. Eger, Thesis, Munich, 1969.
191. E.H. Land, D.C. Farney, and M.M. Morse, Photogr. Sci. Eng. 15, 4 (1971).
192. E. Klein and R. Matejec, Z. Angew. Phys. 12, 26 (1960).
193. E. Eisenmann and W. Jaenicke, see W. Jaenicke, J. Photogr. Sci. 20, 2 (1972).
194. W. Jaenicke and H. Eggenschwiller, Z. Elektrochem. Ber. Bunsenges. Physik. Chem. 64, 391 (1960).
195. P.B. Price, D.A. Vermilyea, and M.B. Webb, Acta Met. 6, 524 (1958).
196. E.J. Perry, Photogr. Sci. Eng. 5, 355 (1961).
197. C.R. Berry, Photogr. Sci. Eng. 13, 65 (1969).
198. W. Jaenicke and B. Schilling, Z. Elektrochem. Ber. Bunsenges. Physik. Chem. 66, 563 (1962); see also C. Wagner, J. Electrochem. Soc. 97, 3 (1950).
199. H. Gerischer and R.P. Tischer, Z. Elektrochem. 58, 819 (1954); W. Jaenicke, H. Gerischer, and R.P. Tischer, Z. Elektrochem., 59, 448 (1955).
200. H.D. Keith and J.W. Mitchell, Phil. Mag. 7 44, 877 (1953).
201. T.H. James, Photogr. Sci. Eng. 9, 121 (1965).
202. G.F. van Veelen, Photogr. Korr. 101, 149 (1965); 103, 137 (1967).

203. R. Matejec, in Mitteilungen aus den Forschungslaboratorien der Agfa-Gevaert AG, Leverkusen-München, Vol. 4, Springer Verlag, Berlin, Heidelberg, New York, 1964, p. 1.
204. F.E. Ford, Photogr. Sci. Eng. 13, 171 (1969); J. Karrer and W.F. Berg, J. Photogr. Sci. 19, 143 (1971); Chimia 26, 641 (1972).
205. J. Tusl, Thesis, Prague, 1969.
206. R.B. Pontius and R.G. Willis, Photogr. Sci. Eng. 17, 21, 157 (1973).
207. R.B. Pontius and R.G. Willis, Photogr. Sci. Eng. 17, 326 (1973).
208. T.H. James, J. Franklin Inst. 240, 83 (1945); Photogr. Sci. Eng. 10, 344 (1966); 14, 371 (1970).
209. T.H. James, SPSE News 11, 4 (1968).
210. R.B. Pontius, Photogr. Sci. Eng. 6, 283 (1961).
211. See J. Heyrowsky and J. Kuta, Principles of Polarography, Publishing House of the Czechoslovak Academy of Sciences, Prague, 1965, p. 126, 181, 205, 221.
212. R. Matejec, Photogr. Korr. 104, 153 (1968).
213. H. Jonker, A. Molenaar, and C.J. Dippel, Photogr. Sci. Eng. 13, 38 (1969).
214. T.H. James, Photogr. Sci. Eng. 1, 141 (1957); 3, 225 (1959).
215. R. Matejec and R. Meyer, Z. Wiss. Phot. Photophysik, Photochem. 57, 45 (1963); Photogr. Sci. Eng. 6, 265 (1963).
216. I. Konstantinov and J. Malinowski, J. Photogr. Sci. 23, 145 (1975); see also P.J. Hillson, J. Photogr. Sci. 22, 31 (1974).
217. Review article: M. Austin, J. Photogr. Sci. 22, 293 (1974).
218. T. Suga, Bull. Chem. Soc. Japan 45, 3464 (1972); 47, 2463 (1974).
219. Review article: G.I.P. Levenson, Photogr. Sci. Eng. 13, 299 (1969).

220. E.P. Honig, Trans. Faraday Soc. 65, 2248 (1969).
221. J.F. Willems, G.F. van Veelen, and A. Vandenberghe, Photogr. Sci. Eng. 13, 312 (1969).
222. J.F. Willems, in R.J. Cox (Ed.), Photographic Processing, Proceedings of the Symposium on Photographic Processing held at the University of Sussex 1971, Academic Press, London, New York, 1973, p. 71.
223. W. Jaenicke, H. Raithel, and M. Brezina, Photogr. Sci. Eng. 15, 230 (1971).
224. H. Raithel, Thesis, Erlangen, 1975.
225. A. Meyer, J. Photogr. Sci. 13, 95 (1965); Photogr. Sci. Eng. 18, 530 (1974).
226. E. Günther and R. Matejec, J. Photogr. Sci. 19, 106 (1971).
227. M. Schellenberger and R. Steinmetz, Helv. Chim. Acta 52, 433 (1969); M. Schellenberger, Helv. Chim. Acta 53, 1151, 1169 (1970).
228. E. Kramp and R. Schaller, Ber. Bunsenges. Physik. Chem. 77, 899 (1973).
229. R. Steiger, Photogr. Sci. Eng. 14, 269 (1970).
230. W. Jaenicke, Z. Elektrochem. 55, 648 (1951); 56, 473 (1952); W. Jaenicke and M. Haase, Z. Elektrochem. Ber. Bunsenges. Physik. Chem. 63, 521 (1959); K.J. Vetter, Elektrochemische Kinetik, Springer Verlag, Berlin, Göttingen, Heidelberg, 1961, p. 575.
231. C. Wagner, unpublished results, see ref. 232.
232. N. Ibl. W. Richarz and H. Wiederkehr, Z. Physik. Chem. NF 98, 123 (1975).
233. H. Eggenschwiller and W. Jaenicke, Photogr. Korr. 95, 149 (1959).
234. W.E. Lee and T.H. James, Photogr. Sci. Eng. 6, 32 (1962).
235. F.A. Kröger, The Chemistry of Imperfect Crystals, 2nd ed., North Holland Publ. Comp., Amsterdam, 1974; see also H.

Rickert, Einführung in die Elektrochemie fester Stoffe, Springer Verlag, Berlin, Heidelberg, New York, 1973, p. 7.
236. L.M. Kellog, Photogr. Sci. Eng. 18, 378 (1974).
237. S.G. Bogdanov, Zh. Nauch. Prikl. Fotogr. Kinematogr. 13, 62 (1968); M. Brezina, W. Jaenicke and H. Raithel, Angew. Chem. 89, 293 (1969).
238. J. Teltow, Z. Phys. Chem. (Leipzig) 195, 213 (1950).
239. See L.F. Santacruz, S.F. da Cunha, and J.A. de Saja, Bull. Soc. Photogr. Sci. Technol. Japan 57 (1972). Chem. Abstr. 80, 42905 (1974).
240. W.E. Mueller, J. Photogr. Sci. 19, 132 (1971); Photogr. Sci. Eng. 15, 369 (1971); 17, 94 (1973); W. Gross, J. Photogr. Sci. 19, 207 (1971); Discussion: Photogr. Sci. Eng. 17, 99 (1973).
241. C.J. Battaglia, Photogr. Sci. Eng. 14, 275 (1970). See also: M.R.V. Sahyun, Photogr. Sci. Eng. 18, 383 (1974).
242. P.J. Hillson and H.M. Adam, J. Photogr. Sci. 23, 104 (1975).
243. J. Malinowski, Photogr. Sci. Eng. 17, 86 (1973).
244. D.O. Raleigh, J. Phys. Chem. Solids 25, 329 (1965); F.A. Kröger, J. Elektrochem. Soc. 117, 69 (1970).
245. C.R. Berry, Photogr. Sci. Eng. 16, 269 (1972).
246. E.R. Brown and L.K.J. Tong, Photogr. Sci. Eng. 19, 314 (1975).
247. J.M. Harbison and J.F. Hamilton, Photogr. Sci. Eng. 19, 322 (1975).
248. J. Eggers, Photogr. Sci. Eng. 15, 128 (1971); J. Charkoudian, A. Ames, and A. Hoffman, Photogr. Sci. Eng. 17, 456 (1973).
249. H.J. Hefter, Photogr. Sci. Eng. 19, 179 (1975).
250. See J. Heyrowský and J. Kuta, Principles of Polarography, Publishing House of the Czechoslovak Academy of Sciences, Prague, 1965, p. 181.

251. R. Matejec, Ger. Offen. 1 813 920 (see Chem. Abstr. 73, 72090 (1970).
252. R. Matejec, R. Meyer, and E. Ranz, Ger. Offen. 2 044 993 (see Chem. Abstr. 77, 120785 (1972).
253. K. Gossner, H. Heidrich, D. Körner, Z. Physik. Chem. 67, 220 (1969), 68, 293 (1969).
254. V.L. Bissonette, Ger. Offen. 2 226 770 (see Chem. Abstr. 78, 117576 (1973); V.L. Bissonette and W.B. Travis, Ger. Offen. 2 226 771 (see Chem. Abstr. 78, 90972 (1973).
255. V.L. Bissonette, Ger. Offen. 2 360 326 (see Chem. Abstr. 81, 113630 (1974).
256. I. Konstantinov and J. Malinowski, J. Photogr. Sci. 23, 1 (1975).
257. A. Milchev and J. Malinowski, J. Photogr. Sci. 23, 12 (1975).

The Work Function in Electrochemistry
SERGIO TRASATTI
University of Milan, Milan, Italy

1	Introduction	215
2	The Work Function	217
	2.1 Physical Definition	217
	2.2 Chemical Definition	219
	2.3 The Chemical Potential of Electrons	221
	2.4 Experimental Values of Work Function	223
	2.5 The Concept of Electrochemical Work Function	226
	2.6 Relation of Work Function to Properties of Metals	228
	2.6.1 Electronegativity of Elements	228
	2.6.2 Heat of Adsorption	230
3	Electrochemical Thermodynamics	235
	3.1 The Surface Structure of a Phase	236
	3.2 Work Function for Electrons in Solution	237
	3.3 Work Function of Metals in Solution	238
	3.4 The Absolute Electrode Potential	239

Work supported by the National Research Council (C.N.R., Rome).

3.5	Operative and Thermodynamic Electrode Potentials	240
3.5.1	Further Developments	243
3.6	Calculation of the Operative Potential	245
3.7	Derivation of Electrochemical Work Function	246
3.8	Underpotential Deposition of Metals	251
3.8.1	Other Systems	255
4	Double-Layer Structure	258
4.1	Relationship between Work Function and Potential of Zero Charge	258
4.2	The Meaning of the Interaction Term $\delta\chi^{M}_{(H_2O)}$	261
4.3	The Surface Potential of Water on Metals	264
4.4	The Role of Water Molecules in the Double Layer	266
4.5	The Inner Layer Capacity	270
4.6	Effect of Single Crystal Faces	273
4.7	Operative Zero-Charge Potentials	274
4.8	A Model for Water Molecules in the Double Layer	276
5	Electrochemical Kinetics	279
5.1	The Work Function in Kinetic Equations	280
5.2	Simple Electron Exchange Reactions	281
5.3	Reactions with Strong Metal-Particle Interactions	290
6	Conclusions	299
7	List of Symbols	303
8	References	305

1 INTRODUCTION

The work function (Φ) is defined (1) as the minimum work to extract an electron from a metal. According to the electronic theory of metals (2), the electron is extracted from the Fermi level at the top of the energetic distribution. Hence the work function taken with the reverse sign represents precisely the energy of electrons in the Fermi level of the metal. Experimentally, the work function is typically a physical quantity (3,4). Conceptually, however, it is as relevant to chemistry as the electronegativity (5) of atoms is. As electrons of metals are exchanged with the environment, no matter of the detailed mechanism, the work function should at least conceptually be involved.

In electrochemistry, electrons are, by definition, exchanged with the environment. It has long been known (6) that the rate of electrochemical reactions depends on the nature of the metal constituting the electrode. Many properties of metals have been employed (7,8) in attempts at correlating electrochemical behavior with the intimate physical structure of the substrate. The work function is expected to be a most relevant physical property for two reasons: (1) it accounts for metal surface orientation effects; (2) it is involved in all electrochemical processes. This means that sound theoretical equations exist between work function and electrochemical parameters that are expected to possess quantitative validity.

Work function concepts were used long ago (9) to discuss thermodynamic aspects of electrochemistry. Later, following the slower development of kinetic concepts (10), the work function was also introduced (11) into electrochemical kinetics as a parameter to characterize the correlation between the nature of metals and the hydrogen overvoltage. Afterwards, the work function has been used many times in correlations with the potential of zero charge of metals (12-18), the rate of hydrogen evolution (19-27), and the rate of simple redox reactions (27-29). Analysis of all of these results reveals that correlations have never gone beyond a mere qualitative aspect. The main reason for this seems to be the use by different authors of different Φ values to characterize the same metal surface. Rationalization in the use of Φ values is thus thought to be a prerequisite to attaining consistent results.

The main purpose of this chapter is not to review extensively and comprehensively the literature in the field of correlations with the work function. Rather, after an appropriate conceptual assessment of the field and rationalization of experimental values of work function, we show that the use of Φ in correlations may lead to consistent and very often quantitative results. This chapter is intended to illustrate ideas and developments that may arise through handling electrochemistry with tools and concepts of solid-state physics and surface science. The basic idea is that a metal interacts with the environment essentially through its valence electrons, and the electronic energy must therefore be directly responsible for phenomena like chemisorption, electrode potential, contact potential, reaction rate, and so on. A metal under quite different circumstances should be identifiable with an identity card, and that most directly related to the energy of electrons is certainly the work function, although on a very general basis other parameters may be needed to characterize all the complex properties of metals. The behavior of the metals is thus expected to be largely dependent on the values of their work functions. There is extensive evidence (24, 3034) that the reactivity scale of metals does not change substantially from catalysis to electrochemistry up to chemistry of elements. The general picture of the behavior of metals is thus expected to show a high degree of internal consistency with a large possibility of exchange of information between adjoining fields. In this light, electrochemistry becomes simply an aspect of surface science, even though it is likely to be the most complicated aspect both conceptually and experimentally. The work that follows has developed in the context of the above ideas. It is anticipated that detailed discussion to corroborate single arguments cannot be given here.

2 THE WORK FUNCTION

The operative definition of work function as the minimum work to extract an electron from a metal is based on practical concepts connected with its experimental determination (1,4). Theoretically, however, the energy of electrons is split into various contributions (2). Splitting of energy forms is present in both the physical (3) and the chemical (13) definition of work function. However, the two definitions do not coincide as regards the mode of separation into contributions, and some conceptual difficulty can also be found. This fact may be misleading and a discussion on this point is needed to assess the situation.

2.1 Physical Definition

Figure 1 depicts schematically the usual way in physics (4, 35 - 39) to represent the various terms of energy contribution to the work function. If we resort to the simplified Sommerfeld model (40,41) largely known to chemists, the work function of a metal is given by

$$\phi^M = V - \varepsilon_F \qquad (1)$$

where V is the difference in potential energy of electrons between a point inside and a point outside the metal, and ε_F is the kinetic energy of electrons at the Fermi level. As shown in Fig. 1, V is measured with respect to an electron at rest at infinity in a vacuum, while ε_F is measured with respect to an electron at rest inside the metal taking V = 0.

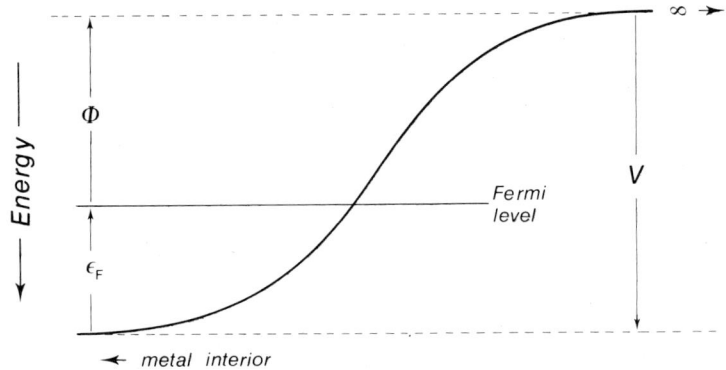

Fig. 1. Physical definition of the work function as the sum of the potential energy V and the kinetic energy ε_F. Arrows indicate sign and zero position of the given quantities.

Sommerfeld's model is conceptually quite simple, since a sharp separation is implied between potential and kinetic energy. However, such simplicity in concepts is hardly found in specialized papers. It is a practice among physicists (2 - 4, 38,42) to call ε_F in Fig. 1 the chemical potential of electrons, but this quantity actually corresponds in some papers (4,37,43) to the pure kinetic energy, whereas in other works (38,42) it represents the kinetic energy <u>inside the metal</u>; that is, it also includes a well-defined potential energy. It is this unclear concept of chemical potential that has led to misleading conclusions in the electrochemical literature (27,44).

According to the usual Sommerfeld representation, V is the total energy barrier at the surface, that is, the energy of interaction of electrons with the ionic cores plus surface effects. ε_F is the sole kinetic energy. It must thus be understood that the Fermi energy is the total energy of electrons at the Fermi level, whereas ε_F should be termed the kinetic Fermi energy. The practice among physicists to call ε_F either Fermi energy or chemical potential may lead one to consider that the chemical potential of electrons coincides with the kinetic Fermi energy, which is incorrect.

2.2 Chemical Definition

Consider an uncharged metal. The change in free energy of the system at constant T and P as an electron is added to the metal defines unambigiously the partial molar free energy of electrons in the metal (3,13). This is the energy binding electrons to the interior of the metal. This energy, taken with the reverse sign, is the work to extract an electron from the metal; it coincides with the work function. From a chemical point of view, it is usual to separate (13) so-called chemical effects (short-range interactions) from electrical effects due to dipoles or free charges. For uncharged metals, the chemical part of the total energy defines the chemical potential of electrons, and the electrical part defines the surface contribution due to dipoles. With Lange's (45, 46) notations,

$$\phi^M = - \mu_e^M + e\chi^M \qquad (2)$$

Conceptually, μ_e^M includes both the potential energy, namely, the interaction of electrons with ion cores inside the metallic phase, which is a typical binding energy according to chemical concepts, and the kinetic energy. χ^M includes only a part of the potential energy V in the Sommerfeld model. Comparison of Eqs. 1 and 2 clearly shows that the chemical potential μ_e^M cannot coincide with ε_F. The correct physical definition of chemical potential of electrons is thus the energy at the Fermi level measured with respect to the energy of an electron at rest inside the metal. However, in this case a representation like that in Fig. 1 may be misleading and inadequate.

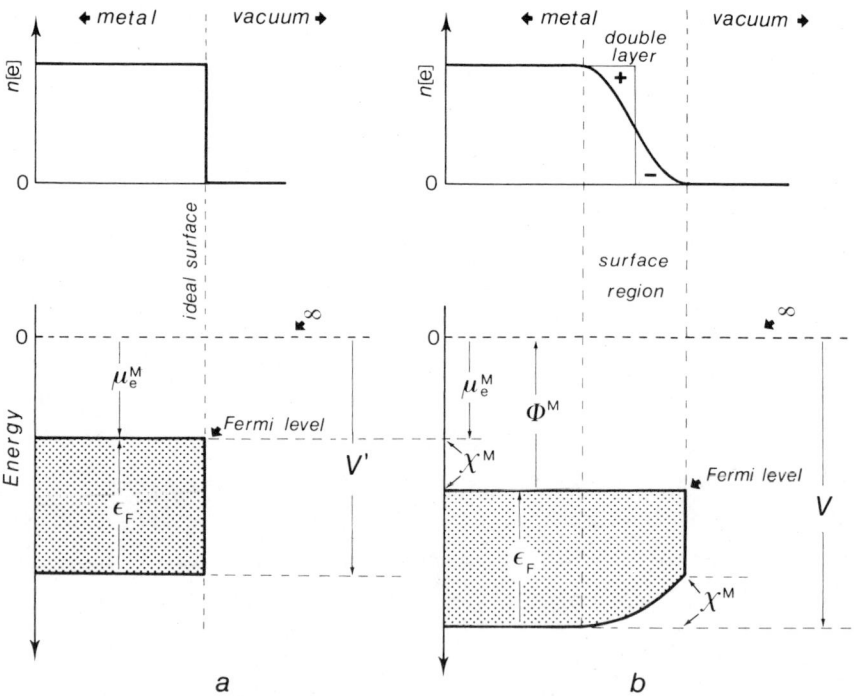

Fig. 2. Simplified electron density and energy diagrams for ideal (a) and real (b) metals, obtained by applying double-layer concepts to the original Sommerfeld model. Symbols as in the text.

Figure 2 illustrates the energetic distribution of electrons in a metal according to the usual Sommerfeld model modified to take into account surface effects. In fact, in the original model no surface effects were present; the barrier at the surface was considered to be infinite so that no overspilling of electrons was possible (40,41). In an ideal metal the distance between the Fermi level and the energy of electrons at infinity measures the chemical potential μ_e^M, whereas ε_F and V retain their usual meanings. One accounts for surface effects due to the presence of surface dipoles by introducing a curvature at the bottom of the band. χ^M results in fact in a shift of the potential energy.

2.3 The Chemical Potential of Electrons

The correct concept of chemical potential of electrons in a metal is that of bulk contribution to the work function (47). If a metal had no surface effects, its work function taken with the reverse sign would correspond to the chemical potential of electrons. Although μ_e^M is a quantity extremely relevant to electrochemistry, electrochemists can only resort to calculations by physicists in theories of metals.

Theoretic calculations for energies of electrons are reliable in the case of alkali metals and few other sp metals. The theory is far from giving satisfactory results for d metals. Therefore, calculations of μ_e^M must be necessarily limited to very few metals. A critical analysis of theoretic calculations is far from the scope of this work, so that results by different authors (3, 35, 37, 38, 43, 48, 49) are given the same weight. Table 1 shows that the agreement among various calculations is moderately satisfactory. Since the theoretic calculations of μ_e^M are completely separated from the calculation of χ^M, therefore the uncertainty

TABLE I

Chemical Potential of Electrons in Metals, $-\mu_e^M/\text{eV}$

Li	Na	K	Rb	Cs	Zn	Cd	In	Ref.
1.96	1.96	1.95	1.96	1.86	–	–	–	3
1.5	1.9	2.1	2.0	2.1	1.4	2.3	2.6	42
1.55	2.14	2.37	2.38	2.38	–	–	–	37
1.55	1.91	1.91	1.87	1.64	–	–	–	37[a]
2.02	1.92	1.92	1.92	–	1.16	2.26	2.64	48
1.74	1.98	2.00	2.00	1.72	–	–	–	43
2.06	2.68	2.32	2.44	–	–	–	–	49
1.57	2.15	2.39	2.38	2.38	–	–	–	38
1.30	1.79	1.95	1.95	1.79	–	–	–	38[a]
2.62	2.35	2.28	2.30	–	1.97	2.57	3.35	35
2.27	2.20	2.03	1.96	–	2.15	1.53	2.55	35[a]
1.80	2.06	2.10	2.09	1.97	1.7	2.2	2.8	probable
±0.3	±0.1	±0.1	±0.1	±0.2	±0.4	±0.3	±0.3	value

[a] From $(\Phi_{exp} - e\chi^M)$.

in the knowledge in μ_e^M is independent of the uncertainty in χ^M and it may even be higher in principle. For this reason, for the sake of comparison, Table 1 reports also values of μ_e^M obtained by subtracting χ^M (35,37,38) from the experimental work function according to Eq. 2.

It should be noted that χ^M, as usually calculated, does not take into account the anisotropy of surface orientation. It is the intrinsic dipole arising from overspilling of electrons. Since the spread of electrons depends only on ε_F and V, it is independent of surface orientation. The latter can actually change χ^M as a result of some redistribution in the electron density at the surface. However, this is extremely difficult to be accounted for in calculations. A few examples (38,43) can be found, but the success is doubtful. If χ^M as calculated is isotropic, it is likely to be a maximum value because it refers to an ideally smooth surface. In fact, the surface potential is usually derived from the "jellium" model (4), which ignores the detailed structure of the surface. Other calculations (35) take into account the possibility for discrete structure of the surface even though no surface orientation is explicitly accounted for. For these reasons, it is thought that χ^M may introduce an uncontrollable degree of uncertainty, and in any case it may give low values for the chemical potential of electrons when subtracted from the experimental value of the work function.

The last line in Table 1 reports the mean value of μ_e^M. The uncertainty is seen to be kept within acceptable limits, at least for a fist practical introduction of this quantity into electrochemistry.

2.4 Experimental Values of Work Function

This topic constitutes the main point of the present approach. Rationalization of criteria for the choice of ϕ^M values is a prerequisite to attaining meaningful results in correlations. These criteria are not simply connected with possible experimental irreproducibility resulting from contamination of surfaces. Different ϕ^M values for the same metal may well be meaningful (1). Equation 2 shows that since ϕ^M depends on χ^M, it depends on the discrete structure of the surface. In terms of Smoluchowski's concepts (50), now widely accepted, a close-packed surface presents a higher work function than a defective surface. In other words, low index planes have higher ϕ^M values than high index planes. Thus, only well-defined planes should possess well-defined work function values.

Polycrystalline surfaces, as mostly implied in electrochemistry, may present different work function values if different proportions of facets are present. However, it is thought that, energetically, a polycrystalline surface should also be in some way reproducible. Evaporated films deposited at low T show increasing ϕ^M values with increasing sintering temperature (51) (see Fig. 3). ϕ^M increases with the degree of packing of atoms on the surface (1). A polycrystalline surface may be defined as that of an evaporated film deposited at low T and annealed at

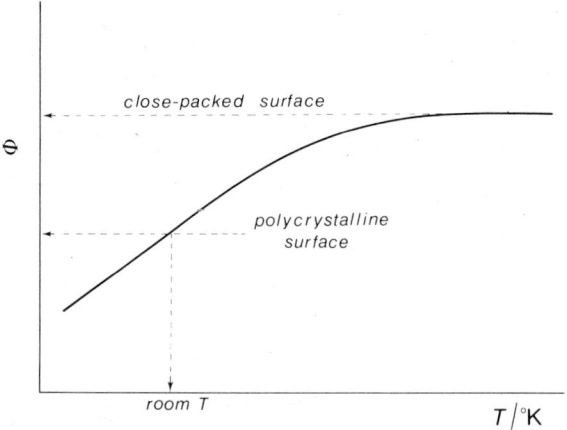

Fig. 3. Sketch of change in work function of evaporated metal films with sintering temperature.

room T. This surface is reproducible and exhibits consistent work function values. This is shown in Table 2 where ϕ^M values for polycrystalline surfaces are seen to locate in definite positions in the series of ϕ^M values for single crystal faces depending on the crystal system. Data of different authors (1, 51 - 58) for evaporated films and for bulk polycrystalline metals (32,59) prove that this approach is reasonable.

TABLE II

Work Functions of Single Crystal Faces, ϕ/eV

	Ta (BCC)	W (BCC)	Mo (BCC)	Cu (FCC)	Ni (FCC)
	4.82 (110)	5.25 (110)	5.10 (110)	4.94 (111)	5.22 (111)
	4.17 (100)	4.63 (100)	4.40 (100)	4.59 (100)	4.89 (100)
	4.15 (poly)	4.55 (poly)	4.30 (poly)	4.55 (poly)	4.80 (poly)
	4.02 (111)	4.47 (111)	4.15 (111)	4.48 (110)	4.64 (110)
Ref.	310	311	312	313	56

Some difference in behavior may be expected between sp and d metals. The former are generally low-melting metals so that packing of the surface occurs at low T. At room T these metals usually show the maximum value of ϕ^M. The spread of data may thus be lower than in the case of d metals. However, in contrast to d metals, sp metals are comparatively little studied as regards experimental ϕ^M values (1,4). Most values refer to old measurements (59) when vacuum techniques were likely to be inadequate

(1). It is thought that for this reason values of ϕ^M for some of the sp metals are suspiciously close to values expected for oxides.

Clean metal surfaces in contact with a vacuum are thought to be equivalent to clean metal surfaces in a solution. If the surface does not react with the solvent and account is made for possible redistribution of electrons as a result of the contact with the liquid phase, then a metal surface in solution can be treated with the same concepts as a surface in a vacuum. The only difference is in the number of parameters involved.

2.5 The Concept of Electrochemical Work Function

The main idea behind this work is that the work function may represent an "identity card" for metals that is much more meaningful on a quantitative basis than the atomic number is. The reactivity scale for metals in electrochemistry, catalysis, and surface science is thus expected to be closely the same. If this is the case, then from the behavior of metals in a given field it should be possible to predict their behavior in an adjoining field, and to make cross-examinations. It is thought that the behavior of metals under different circumstances should present strong internal consistency. This idea suggests that the work function may be used, on the one hand, to locate a metal in the reactivity scale, but on the other hand, the behavior of a metal should in turn enable one to test the reliability of the work function used. Along these lines, the idea may be derived of an operative work function, that is, of a work function not experimentally measured but estimated indirectly from the general behavior of a given metal. It is shown later that in electrochemistry this is possible both through a rigorous route and on a more empirical basis. Work functions derived in this way are called *electrochemical* (16) in that they are estimated from electrochemical properties of metals. They may in turn be tested with other correlations in a search for internal consistency.

TABLE III

Preferred Values of Work Function
for Polycrystalline Surfaces, Φ/eV (32,59)

Ag	4.30	Ga	4.25[a]	Ni	4.80	Sn	4.40[a]
Al	4.20	Hf	3.65	Os	4.85	Sr	2.75
Au	4.80[a]	Hg	4.50	Pb	4.24[a]	Ta	4.15
Ba	2.35	K	2.30	Pd	5.00	Tc	4.90[a]
Be	5.10	In	4.16[a]	Po	4.6	Te	4.70
Bi	4.35[a]	Ir	4.97[a]	Pr	2.7	Th	3.7
Ca	2.70	La	3.4	Pt	5.03[a]	Ti	4.10
Cd	4.12[a]	Li	3.10	Rb	2.20	Tl	4.10[a]
Ce	2.80	Mg	3.65	Re	4.95	U	3.5
Co	4.70	Mn	3.90	Rh	4.98	V	4.45
Cr	4.40	Mo	4.30	Ru	4.80	W	4.55
Cs	1.90	Na	2.70	Sb	4.55	Zn	4.30
Cu	4.55	Nb	4.20	Sc	3.5	Zr	4.00
Fe	4.65	Nd	3.1	Sm	2.95		

[a] Electrochemical work function.

It is curious that, whereas reliable experimental work function values exist for d metals and less satisfactory values for sp metals, rigorous electrochemical data are largely available for sp metals but not for d metals. An essential reason for this may be the possibility of interaction of the solvent with d metals, which would complicate the understanding of experimental measurements and the theoretic treatment. As a consequence, along the above lines, ϕ^M values could lend themselves especially well to predicting electrochemical properties in the case of d metals, whereas the situation for sp metals may be favorable for the rigorous introduction of ϕ^M into equations and therefore for an electrochemical refinement of ϕ^M values.

Table 3 summarizes suggested work function values for metals as used in this work. The reader is referred to the original papers (16,32,59) for a detailed discussion of the criteria for choosing. The electrochemical work functions are discussed in detail later, but their values must be introduced at this stage to make the presentation of the results self-consistent.

2.6 Relation of Work Function to Properties of Metals

An a posteriori proof that ϕ^M can profitably be used as an identity card for metals may be constituted by the observation that the work function bears simple relationships to properties of metals (24,32). If ϕ^M is expected to be involved in all other properties, these must be in some way related to ϕ^M. Since no theoretic equations can often be found to support this, it is necessary as a first step to rely on simple correlations.

2.6.1 ELECTRONEGATIVITY OF ELEMENTS Since a solid phase is formed by interaction of isolated atoms, properties of the latter are obviously expected to be changed but at the same time to govern the bulk properties of the resulting phase to some extent. If Mulliken's concepts (60) for electronegativity of an isolated atom (x_e^M) are applied to a metal (61), the electronegativity of the solid surface (x_s^M) may be identified with the work function. The value of x_s^M on the Pauling scale (5) can be derived from the experimental value of ϕ^M using an appropriate scaling factor (62). Hence a correlation is expected to exist between x_e^M and ϕ^M. Figure 4 shows that this is in fact the case, and simple linear

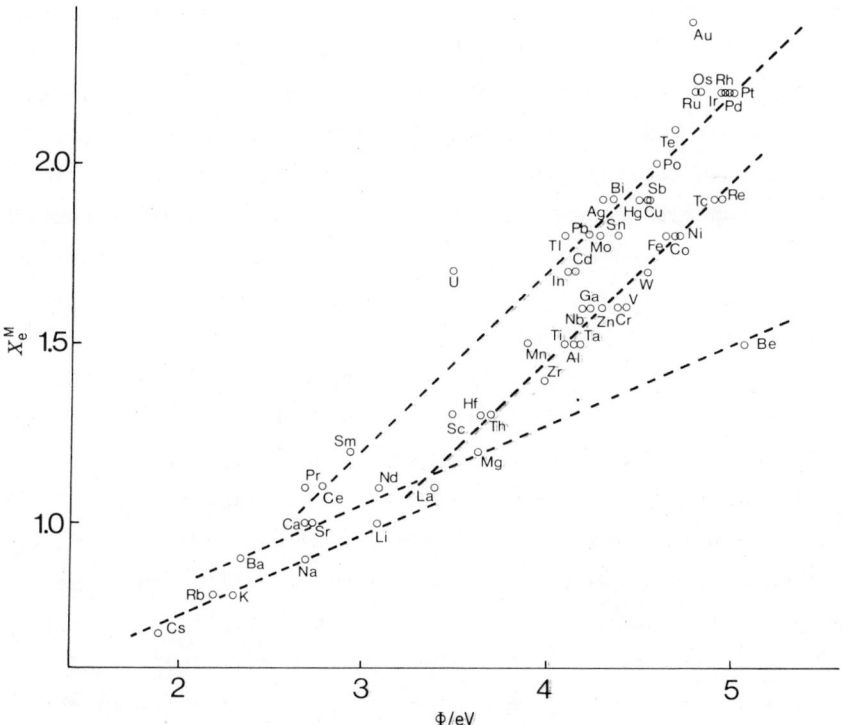

Fig. 4. Correlation between Pauling electronegativity for elements and work function for bulk metals.

relationships of 0.5 slope are observed for d and sp metals separately. Previous correlations (21,63) suggested that metals were probably lumped in a single group. The higher selectivity of the plot in Fig. 4 is thought to be a first effective result due to the more consistent group of ϕ^M values used here. The separation of the metals into groups is thought to have a definite physical meaning. It is likely to be a result of different screening powers of inner electrons (32).

2.6.2 HEAT OF ADSORPTION Adsorption processes are commonly encountered in electrochemistry and they are particularly important in kinetics. The work function can be shown to be conceptually involved. According to simple quantum-mechanical concepts, if adsorption is regarded as an interaction between electrons in the particle and electrons in the metal, the energy of electrons in the resulting surface bond is expected to be related in some way to the work function.

Although a sound theoretic equation expressing the above interdependence cannot be established, it is nevertheless possible to show that the heat of adsorption is generally related in a simple way to the work function. With the purpose of making the evidence compelling, in Figs. 5 to 8 the heat of the adsorption of a number of gases on bare metals (64) is plotted as a function of the work function of metals. Although some scattered points can be seen and the data are scanty in some cases, it is still possible to infer that the simple linear relationship between ΔH_{ads} and ϕ^M may be a quite general rule. Conversely, the linearity in the plots is regarded as an indirect support to the reliability of the work function values used.

Plots in Figs. 5 to 8 refer only to d metals. Experimental adsorption heats are available almost exclusively for these metals. ΔH_{ads} is seen to decrease linearly as ϕ^M is increased. Practically no experimental data useful for correlations are available for sp metals. In this case, heats of formation of

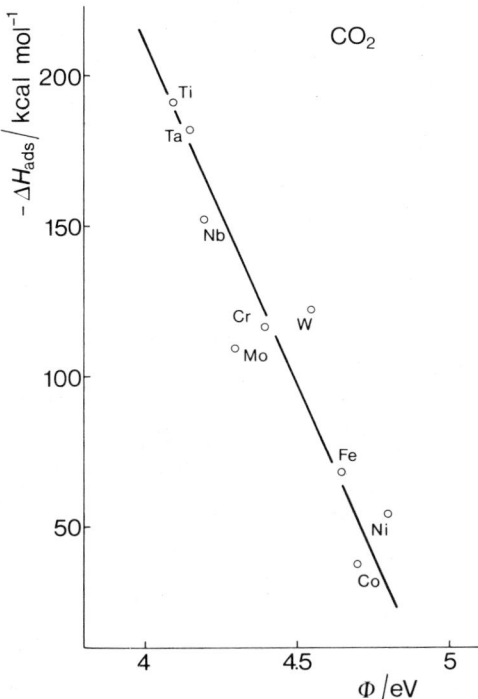

Fig. 5. Heat of adsorption of CO_2 from the gas phase on transition metals (64) as a function of substrate work function.

bulk compounds may be used in the place of ΔH_{ads}. This is not quantitatively correct, but if the general idea is followed that the general properties of a metal are retained under any circumstances, the scale of ΔH_f may be expected to parallel ΔH_{ads}, that is, the bulk reactivity scale should parallel the surface reactivity scale (65,66). Similar concepts can be found in catalysis (67) in connection with Balandin's volcano curves (68) where the catalytic activity is plotted as a function of the formation heat of the bulk compound if properties for the surface compound are unknown (69). An alternative possibility is usually suggested (21,70) to be the calculation of ΔH_{ads} with the Pauling equation

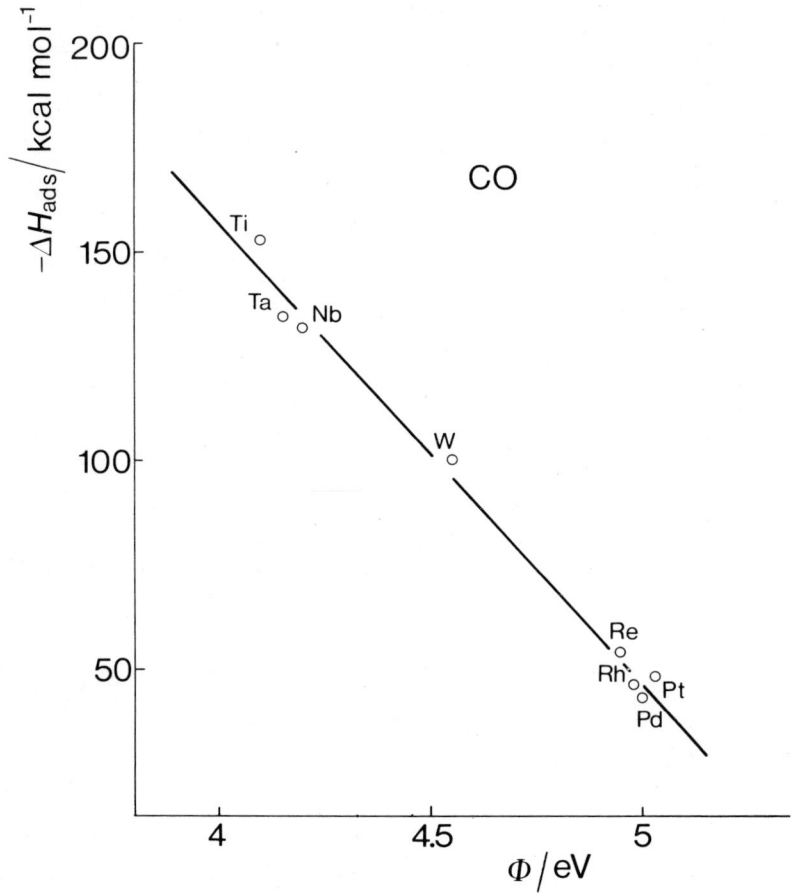

Fig. 6. Heat of adsorption of CO from the gas phase on transition metals (64) as a function of substrate work function.

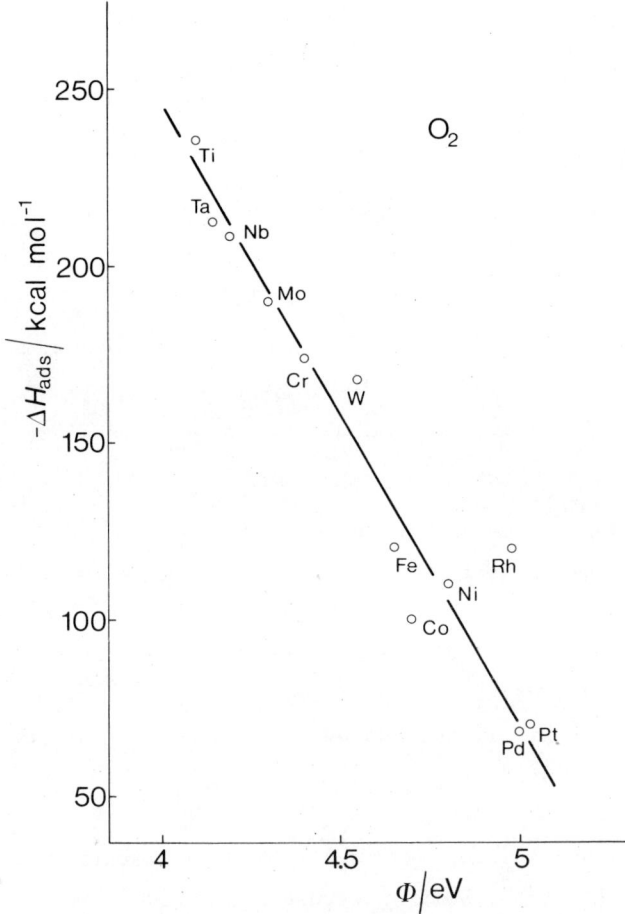

Fig. 7. Heat of adsorption of O_2 from the gas phase on transition metals as a function of substrate work function.

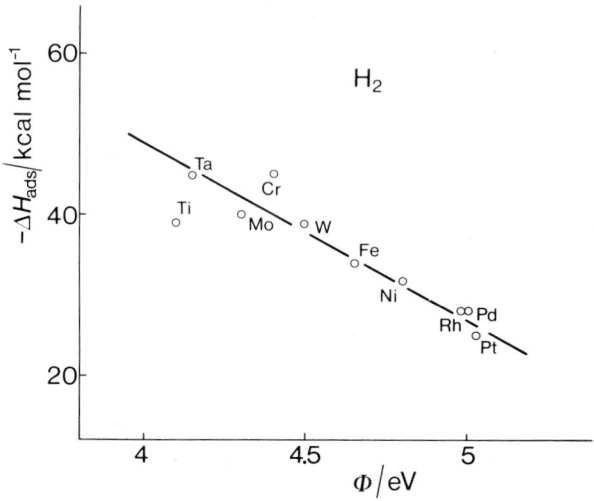

Fig. 8. Heat of adsorption of H_2 from the gas phase on transition metals as a function of substrate work function.

(5) for bond energy. Eley's modification (61,71) of the Pauling equation indeed gives good results with d metals if some refinements are appropriately introduced (32). However, this equation is known (72) to predict unreasonable ΔH_{ads} in the case of hydrogen adsorption on sp metals.

Figure 9 shows ΔH_f for hydrides (73) of sp metals as a function of work function. Some spectroscopic dissociation energies (21) have also been introduced into the plot. ΔH_f can be seen to increase linearly with Φ^M. Thus in spite of the difficulty of establishing theoretic relationships among elemental, collective, bulk, and surface properties of a metal, it appears that simple linear interrelationships exist; this indicates that the reactivity scale is unchanged under any circumstances. These correlations, if not useful for quantitative calculations for a single metal, are extremely useful for quantitative intercomparison of groups of metals. It should, however, be noted that

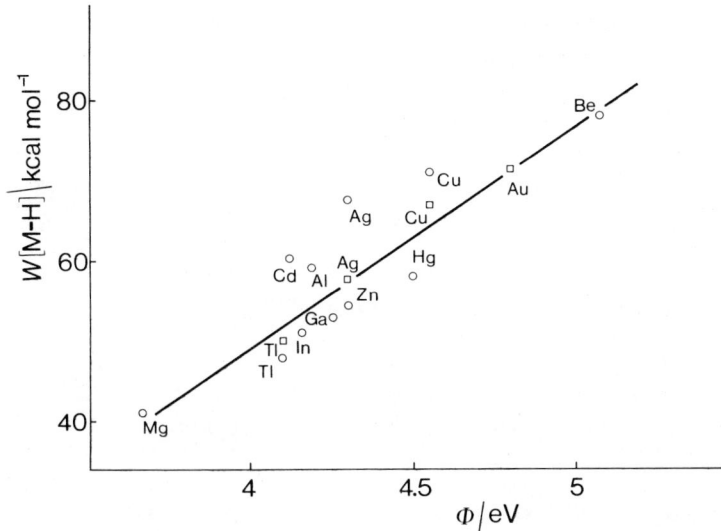

Fig. 9. M-H bond strength for sp metals derived from hydride formation heats (73) as a function of work function. Open square, spectroscopic dissociation values.

the use of ΔH_f proves unfortunately not to be as general an approach for sp metals as it appears to be for d metals. In the case of oxides of sp metals, ΔH_f decreases with increasing ϕ^M as d metals do, although with different slope, but the existence of a sharp, unique correlation gathering all sp metals is doubtful. Simple correlations can, however, be found within subgroups.

3 ELECTROCHEMICAL THERMODYNAMICS

The work function is contained in a rigorous way in thermodynamic equations. In this case, the relation between physics and electrochemistry is expected to be strictly quantitative. A possible objection by physicists is the strictly thermodynamic significance given here to ϕ^M values. The work function is very often measured with nonthermodynamic means. The energy to extract an electron from a metal may be anisotropic on the same surface (74). Only contact potential difference measurements have strict thermodynamic significance (1). However, these are problems regarding particular criteria for choosing ϕ^M values and should not influence directly developments of this approach. All ϕ^M values used here are very likely to be negligibly affected by similar problems.

3.1 The Surface Structure of a Phase

Ordered and orientated particles are likely to exist in the surface region of a phase because of the presence of unbalanced forces. As a result, modifications in the electron distribution give rise to the surface potential of metals (4,39), χ^M. Similarly, ordered layers of solvent dipolar molecules produce a potential drop at the free surface of the solution (75,76), χ^S. As the two phases are brought into contact and if no charge transfer can occur, the electron distribution in the surface of the metal is likely to be modified,

$$g^M_{(S)}(\text{dipole}) = \chi^M + \delta\chi^M_{(S)} \qquad (3)$$

and the orientation of dipoles on the liquid phase is expected to be changed,

$$g^S_{(M)}(\text{dipole}) = \chi^S + \delta\chi^S_{(M)} \qquad (4)$$

Although the interaction terms cannot be measured separately, their sum is capable of experimental measurement. An uncharged metal in contact with a solution does not show a zero contact potential difference (cpd) (18,44). In the case of Hg in contact with water, the cpd has been measured to amount (77,78) to 0.26 V. Thus

$$\delta\chi_{(Hg)}^{H_2O} - \delta\chi_{(H_2O)}^{Hg} = 0.26 \qquad (5)$$

Since neither theoretic nor experimental data are available for $\delta\chi_{(S)}^{M}$, information about this term can be obtained only by estimating $\delta\chi_{(M)}^{S}$, a more accessible quantity.

Equation 3 and 4 show that a potential drop exists at the boundary between an uncharged metal and a liquid phase, given by

$$g_S^M(\text{dipole}) = g_{(S)}^M(\text{dipole}) - g_{(M)}^S(\text{dipole}) \qquad (6)$$

3.2 Work Function for Electrons in Solution

Although the solution phase does not contain electrons intrinsically, work function concepts may be extended to this phase as well. If electrons are injected into a solution containing no electron donors and acceptors, they will be trapped in the structure of the phase, regardless of the next step. Electrons trapped in a solution are called solvated electrons. This concept and also experimental results in this field have widely spread in electrochemistry in the last two years (79 - 82). Reference is here made to electrons in thermodynamic equilibrium with the structure of the solution. In this case solvated electrons may be regarded as F centers in ionic solids and may be thought to lie in localized levels in the solution.

Concepts applied to electrons in metals can be applied to electrons in solution. Conceptually, the energy binding an electron to the interior of a liquid phase is the chemical potential of electrons in that phase, μ_e^S. If surface effects are accounted for, it is possible to introduce the work function of electrons in solution, which is defined by an equation quite similar to that for ϕ^M. Thus

$$\phi^S = -\mu_e^S + e\chi^S \qquad (7)$$

Detailed discussions on energy levels for electrons in solution can be found in the literature (83 - 87). Here we conceptually refer to isolated electrons for which the approximation of localized levels can be applied. In this case, ϕ^S has a single definite value.

3.3 Work Function of Metals in Solution

It is now easy to define the work function of a metal in a solution (82, 88 - 91). The presence of the solution acts like a change in the environment. Figure 10 shows that electrons from

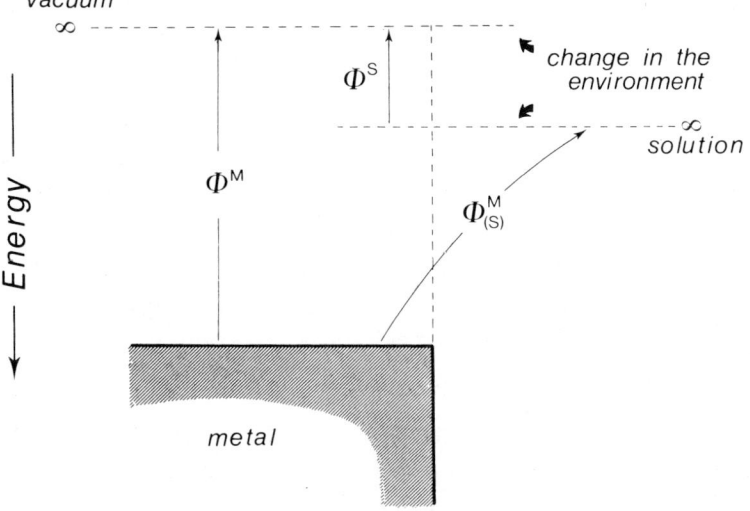

Fig. 10. Sketch of energy levels defining work functions of metals (ϕ^M), of solutions (ϕ^S), and of metals in solution ($\phi^M_{(S)}$).

the metal are no longer taken to infinity in a vacuum, but to infinity in a solution. There is no change in the reference energy level. However, the work to extract an electron from the metal is also different because the surface potential of the phase may change at the interface (interaction term). According to Fig. 10, it results from Eqs. 3, 4, and 6 that

$$\Phi^M_{(S)} = \phi^M + e\delta\chi^M_{(S)} - e\delta\chi^S_{(M)} - \phi^S \qquad (8)$$

If the work function of the solution is split into the volume and the surface contribution, a general equation is obtained where the orientation of the solvent molecules at the metal/solution boundary enters explicitly:

$$\Phi^M_{(S)} = \phi^M + e\delta\chi^M_{(S)} - eg^S_{(M)}(\text{dipole}) + \mu^S_e \qquad (9)$$

3.4 The Absolute Electrode Potential

Consider two uncharged metals in contact with the same solvent (Fig. 11). Let E be the experimental cell potential. The work

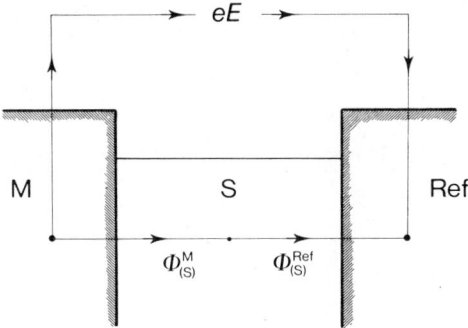

Fig. 11. Scheme showing the equality in the works to transfer one electron from metal M to the reference metal through a vacuum and through the solution.

to take an electron from the metal to the reference electrode through the external circuit must equal that to take an electron from the metal to the reference electrode through the internal circuit, that is, through the solution phase. Therefore,

$$eE = \phi^M_{(S)} - \phi^{Ref}_{(S)} \tag{10}$$

In Eq. 10 each separate work is independent of the nature of any other electrode. Thus, since

$$eE = e(E^M_{\sigma=0} - E^{Ref}_{\sigma=0}) = e(_{abs}E^M_{\sigma=0} - {_{abs}}E^{Ref}_{\sigma=0}) \tag{11}$$

where E^M is the potential of metal M in the solution measured on a conventional scale (e.g., hydrogen electrode scale), and $_{abs}E^M$ is the absolute value of the potential (92,93), it is possible to conclude that

$$_{abs}E^M_{\sigma=0} = \phi^M_{(S)} \tag{12}$$

Thus the work function of metals in solution measures precisely the absolute value of the potential of uncharged metals. A more general way of looking at this problem is that the absolute potential of an electrode is quantitatively measured by the threshold for photoemission of electrons into the solution (82). As a consequence, electrodes at the same E^M have the same value of $_{abs}E^M$ and the same threshold for photoemission (94). It is obvious that $_{abs}E^M$, although independent of the nature of the reference electrode, depends on the nature of the solvent.

3.5 Operative and Thermodynamic Electrode Potentials

Equations 10 to 12 can be developed to derive a general expression for charged and uncharged metals. From Eq. 10, taking into account Eq. 8, it follows that

$$E^M_{\sigma=0} = \frac{\phi^M}{e} + \delta\chi^M_{(S)} - g^S_{(M)}(\text{dipole}) + \frac{\mu^S_e}{e} - \frac{\phi^{Ref}_{(S)}}{e} \qquad (13)$$

from which

$$E^M_{\sigma=0} = \frac{\phi^M}{e} + \delta\chi^M_{(S)} - g^S_{(M)}(\text{dipole}) + \text{const} \qquad (14)$$

where the constant term includes the energy barrier at the reference electrode/solution interface and the chemical potential of electrons in solution. Now, for uncharged metals, with Eqs. 2 and 6,

$$E^M_{\sigma=0} = -\frac{\mu^M_e}{e} + g^M_S(\text{dipole}) + \text{const} \qquad (15)$$

and for charged metals

$$E^M = -\frac{\mu^M_e}{e} + g^M_S(\text{ion}) + g^M_S(\text{dipole}) + \text{const} \qquad (16)$$

where $g^M_S(\text{ion})$ accounts for the potential drop due to free charges on the metals (σ^M). The total potential drop across the interface is now,

$$\Delta^M_S\phi = g^M_S(\text{ion}) + g^M_S(\text{dipole}) \qquad (17)$$

and, from Eq. 16,

$$E^M = -\frac{\mu^M_e}{e} + \Delta^M_S\phi + \text{const} \qquad (18)$$

Equation 18 shows that the experimentally measured potential E^M has nothing to do (44,92,93, 95 - 97) with the actual electric potential drop across the interface, $\Delta^M_S\phi$. Since the latter is the actual potential that operates on charge-transfer steps during an electrochemical reaction and it is the sole term in Eq. 18 changing as E^M is changed, therefore $\Delta^M_S\phi$ may be defined as the operative potential of the electrode. E^M is, on the contrary, related to a thermodynamic view of the electrode process in that its form is a result of the way it is measured. E^M is then defined as the <u>thermodynamic potential</u> of the electrode. It is the sole potential experimentally measurable and, as a consequence, when the term "electrode potential" is used, it is tacitly understood that E^M is being referred to. Since E^M is usually measured with respect to a reference electrode, the concept of absolute electrode potential must also be referred exclusively to E^M and not to $\Delta_S^M \phi$. However, when the electrical state of a metal is changed by polarization, E^M and $\Delta_S^M \phi$ change by the same amount, as Eq. 18 shows. The thermodynamic and the operative potentials differ very much conceptually but they give the same practical result since they are changed by the same amount.

On the basis of equations developed above, it will be realized that the conceptual difference between E^M and $\Delta_S^M \phi$ is to be sought in the form of energy involved. Thus, the operative potential may be defined in terms of the electrical work to transfer one electron from the metal to the solution. Its reference zero is an electrical zero. Conversely, the thermodynamic electrode potential may be defined in terms of the electrochemical (chemical plus electrical) work to transfer one electron across the interface. It is defined on an energetic base and its zero is an energetic zero.

The absolute electrode potential is the absolute value of E^M. From Eqs. 13 to 18,

$$_{abs}E^M = -\frac{\mu_e^M}{e} + \Delta_S^M \phi + \mu_e^S \qquad (19)$$

It is to be noted that, as a matter of fact, as an electron is extracted from the metal, passed through the solution, and taken to the reference electrode, μ_e^S appears twice in Eq. 13 with opposite signs and cancels out. This is equivalent to assuming $\mu_e^S = 0$ and shifting the zero position of the absolute potential.

Thus, a new potential may be defined, not properly absolute but conditional (98):

$$_r E^M = - \frac{\mu_e^M}{e} + \Delta_S^M \phi \qquad (20)$$

$_r E^M$ may be termed the reduced absolute potential. It is easy to show that this is the only type of potential that can be obtained from the application of a Born-Haber cycle to the electrode reaction. The use of $_r E^M$ in the place of $_{abs} E^M$ is sufficient for conceptual discussion involving the electrode potential.

3.5.1 FURTHER DEVELOPMENTS The concept of absolute electrode potential has recently aroused new interest and has given rise to some controversy (315 - 317). No consensus has apparently been achieved, but it is now thought that opposite views may in fact be reconciled ultimately. Since the cell potential E is actually a measure of the difference in electrochemical potential of electrons at the two ends of the system (13), the contribution each of the single half-cells actually gives to E may be best recognized by examining the profile of electrochemical potential of electrons inside the cell (317). This leads to the necessity of characterizing the electrochemical potential in solution. Two alternative routes may be followed within the framework of accepted models for electrodes. If polarizable electrodes are dealt with (317) then the situation is that discussed in Section 3.4. In this case, Eq. 10 holds and the contribution each single electrode gives to E is definable in terms of the work function of the metal in the solution. If nonpolarizable electrodes are dealt with, then an electronic equilibrium between M and S may also be postulated (315,316). Under these circumstances, the two half-cells must be kept separated (salt bridge) and the energy level for electrons in solution in one of the half-cells is expected to differ from that in the other one. In particular, $\bar{\mu}_e$ is constant within the former half-cell from M to S, and constant within the latter half-cell from S to Ref (Fig. 11). On the whole, the potential of a single electrode is definable in terms of the electrochemical potential of electrons in the solution. Alternatively, it may also be defined in terms of the electrochemical potential of electrons in the metal, since, as a result of electronic equilibrium, $\bar{\mu}_e^M = \bar{\mu}_e^S$.

The two preceding definitions are separately valid in the context of the model chosen for the interface. Thus with polarizable electrodes metal-solution interfaces are treated, as regards electrons, in the way physicists treat metal-vacuum interfaces. With nonpolarizable electrodes, the situation at metal-solution interfaces is treated as rigorous thermodynamics sees it. Neither of the two approaches is concretely valid universally in that there are well characterizable potential ranges where only one of the two may be reasonably applied to electrons to describe the actual situation.

The reduced absolute potential may, however, be shown to be concretely independent of the model chosen. In fact, such a potential results from the expression of E in terms of either of the two above single potentials if μ_e^S is cancelled out, in that it compares twice with opposite signs, and if any liquid junction potential is neglected. This is tantamount to assuming $\mu_e^S = 0$ in both cases. $_rE^M$ thus is the form of potential most useful in practice for explicit parameters contributing to the electrode potential without necessarily making any assumption about the electronic polarizability or nonpolarizability of the interface.

If $_{abs}E^M$ must be quantitatively defined, however, then μ_e^S must also be concretely characterized. With polarizable electrodes, the zero of single (absolute) electrode potentials is measured by the energy of electrons in solution. Since electrons are not present in solution by definition, they could be used as chemical charged probes to pick up the energy states in the phases (317). Thus the energy level in solution may in this case be that of an isolated electron interacting with the structure of the solvent, which can be calculated on the basis of models (85).

With nonpolarizable electrodes, as long as we also postulate the validity of thermodynamic formulas for equilibrium situations involving electrons in cells, μ_e^S is by definition the chemical potential of electrons in the solution in electronic equilibrium with the metal. However, single electrode potentials have been shown above to be measured by $\overline{\mu}_e^S$ (or $\overline{\mu}_e^M$), and the zero level is now represented by electrons in vacuum. The problem of the significance of defining chemical potentials for electrons at such low electronic populations in the liquid phase as expected in usual electrochemical cells is not relevant for this chapter and is treated in detail elsewhere.

3.6 Calculation of the Operative Potential

From Eq. 20 it can be shown (92) that if a standard hydrogen electrode is used as a reference electrode, the standard thermodynamic potential of metal M for the reaction $M^+ + e \rightleftharpoons M$ is given by

$$E^°(M^+/M) = \frac{\mu_e^M}{e} + \Delta_S^M \phi^° - {}_rE^°(H^+/H_2) \tag{21}$$

Rearranging Eq. 21, an equation useful for calculating the operative electrode potential is derived:

$$\Delta_S^M \phi^° = E^°(M^+/M) + \frac{\mu_e^M}{e} + {}_rE^°(H^+/H_2) \tag{22}$$

Since $E^°(M^+/M)$ can be measured experimentally (99), $_rE^°(H^+/H_2)$ can be calculated by applying a Born-Haber cycle to the electrode reaction, and since μ_e^M has been shown to be derivable from the electronic theory of metals, the operative electrode potential is now calculable. The value of $_rE^°(H^+/H_2)$ has been found (92,93) to be 4.31 V.

TABLE IV

Operative Standard Electrode Potential

Electrode	$\Delta_S^M\phi^°/V$
Li^+/Li	$- 0.54 \pm 0.3$
Na^+/Na	$- 0.46 \pm 0.1$
K^+/K	$- 0.72 \pm 0.1$
Rb^+/Rb	$- 0.71 \pm 0.1$
Cs^+/Cs	$- 0.58 \pm 0.2$

Table 4 summarizes the values of $\Delta_S^M\phi^°$ for alkali metals. The entire uncertainty in $\Delta_S^M\phi^°$ derives essentially from the uncertainty in the values of μ_e^M. Unfortunately, the knowledge of a fundamental quantity in electrochemistry like $\Delta_S^M\phi$ cannot be improved by electrochemists directly.

3.7 Derivation of Electrochemical Work Functions

Starting from Eq. 16, the thermodynamic potential of an electrode can be related explicitly to the work function:

$$E^M = \frac{\phi^M}{e} + \delta\chi_{(S)}^M - g_{(M)}^S(\text{dipole}) + g_S^M(\text{ion}) + \text{const} \qquad (23)$$

Since E^M is a function of the charge on the metal, some terms in Eq. 23 must be a function of charge. If $\delta\chi_{(S)}^M$ is assumed to be metal independent, only $g_{(M)}^S$ (dipole) and g_S^M(ion) depend strictly on the charge. For the rationale behind the splitting of $\Delta_S^M\phi$ into dipole and free charge contributions (13), g_S^M(ion) must be metal independent at the same charge. In double-layer models accepted for sp metals, water molecules are thought to rotate under the influence of the electric field in the double layer. $g_{(M)}^S$ (dipole) is metal dependent (16, 100) when $\sigma^M = 0$, but at sufficiently negative charges as the orientation of water is the same on all metals because electrical forces predominate over chemical forces, this term becomes metal independent (101, 102).

The change in E^M with charge can be explored experimentally by deriving charge-potential curves by integration of double-layer capacity data. Figure 12 shows charge-potential curves for Hg and Ga. Data for Hg are taken from Grahame (103). Data for Ga

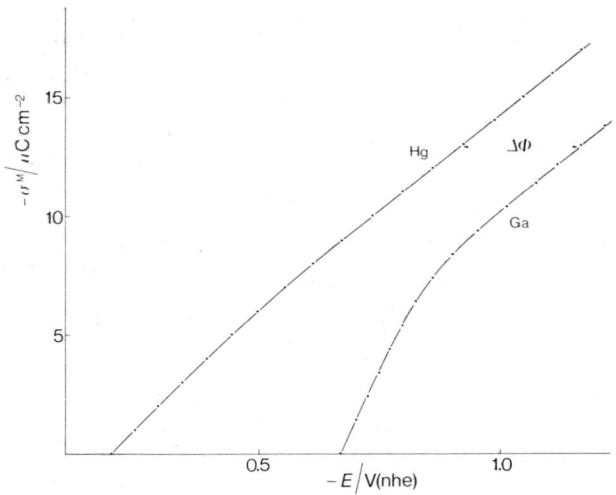

Fig. 12. Charge-potential curves from integration of double-layer capacity data for Hg in 0.1 M NaF (103) and for Ga in 0.1 M NaClO$_4$ (104).

have been obtained by graphical integration of capacity curves for 0.1 N NaClO$_4$ solutions (104). At low charges, the two curves differ because $g_{(M)}^S$ (dipole) differs on the two metals. At high negative charges, the two curves become parallel. At this point, the properties of the solvent are metal independent. The distance on the E axis separating the two curves measures the difference in electronic energy at the Fermi level:

$$\Delta_{M'}^{M} E = \Delta_{M'}^{M} \Phi \tag{24}$$

Equation 24 provides a way to evaluate the work function at the metal-vacuum interface for the given surface starting from electrochemical data. Once the work function for one metal is independently known, the value of Φ for the other is readily obtained. The best reference metal is Hg, for which very reliable capacity data are available and whose work function value of 4.50 eV is known with good accuracy (59). The derived value of the work function for Ga must, of course, be solvent-independent if the present approach is valid. Results in acetonitrile (105) prove that this is in fact the case.

The ideas developed first (101) for Ga can be extended to other metals for which reliable capacity data are now available. For Eq. 24 to be applicable, charge-potential curves are required to be closely parallel at negative σ^M. Nonparallel behavior may be a result of inaccurate capacity data, but may also be a consequence of other factors. In this way spurious results may be easily found and this approach may constitute a new tool for checking the reliability of capacity data or for indicating whether the model for the double layer needs some revision or refinement.

Charge-potential curves can be found in the literature (106-111) for Cd, Pb, Bi, Au, and Sn. Comparison of E^M values have been made at $\sigma^M = -15$ $\mu C/cm^2$ using Hg as the reference metal. Data for

Tl and In have been taken from Frumkin et al. (18). Zn and Ag are ignored for the moment because the interpretation of the relevant data (112,113) requires more study. Table 5 compares electrochemical work functions and ranges of experimentally measured values (32,59). Data in the table refer to polycrystalline surfaces.

TABLE V

Physical and Electrochemical Work Functions, Φ/eV

Metal	Electrochemical	Physical (1,4,32,59)
Au	4.80	5.2 - 5.3
Bi	4.35	4.3 - 4.4
Cd	4.12	4.0 - 4.2
Ga	4.25	4.2 - 4.4
In	4.10	4.1^a
Pb	4.24	4.0^a
Sn	4.40	4.2 - 4.4
Tl	4.10	3.7 - 3.8

[a] Maximum value.

With the striking exception of Au, electrochemical values fall within the range of uncertainty of experimental values. Electrochemical work functions are operative quantities inasmuch as they refer precisely to the state of surfaces during electrochemical measurements. There are reasons to believe that the electrochemical measurement of the work function, where it applies, is more precise than the direct physical measurement. As an example, the high value for Tl is likely to arise because the electrochemical measurement refers to a really reduced surface often not warranted by inadequate vacuum techniques in old measurements. For this reason, it is thought that these values can be useful for improving the reliability of the list of work function and have been included in Table 3. The validity of electrochemical work functions is closely connected to the validity of the model of the double layer as a simple "water capacitor".

Equation 24 cannot, in principle, be applied to transition metals. Reliable values of capacity are not yet available for these metals. Decomposition of the solvent is likely to occur (114-116). Pseudocapacity is often present (117). Attempts at an elimination of this leads (118,119) to a capacity curve remarkably different from that of Hg. Integration of such a curve gives (120) a σ/E curve that apparently parallels that of Hg and does not show the steep portion observable in that of Ga in Fig. 12. Although reorientation of the solvent is suggested (121), it is doubtful that this occurs. Solvent molecules are likely to be anchored to the surface and would not reorient. The electric field can only govern charge transfer from and to the electrode. Partly empty d levels are thought to be responsible for this behavior. Electrochemical work function values can nevertheless be suggested (16) for some d metals, but on a more empirical basis than Eq. 24. This point is discussed further later.

The case of Au is admittedly puzzling. A possible explanation may be in terms of difference in crystallinity of the surface in the different conditions of measurement. There is evidence (122-125) that for evaporated films in a vacuum some rearrangement of the surface layer is possible for Au favoring close-packed structures. According to some views (122,126), this would be possible only in a vacuum, whereas contamination of the surface would hinder this process. Presence of water may simulate the effect of contaminations. Thus Au in an operative state would present a surface where low index planes would be stabilized, thus giving lower work function values. This explanation may be regarded with little confidence by physicists on the ground that polycrystalline wires or films sintered at room or lower T never show (51,127) values close to 4.8 eV. However, quite recent measurements (128) of work function for Au deposited on W layer by layer show that the second layer may have values close to 4.8 eV if the deposition T is below 300°K. This structure is apparently stable for eight layers; at heavier coverages, crystallites form. Although the present view for explaining the case of Au may leave a number of questions still open, no other conclusive description can be offered at the moment. This point is further discussed later.

3.8 Underpotential Deposition of Metals

Although reactions of discharge of particles with deposition of products on the electrode surface are more relevant to the field of kinetics, the underpotential deposition is a special case that is most profitably treated in the field of thermodynamics. This is a particular phenomenon where an electrode reaction occurs before the reversible potential. This is termed underpotential in contrast to the usual concept of overpotential. The underpotential deposition offers the possibility for a discussion in terms of concepts outlined above.

Underpotential deposition of metals is observed for the discharge of the first monolayer of metal M on a foreign metallic substrate M'. Application of a Born-Haber cycle to this reaction suggests (129) that the underpotential deposition is due to the excess binding energy (130) of metal M on M' with respect to M atoms on bulk M. This phenomenon is paralleled by the adsorption of a metal monolayer on a metallic substrate from the gas phase. Desorption measurements applied to alkali metals deposited on a foreign substrate show (131) that the binding energy is in excess in the first monolayer with respect to any other subsequent layer. The latter possess the same binding energy as the bulk metal. The work function of the substrate changes from the value of bare M' to the value pertaining to bulk M upon completion of the first monolayer (131-135).

A physical model for the underpotential deposition of metals has been only recently proposed by Gerischer et al. (130,136) in terms of contact potential difference. Starting from the observation that the monolayer presents bulk metallic properties in the case of alkali metal adsorption from the gas phase (131), the excess binding energy is suggested by Gerischer et al. to be related to a partial ionic bond expressed in terms of difference in work function between M and M'. A linear correlation is indeed obtained as $\Delta_M^{M'} E_d$ (underpotential deposition for M on M') is plotted as a function of $\Delta_M^{M'} \Phi$. A slope of 0.5 found for this relationship has been interpreted on the basis of the 0.5 slope for the correlation between the electronegativity and the work function (32)(see Fig. 4).

An alternative way to calculate the excess binding energy by means of the Pauling equation (5,61,71) has been suggested (137). The Pauling equation has, however, been used as such without any correction to take into account that not isolated molecules in the gas phase but bonds on metal surfaces are dealt with (32). Nevertheless, a linear correlation of unity slope is claimed (137) between $\Delta_M^{M'} E_d$ and excess binding energy. However, the plot reveals a broad trend rather than a definite quantitative relationship and confirms that the Pauling equation cannot be applied indiscriminately to d and sp metals. Account for the presence of a metal surface can be attempted, but this does not seem to improve the situation very much.

The physical description of Gerischer et al. (130,136) seems to be the most adequate. However, a consistent analysis should not be made at the peak potential where the transformation from nonmetallic to the metallic state of the adsorbed layer (138) is likely to be incomplete (139). Furthermore, the monolayer does not show metallic properties in all cases (140). Some metals attain the metallic state only after deposition of a number of layers (141). Thus the best approach may be to refer to the potential of incipient deposition where the surface bond is very likely to be fully ionic as proved by the steep slope for the change in work function of the substrate at low coverages. In this case the process can best be rationalized on a modellistic basis. For the deposition of a metal atom M on bulk metal M', the energy of adsorption at $\theta \to 0$ can be written (142,143) as

$$W(M'-M) = -I^M + \phi^{M'} + \frac{e^2}{4x} \qquad (25)$$

where I^M is the energy of ionization of metal atom M and the last term is the image energy. For the deposition of the metal atom M on another substrate M'', the energy is

$$W(M''-M) = -I^M + \phi^{M''} + \frac{e^2}{4x} \qquad (26)$$

The difference between Eqs. 25 and 26 represents the difference in underpotential deposition on the two substrates:

$$\Delta_M^{M'} E_d - \Delta_M^{M''} E_d = \frac{\phi^{M'} - \phi^{M''}}{e} \qquad (27)$$

This concept, developed in the case of underpotential deposition on a single substrate requires, that the depositon of a metal atom M on bulk M could be written as

$$W(M-M) = -I^M + \phi^M + \frac{e^2}{4x} \qquad (28)$$

Difference between Eq. 26 and Eq. 28 gives

$$W(M'-M) - W(M-M) = e\Delta_M^{M'} E_d = \Delta_M^{M'} \Phi \qquad (29)$$

Equation 29 predicts a linear relationship of unity slope between ΔE_d and $\Delta \Phi$ without it invoking the metallic nature of the adsorbed monolayer and without resorting to Pauling's electronegativity for isolated atoms. Figure 13 shows that, if ΔE_d is appropriately corrected to $\theta \rightarrow 0$ by adding halfwidth values for the peaks (130) and Φ^M values from Table 3 including electrochemical values are

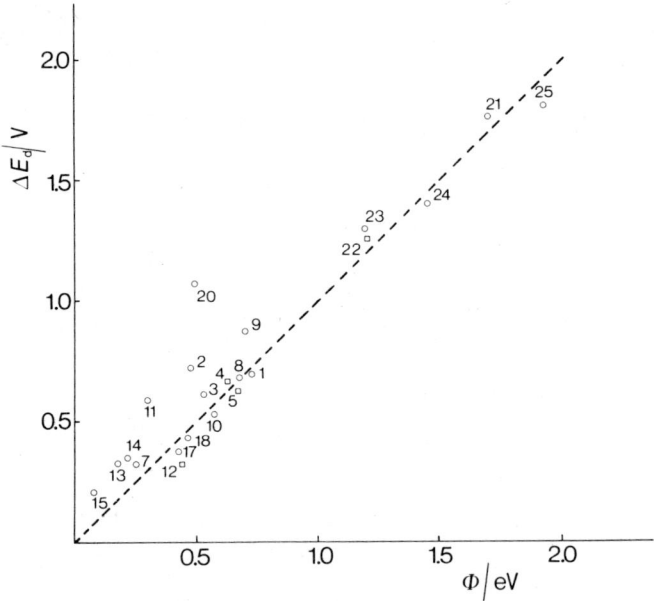

Fig. 13. Underpotential, corrected to $\theta = 0$, for deposition of metal M on foreign metallic substrate M' plotted as a function of difference in work function between bulk M and bulk M'. Numbers refer to data in Table 1 of Ref. 130 arranged in sequential order. Open square, correction to $\theta = 0$ estimated from Fig. 7 of Ref. 130.

used, then the points representing the various systems fit a
linear relationship of unity slope fairly well.

The preceding approach introduces Eq. 28, which seems questionable in that it is based on the assumption that a fully ionic bond is formed as a metal atom M is deposited on the surface of bulk M. This is, however, not an unreasonable possibility in view of results (144) showing that deposition of W atoms on bulk W from the gas phase produces a large drop in the work function of the metal. A metal atom isolated on a surface is presumably deprived of its valence electrons so that it behaves as an ionized particle.

3.8.1 OTHER SYSTEMS Ideas developed for the deposition of metals may be extended to systems to which the concept of underpotential has never been explicitly applied. Prewaves in polarographic studies of redox reactions (145) have been attributed (146) to species in the adsorbed state. Along the above lines, this may be defined as an underpotential charge transfer governed by the excess energy of species in the adsorbed state with respect to species in solution. If prewaves for a given redox couple could be available on different substrates, the prewave potentials should be linearly related to the work function of the substrate.

Hydrogen and oxygen adsorption on electrodes may be treated as underpotential deposition if the evolution reaction is regarded as the deposition at the reversible potential. Thus, substrates forming surface bonds with hydrogen stronger than the H-H bond in the molecule should show underpotential deposition of hydrogen, that is, adsorption waves in voltammetric curves. This is indeed true in the case of metals of Pt group (147), but the potential for incipient adsorption becomes more cathodic rather than more anodic (148,149) as it should do since the M-H bond strength, as measured in adsorption from the gas phase, increases. Furthermore, metals known to adsorb hydrogen very strongly from the gas phase (64), for example, W, Ta, Nb, Ti, do not show any underpotential deposition of hydrogen from solutions. This may be related to Barclay's concept (151) of a weakening of the M-H bond strength on these metals in solution with respect to the gas phase. A possible explanation may be a competition with water molecules whose energy of interaction with the surface of metals (100) is likely to increase more rapidly than the M-H bond energy does, as Figs. 7 and 8 suggest. This point is discussed further later in connection with the hydrogen evolution reaction.

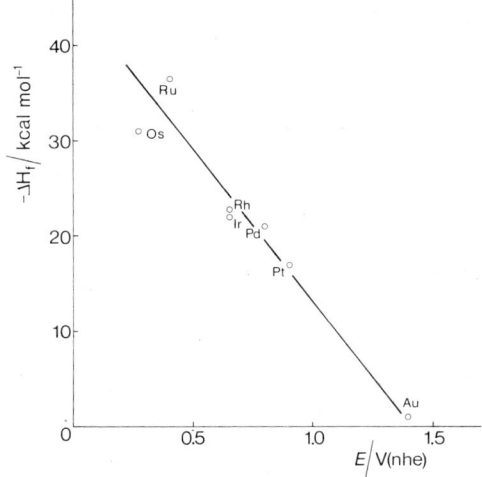

Fig. 14. Potential for incipient oxygen adsorption from acid solutions (152) plotted as a function of substrate oxide formation heat (187).

Adsorption of oxygen lends itself to some quantitative considerations. Figure 14 shows a good linear dependence between the potential of incipient oxygen adsorption on various metals (152) and the enthalpy of formation (73) of the oxide MO assumed to simulate to a first approximation the M-O surface bond strength. In this case there is clearly no reversal in the order of reactivity with respect to the gas phase (64). This is probably because there is no competition with the solvent. Water molecules simply evolve to adsorbed oxygen. According to concepts developed for the underpotential deposition of metals, the potential of incipient adsorption of oxygen should be linearly related to the work function of the substrate. Although a wide range of Φ^M values cannot be explored, Fig. 15 shows that this is probably the case for transition metals, but Au falls far from the common line if the electrochemical work function is used. This may be an indication that since covalent bonds are here prevalently involved, splitting of metals into d and sp metals is probably to be expected. However, the behavior of gold may in a general way be

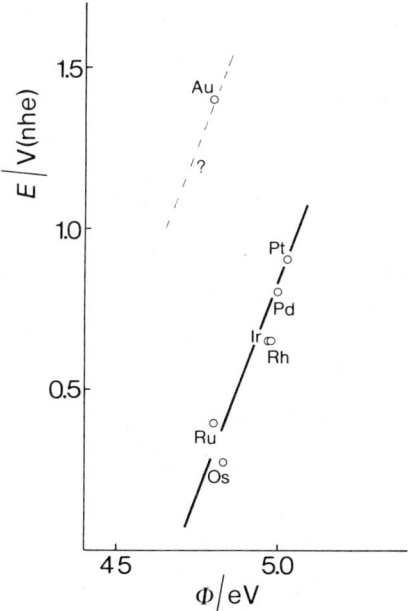

Fig. 15. Potential for incipient oxygen adsorption from acid solutions (152) plotted as a function of substrate work function.

open to questions, so that further study is needed to settle the situation.

4 DOUBLE-LAYER STRUCTURE

The work function is strictly involved in double-layer concepts. Since the structure of the double layer is essentially governed by the position of the potential of zero charge, relation to the energy of electrons at the Fermi level in metals is immediate. As a matter of fact, work function concepts were first introduced into double-layer problems (9). Various approaches using correlations with work functions (12 - 15, 17, 153) always led to a cautious suggestion that the structure of the double layer should not be affected by the nature of the metal. It is believed once more that these conclusions are not due to an inadequacy in the electrochemical data but to the paucity and inaccuracy of work function values used. Recent approaches (16, 18, 93, 100) clearly suggest that the nature of the metal can profoundly affect the structure of the double layer. The situation is now such that double layer and work function data are likely to be useful for reciprocal clarification.

4.1 Relationship between Work Function and Potential of Zero Charge

A peculiar aspect of the work function approach to electrochemical interfaces is the investigation of the relationship between work functions and potentials of zero charge. In Eq. 14, although the potential of zero charge and the work function are measurable quantities, the other terms are not capable of direct experimental measurement (13). A graphical correlation is thus expected to be of help in gaining insight into this matter. Figure 16 shows a plot of ϕ^M versus $E^M_{\sigma=0}$ measured in the standard

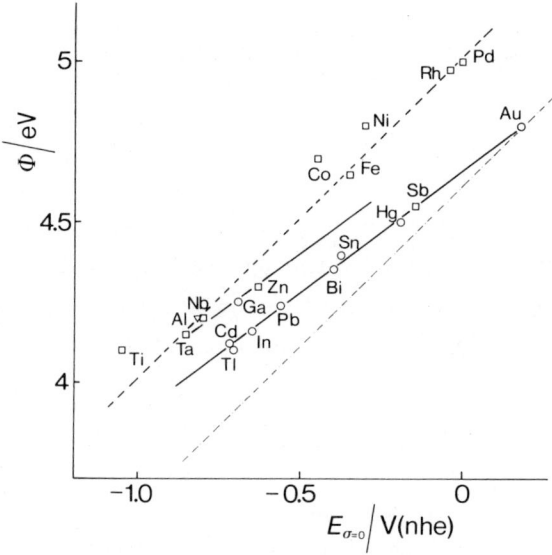

Fig. 16. Dependence of potential of zero charge of metals on work function. Open square, physical work functions; open circle, electrochemical work functions; open triangle, value of potential of zero charge unavailable experimentally.

hydrogen electrode scale. Work functions are taken from Table 3 so that they include electrochemical values. However, Fig. 12 shows that, in principle, $\Delta^{M'}_M E$ at constant charge is not expected to bear any definite relationship to $\Delta^{M'}_M E_{\sigma=0}$. Thus, the results from Fig. 16 are thought not to be an artifact. If definite relationships are obtained in the ϕ^M versus $E^M_{\sigma=0}$ plot, this is simply a consequence of the double-layer conforming to a well-defined model.

The values of $E_{\sigma=0}^M$ in Fig. 16 are taken for most of the sp metals from those suggested by Frumkin et al. (154,155). The potential of zero charge for Al is not an experimental value but merely suggested to test the model. The value for Zn (16) is not from capacity data in that the metal starts dissolving at negative rational potentials (156). $E_{\sigma=0}^M$ values for transition metals are not so reliable as for sp metals. They are introduced with the hope of finding out some definite trend in order to build up a model and test its validity a posteriori. Accurate values are available (157,158) for the group of Pt. However, differences exist between data obtained by the open-circuit scrape technique (158) and data from adsorption measurements (157). Reasons for the preference given here to the former are discussed elsewhere (16) in terms of some contribution from adsorbed hydrogen to the interfacial potential drop in the latter (159,160). The opposite preference has been justified (18) in terms of the significance of the data obtained with the two techniques. The role of adsorbed hydrogen atoms seems, however, to be undisputable (161).

It is to be noted that in Table 3 the work functions of Rh and Pd are experimental values (52,59), whereas experimental values for Pt and Ir would not fit in correlations. For this reason, since the values of $E_{\sigma=0}^M$ are fairly accurate for these metals and because Eq. 14 is an exact equation, Φ^M values for these metals may be derived by fitting $E_{\sigma=0}^M$ values to the graphical correlation for transition metals. The derived values are electrochemical work functions and are included in Table 3. These values are much lower than those experimentally measured. The case is quite similar to that of Au. Although a value of about 5.0 eV is, however, found (162) for Ir around the (110) face, no value close to this figure can be experimentally measured for Pt. It is quite interesting to note that these are the only metals which, together with Au, undergo a surface rearrangement (126) toward a close-packed structure when exposed to a vacuum. An explanation similar to that given for Au may be invoked here.

Ag and Cu are ignored in this plot, the former for reasons given in Section 3.7, and the latter because the potentials of zero charge observed with single crystal faces (163) are in disagreement with that for the polycrystalline surface (164).

Figure 16 shows that metals divide into two groups. Transition metals are gathered in a group for which the linear dependence has unity slope:

$$E_{\sigma=0}^{M} = \phi^{M} - 5.01 \qquad (30)$$

With reference to Eq. 14, this means that the sum of $\delta\chi_{(H_2O)}^{M}$ and $g_{(M)}^{H_2O}(\text{dipole})$ is the same for all metals. No separate information can be directly obtained on these two quantities, but in the absence of an unlikely compensation effect, assuming as above that $\delta\chi_{(H_2O)}^{M}$ is the same for all metals, unity slope means that the orientation of water on transition metals should be metal independent. On the other hand, a slope different from unity in the case of sp metals suggests that, provided the interaction term is the same for all metals (and possibly the same as for transition metals), the orientation of water on sp metals should be metal dependent increasing from Au to Al.

4.2 The Meaning of $\delta\chi_{(H_2O)}^{M}$

Although the values of $\delta\chi_{(H_2O)}^{M}$, $g_{(M)}^{H_2O}(\text{dipole})$, and of the constant in Eq. 14 cannot be determined experimentally, it is necessary to attempt to estimate them separately. Hg lends itself especially well to this purpose in that a number of quantities are accurately known for it. Thus, from the values of ϕ^{Hg} (59)

and $E^{Hg}_{\sigma=0}$ (103), it follows that

$$\delta\chi^{Hg}_{(H_2O)} - g^{H_2O}_{(Hg)}(\text{dipole}) + \text{const} = -4.69 \qquad (31)$$

In Eq. 31, the surface potential of water at Hg can be estimated from the shift of the electrode potential upon adsorption of organic substances (165). The most reasonable value has been found (166) to be 0.08 V. Furthermore, the value of the constant, which is simply the reduced absolute potential of the standard hydrogen electrode, that is, $_rE^{\circ}(H^+/H_2)$ as defined by Eq. 20, can be computed from a Born-Haber cycle applied to the hydrogen electrode reaction. The most reliable value has been found (92,93) to be 4.31 V. Thus from Eq. 31 it is possible to derive for $\delta\chi^{Hg}_{(H_2O)}$ the value of −0.30 V. This is assumed to equal $\delta\chi^{M}_{(H_2O)}$.

The sign of $\delta\chi^{M}_{(H_2O)}$ corresponds to a regression in the surface redistribution of electrons. This may be conceivable if it is considered that electrons of the metal may be pushed back by the electron cloud of the water molecule. The sign of $\delta\chi^{M}_{(H_2O)}$ agrees with an earlier suggestion by Frumkin et al. (75), although in a later treatment (78) this quantity was neglected.

The possibility of understanding qualitatively the physical meaning of the interaction term $\delta\chi^{M}_{(H_2O)}$ on the basis of data for adsorption from the gas phase may exist. It is necessary to resort to processes where the interaction term may be conceptually involved. There is some evidence that the strength of interaction of water molecules with Au (167) is similar to that on Hg (168) and probably less than that on Pt (169). Also, work function drops upon water adsorption on Ga (170) seem to indicate stronger orientation on this metal than on Hg, in agreement with conclusions from analysis of electrocapillary and capacity data (101). A good support for the idea of constancy in $\delta\chi^{M}_{(H_2O)}$ on all metals and in $g^{H_2O}_{(M)}$(dipole) on transition metals, respectively, is given by measurements (171-173) of work function drop upon water adsorption on Fe, Ni, and Al. The same value of $\Delta\phi^M$ supports the suggestion that Al may locate in the group of transition metals at the extreme end of the sp metal group. Evidence exists (174) that water may be regarded as integral molecules on these surfaces.

Adsorption of rare gases from the gas phase could be a way to observe directly the interaction term through the experimental surface potential, since rare gases have no permanent dipoles. Experimental results are not clearcut and interpretations are in conflict. Explanations for work function changes upon rare gas adsorption are based on two possible models (175,176), polarization (177), and charge-transfer-no-bond (178), respectively; the relative validities of the models are debated (179-182). Surface positive fields are invoked (30) to support the polarization model. However, double-layer results in electrochemistry do not appear to require postulating the existence of such a field. Local fields, however, seem possible at asperities on the surface (179). In this case, roughness on an atomic scale would simulate different $\delta\chi^M$ values through different experimental surface potentials for Xe adsorption on different metals.

It should be stressed that a definite difference between the electrochemical situation and adsorption from the gas phase resides in the fact that in the former case metals are under electrical control, which enables the Fermi energy to be kept constant. This is obviously not feasible in the latter case. Thus adsorption of water on Pt and W can occur (115,116) with hydrogen evolution and increase in ϕ^M as a result of oxidation of the surface upon extensive charge transfer. For the same reason, a direct comparison of electrochemical results with data for Xe adsorption might require a particular caution. However, to a first approximation, the same work function drop upon Xe adsorption (52,180,183-186) on Cu, Au, and Ag possessing almost the same melting point, and then (51) the same surface structure at the same T on evaporated films [and presumably the same local field (179), if any] but different work functions, and then different charge-transfer powers, may be a good support for the view that $\delta\chi^M$ may substantially be a metal-independent quantity.

4.3 The Surface Potential of Water on Metals

Once the surface potential of water on Hg is known, a straight line with unity slope can be drawn in Fig. 16, 0.08 V on the right of the point for Hg. This line would represent the place where the surface potential of water is zero. Now, the horizontal distance between the point for a given metal and such a line would precisely measure the surface potential of water at that interface. This procedure is equivalent, if the model of the "water capacitor" holds for all metals, to subtract the difference in potential at the same negative charge from the difference in potential of zero charge for two metals (Fig. 12). In fact, from Eq. 14, we have

$$\Delta_M^{M'} E_{\sigma=0} = \Delta_M^{M'} \phi + \Delta_M^{M'} g^{H_2O}(\text{dipole}) \qquad (32)$$

Recalling Eq. 24, it follows that

$$\Delta_M^{M'} E_{\sigma=0} - \Delta_M^{M'} E = \Delta_{M'}^{M} g^{H_2O}(\text{dipole}) \qquad (33)$$

It is evident that the two procedures would give the same result anyway. But the fact that the points for the metals fit a well-defined relationship is an indication that a definite model is obeyed. Whatever the approach, it must lead to consistent results.

TABLE VI

Surface Potential of Water at Metal-Solution Interfaces[a]

Metal	$g_{(M)}^{H_2O}$(dipole)/V
Au	0.0
Sb	0.07
Hg	0.08
Bi	0.13
Sn	0.14
Pb	0.19
Tl	0.20
In	0.20
Cd	0.23
Zn	0.32
Ga	0.33
d metals	0.40

[a] Estimated from Fig. 16.

Table 6 summarizes the values of the surface potential of water on various metals. In the table a value valid for all transition metals has also been included. This is not an experimental value, because charge-potential curves for transition metals obtained from capacity curves do not possess as straightforward a meaning as in the case of sp metals.

The surface orientation of water changes from zero on Au to

0.4 V on transition metals. The sign corresponds to a dipole
with the negative end toward the exterior of the liquid phase,
that is, with the oxygen atom pointing to the metal surface.
The presence of such an oriented dipolar layer corresponds to a
decrease of the work function of the metal. This is in fact observed upon adsorption of water from the gas phase on metals (171-
173). However, the experimental drop in work function exceeds the
maximum value of 0.4 V even though the effect of the interaction
term is accounted for. A possible reason for this may be that in
the adsorption from the gas phase the monolayer of water is also
subjected to forces at the boundary with the air. The exact reason for the quantitative discrepancy, however, is unclear at the
moment. An important point is that the maximum value of $g_{(M)}^{H_2O}$ (dipole) is very likely to be related to a definite position of water
molecules on the surface.

4.4 The Role of Water Molecules in the Double Layer

The group of sp metals is seen in Fig. 16 to split into two subgroups within which the orientation of water increases as the
electronegativity of the metal decreases. In Fig. 3, Ga, Zn, and
Al are seen to be grouped with transition metals. This indicates
that the splitting of sp metals into two subgroups should be
closely related to the intimate electronic structure of these
metals.

The importance of the concept of electronegativity in this context is evidenced by an empirical equation (16) lumping all metals
together:

$$E_{\sigma=0}^{M} = \phi^{M} - 4.61 - 0.40\,\alpha \qquad (34)$$

where

$$\alpha = \frac{2.10 - x_e^M}{0.6} \tag{35}$$

α is a parameter changing from 0 to 1, and it is related to the degree of orientation of water. It ensues from Eq. 34 that as $\alpha = 0$, water molecules contribute 0 V to the electrode/solution potential drop. As $\alpha = 1$, water dipoles contribute some -0.40 V to the potential of zero charge. Thus, $\alpha = 0$ for Au and $\alpha = 1$ for transition metals.

A model may be proposed for metal-water interactions. Metals with low work functions are electropositive in nature and tend to interact with the electronegative end of the water dipole. A strong orientation of water is expected on these metals. Metals with high work functions are electronegative in nature and interact weakly with the electronegative end of the dipole of water. A weak orientation for water may be expected on these metals. Interaction of water with transition metals is no longer simple. A strongly oriented chemibond may form between the oxygen atom of water and the partly empty d bands of the metals. The strength of this bond should always be enough to force water molecules to the maximum in orientation. Thus, the orientation of water on d metals is expected to be metal independent. Alternatively, along the previous lines, transition metals behave as strongly electropositive metals. However, the nature of electrons at the Fermi level and not simply their energy should be responsible for the splitting of metals into the d and sp groups. The position of maximum orientation must thus be strongly related to the molecular structure of water.

The strength of the metal-water interaction should be a function of the affinity of metals for oxygen (93,100). No direct experimental value is available for oxygen adsorption from the gas phase on most sp metals. In view of the linearity generally observable for the relationships between the heat of formation for the bulk compound and that for the surface compound, thermodynamic data (73,187) for the enthalpy of formation of the oxide MO may be regarded as a measure of the affinity of metals for oxygen. Figure 17 shows that the orientation of water on metals is related in a simple way to the affinity of surfaces for oxygen. Points for some of the metals, however, fall far from the expected plot in a way that requires some explanation.

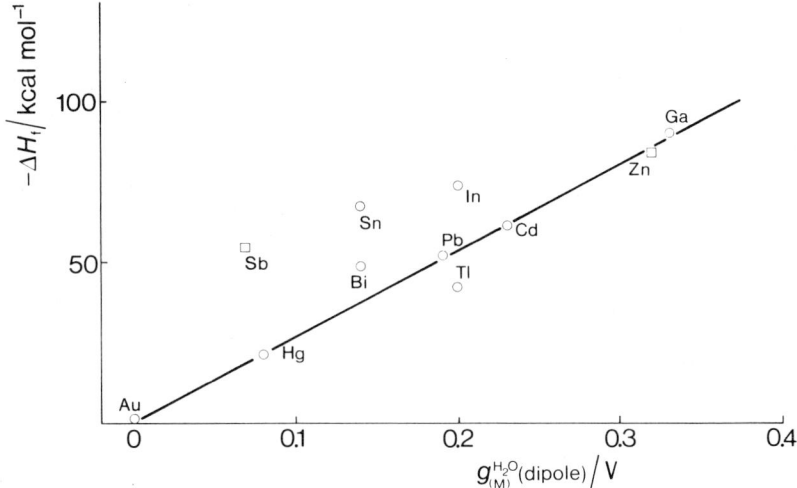

Fig. 17. Correlation between surface potential due to oriented water molecules in the double layer and substrate oxide formation heat.

The value of the strength of the metal-water interaction, which may be termed (154) hydrophilicity of metals, may be estimated in an independent way (188) from the potential at which a given organic substance is expelled from the double layer (189). Since the occurrence of this expulsion is essentially a function of charge, the potential of expulsion should correspond at the same charge, on the various metals (154). Figure 12 suggests that if this model is correct and the charge for expulsion of the organic substance is located on the linear portion of the charge-potential curve, then the peak potential for desorption measured with respect to

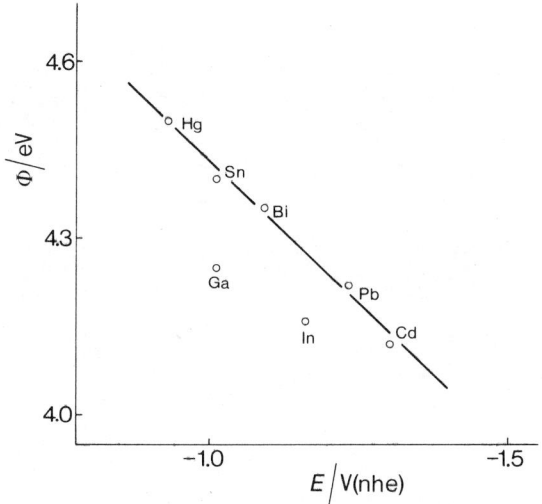

Fig. 18. Change in peak potential for amyl alcohol desorption (154) as a function of electrode work function.

the same reference electrode should be linearly related to the work function. Figure 18 shows that this is indeed the case, but In and Ga apparently form a separate group. This may be an indication that reorientation of water is incomplete on these metals at the given charge. This suggests that the hydrophilicity of In may be higher than expected from the value of the electrochemical work function.

The same charge is attained on different metals at potentials on the rational scale that should be a function of the hydrophilicity of the various surfaces (154) in that the capacity of the double layer is fully governed by the properties of the adsorbed water layer. Figure 19 shows that the peak potential in the ra-

tional scale for desorption of amyl alcohol [the result for Zn
(112) has been included using the consistently estimated (112)
value for $E^{Zn}_{\sigma=0}$] is a definite, though not simple, function of the
hydrophilicity of metals expressed as affinity for oxygen.

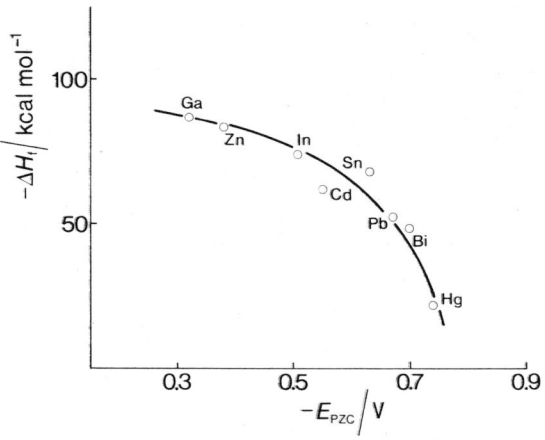

Fig. 19. Peak potential on the rational scale for amyl alcohol desorption (154) as a function of substrate oxide formation heat (73,187).

Comparison of Figs. 17 and 19 suggests that the hydrophilicity of In and Sn as expressed by $g^{H_2O}_{(M)}$(dipole) may not correspond to the actual situation. In is definitely more hydrophilic than Cd, judging from the peak potential for organic desorption and the heat of oxide formation, and Sn seems to be more hydrophilic than Pb. Possible reasons for this are discussed later.

4.5 The Inner Layer Capacity

A further way to investigate the hydrophilicity scale of metals is to resort to inner layer capacity values. The inner layer capacity increase observed (101,104) with Ga as the negative charge is decreased may be described (101,102,190) in terms of water orientation associated with a strong metal-water interaction. Thus an increase in C^i with an increase in $g_{(M)}^{H_2O}$(dipole) may be expected.

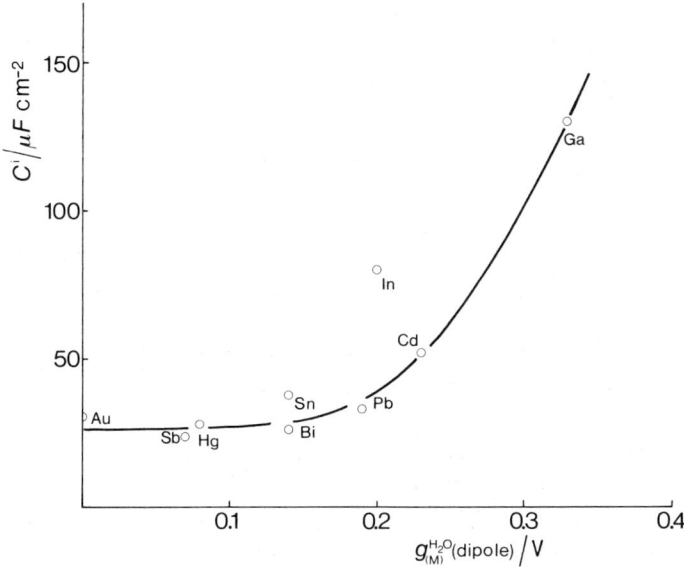

Fig. 20. Dependence of the inner layer capacity on the preferential orientation of water on the metal surface.

Figure 20 shows a plot of C^i at $\sigma^M = 0$ (101,103,107,191-196) as a function of the surface potential of water. C^i depends on $g_{(M)}^{H_2O}$(dipole) in a complex but definite way. In and Sn again fall far from the expected line. Sb is on the line contrary to the expectation on the basis of Fig. 17. Sn and In are indicated by C^i values and peak potentials for organic desorption as being probably more hydrophilic than the value of $g_{(M)}^{H_2O}$(dipole) would suggest. The position of Sb is less sensitive to the value of the orientation of water since the point for this metal lies on the plateau.

A more sensitive plot is C^i versus the affinity for oxygen;

Fig. 21 shows that while In now fits on the line, Sn and Sb seem to fall out. This confirms that the hydrophilicity of In is probably best represented by ΔH_f for the oxide, whereas for Sn the value of ΔH_f is apparently higher than expected from electrochemical parameters. The reason may be sought in the amphoteric nature of SnO_2 so that the value of ΔH_f may not be referable reliably to the surface situation. In the case of Sb, the value of ΔH_f places this metal between Pb and Cd close to Pb. This is in agreement with the suggested (154) position from the value of the peak potential for organic desorption. If this is the case, the value of $g_{(M)}^{H_2O}$(dipole) would actually be higher than expected from capacity data. A complete quantitative explanation requires further study.

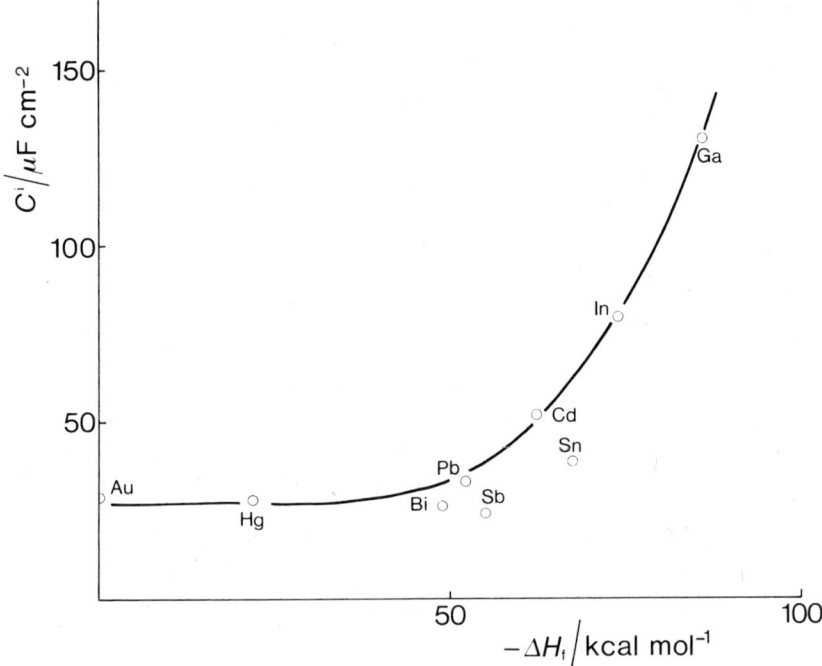

Fig. 21. Inner layer capacity plotted as a function of substrate oxide formation heat (73,187).

Adsorption of ions and organic molecules should offer a further way to test the hydrophilicity scale. However, data are available only for those metals for which the situation is unquestionable. Adsorption should decrease in a general way with increasing strength of the metal-water interaction because this is a replacement reaction (197). This is observed experimentally. Adsorption of ions increases in the sequence Ga<Cd<Pb<Bi<Hg<Au (198-203), and the same occurs for the case of organic adsorption (108, 111,204-207). However, a quantitative assessment of the situation should take into account (154) a number of other parameters involved in the free energy change for the adsorption reaction (189), but the general trend seems to be unquestionable.

The shape of capacity curves for transition metals is complex (117,118,208-210) and does not allow a discussion along the lines seen for sp metals. The fundamental point is that C^i on sp metals is determined by the ability of water molecules to reorient. C^i would thus be expected to be even higher on transition metals that lie beyond Ga in Fig. 16. This is indeed not the case. Chemisorption of water on strongly directional d orbitals preventing reorientation is likely to be the reason for this behavior.

4.6 Effect of Single Crystal Faces

The best approach for gaining definite insight into the matter of water orientation would be to use both electrochemical and work function data relative to single crystal faces. However, such work function data are available for metals (1,4,59) for which no reliable electrochemical data can be found. Conversely, potentials of zero charge for single crystal faces are available for metals (101,155) that have been very little studied from this point of view as regards work function values. Table 7 shows that for Ag (155) and Au (211), existing reliable $E^M_{\sigma=0}$ values for single crystal faces indicate that the potentials of zero charge are ordered as regards planes in the same sequence as work function values for the same crystal system. In particular the location of the polycrystalline sample is the same.

The situation depicted in Table 7 suggests that the values of $g_{(M)}^{H_2O}$(dipole) given in Table 6 actually are mean values determined by the proportions of the various planes present on a polycrystalline surface. As a general rule, the orientation of water molecules on a given metal is expected to increase as the work function decreases. ΔH_f values used to simulate affinity for oxygen are obviously isotropic in nature and can be used only with reference to a polycrystalline surface. Single crystal faces would require that oxygen affinities specific for the particular planes could be used.

TABLE VII

Potentials of Zero Charge of Single Crystal Faces, $E_{\sigma=0}$/V(nhe)

	(111)	(100)	(poly)	(110)	Ref.
Ag	-0.46	-0.61	-0.7	-0.77	154
Au	0.50	0.38	0.25	0.19	211

4.7 Operative Zero-Charge Potentials

The actual potential drop across the double layer at the potential of zero charge can be tentatively calculated, following the model depicted above, for alkali metals and very few other sp metals. From Eq. 22,

$$\Delta_S^M \phi_{\sigma=0} = E^M_{\sigma=0} + \frac{\mu_e^M}{e} + {}_rE^\circ(H^+/H_2) \qquad (36)$$

In the case of alkali metals it is necessary to postulate the same relationship between $E_{\sigma=0}$ and work function as for transition metals. This may be reasonable in view of the strong affinity for oxygen. From Eqs. 21 and 36 one also obtains

$$\Delta_S^M \phi_{\sigma=0} = \chi^M + \delta\chi^M_{(H_2O)} - g^{H_2O}_{(M)}(\text{dipole}) \qquad (37)$$

which, though equivalent to Eq. 36, evidences double-layer parameters. Table 8 shows that the potential drop across the double-layer is definitely far from being zero as the charge on the metal

TABLE VIII

Operative Zero-Charge Potential (from Eq. 36) and Surface Potential (from Eq. 37) of Metals

	Li	Na	K	Rb	Cs	Zn	Cd	In
$E^M_{\sigma=0}/V$	−1.91	−2.31	−2.71	−2.81	−3.11	−0.63	−0.72	−0.65
χ^M/V	1.3 ±0.3	0.6 ±0.1	0.2 ±0.1	0.1 ±0.1	−0.1 ±0.2	2.3 ±0.3	1.9 ±0.3	1.4 ±0.3
χ^M/V [a]	1.45 ±0.2	0.75 ±0.1	0.30 ±0.05	0.26 ±0.05	0.10 ±0.1	2.0	1.7	1.5
$\Delta_S^M \phi_{\sigma=0}/V$	0.06 ±0.3	−0.06 ±0.1	−0.50 ±0.1	−0.59 ±0.1	−0.77 ±0.2	2.0 ±0.3	1.4 ±0.3	0.9 ±0.3

[a] As calculated in theories of metals (35,37,38).

vanishes. Many-electron metals exhibit large positive values for g_S^M =0 in that M for these metals is a large fraction of the total work function.

4.8 A Model for Water Molecules in the Double Layer

A tentative model on a molecular basis can now be given for water molecules in the double layer. This model, which cannot be described in detail here, is based on the reasonable assumption (212 - 217) that the ion-free layer is one molecule thick. The model can account for the "chemical" part of the potential drop due to water dipoles in that it is derived from data at the potential of zero charge.

The minimum value of $g_{(M)}^{H_2O}$(dipole) can obviously be described with a flat position of water molecules giving rise to a zero net dipole moment perpendicular to the surface. This is illustrated in Fig. 22a. The maximum value of $g_{(M)}^{H_2O}$(dipole) is likely to be connected with a well-defined position of the molecule permitting a strong directional bond with the surface and a steric hindrance to further orientation. Possible positions are outlined in Figs.

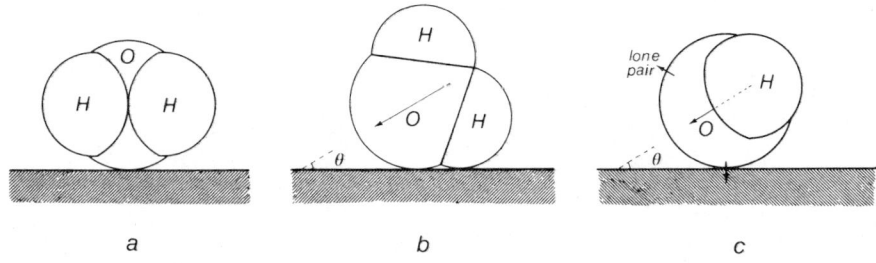

Fig. 22. Suggested positions of adsorbed water molecules on metal surfaces. (a) Minimum interaction; (b) and (c) maximum interaction.

22b and 22c. In the first case the water molecule lies naturally sideways. The alternative position is with one of the lone pair perpendicular to the surface. Lower orientation would correspond to a bent bond and thus to lower strength of interaction.

It is interesting to note that in both positions the molecular dipole forms an angle of about $35°$ to the surface. Simple calculations using the usual formula for the potential drop across a dipolar layer (13,218) show that if for the dielectric constant the value corresponding to atomic and electronic polarizabilities (219) is used, the positions outlined in Fig. 22 can account for the maximum value of 0.4 V. This requires that the value of about 20 $Å^2$ should be used for the area occupied by an adsorbed molecule on the surface. This definitely does not correspond to a close-packed hexagonal structure but to a square lattice. Adsorption from the gas phase (171,172,220) on Ni, Fe, Cu, and Pd seems to suggest that this value may correspond to a definite experimental situation. Since the dipoles are not perpendicular, their positions can rotate around a vertical axis. A square lattice with antisymmetric projected dipoles does not give rise to particle-particle repulsion but rather to small attraction (221), which may conform to the S-shaped isotherm found (222) for water adsorption on Hg. According to the present model, hydrogen bonds can be formed in the surface and toward the bulk of the liquid phase. Under the action of an electric field, molecules are expected to rotate (215,126,223) between minimum and maximum orientation, beyond which polarization occurs. Further work is needed to test this model quantitatively, but it is quite interesting that adsorbance measurements (224) for water on the hydroxylated layer of certain oxides suggest that the plane of the water molecule possibly forms an angel of less than $45°$ to the surface.

A definite model for water at interfaces conceptually based

on statistical thermodynamics has been proposed by Damaskin and
Frumkin (216). Their calculations give the correct shapes of
the experimental capacity curves for Cd and Hg. Values of $g_{(M)}^{H_2O}$
(dipole) obtained by the above authors by fitting calculated
C^i versus E curves to the experimental results agree well for
Hg and satisfactorily for Cd with those derived here. However,
the physical basis of the model is not immediate. Water molecules
are thought to be present in a mixture of chemisorbed molecules
and physically adsorbed clusters. No more detailed molecular
model is given. It seems reasonable to infer that clusters are
weakly bound to the surface and on average not orientated at
$\sigma^M = 0$. Thus adsorption of organic substances is expected to
occur (165,225,226) with selective replacement of the two types
of water molecules. There does not seem to be experimental evidence for this. Nevertheless, the combined electrical properties
of chemisorbed molecules (C^i increases exponentially with charge)
and clusters (C^i shows a maximum at the potential of zero charge)
predicts correctly the hump of the capacity curve at small positive charges although the preferential orientation of water
molecules at the potential of zero charge is definitely (227,228)
with the oxygen atom toward the metal. In Damaskin and Frumkin's
model, which certainly gives the basis for a better understanding
of the often-debated (223, 229 - 237) nature of the hump in
capacity curves, the change in orientation of water molecules
on different metals is replaced by a change in the relative
coverage of the surface with the two types of interfacial water.

A paper by Parsons (314) has recently appeared dealing with a
four-state model for a solvent at the electrode-solution interface. Parsons starts from the idea of Damaskin and Frumkin (216)
that two types of water molecules are present at interfaces, each
of which is thought to exist in two additional possible opposite
positions depending on the field. The physical picture of this

model differs from Damaskin and Frumkin's model in that the isolated molecules are not chemisorbed in a fixed position, although appropriate allowance is made for metal-water specific interactions. Furthermore, the total number of molecules in the monolayer is kept constant by accounting for changes in the number of water molecules in the different states.

The model is successful as regards the shape of the capacity curve over all the range of charges and may even be promising for treating nonaqueous solvents. Quantitatively, the result seems less satisfactory. The value of $g_{(Hg)}^{H_2O}$(dip) is calculated to be only 8 mV, though the sign is that expected. Further, the calculated temperature coefficient of the surface potential of water is opposite to that found experimentally.

It seems that both of the above models stress that capacity curves can be theoretically reproduced only by a two-term equation describing two types of water molecules in the double layer. However, the physical description of the two types of molecules seems to be somewhat elastic and may be open to further change.

5 ELECTROCHEMICAL KINETICS

The concept of work function has found a place in kinetics later than in thermodynamics; probably because of the lack of straightforward theoretical relationships (29, 88, 238) linking the work function with kinetic parameters. Nevertheless, the operative nature of the work function has made its efficacy in kinetics (11, 19 - 29) much greater than in thermodynamics, although kinetic correlations have always retained an empirical character. It is, however, possible to show that the work function approach to kinetics is less empirical than it appears at first sight.

5.1 The Work Function in Kinetic Equations

Consider a general reaction

$$M^z + e(M) \rightleftarrows M^{(z-1)} \qquad (38)$$

At the equilibrium, the absolute value of the rate of reduction is given by

$$\vec{i} = \kappa \frac{ekT}{h} c_{M^z} \exp \frac{\Delta \vec{G}^{\neq}}{kT} \qquad (39)$$

where κ is the transmission coefficient and $\Delta\vec{G}^{\neq}$ is the electrochemical free energy of activation. $\Delta\vec{G}^{\neq}$ consists of a chemical term and an electrical term:

$$\Delta\vec{G}^{\neq} = \Delta\vec{G}^{\neq}_{chem} + \alpha e \Delta^M_S \phi_e \qquad (40)$$

where the electrical term is the operative electrode potential at the equilibrium. It has been pointed out (88,89,238) that the work function is not explicitly involved in kinetic equations. With reference to Eq. 40, it is sufficient to consider that there is a compensation effect (70, 239 - 241) between the chemical and the electrical term so that the term expressing the energy of electrons in the metal cancels out (238) ultimately. However, the work function may be implied in kinetic equations in an indirect but definite way. The discussion of this topic requires reactions to be split into two main groups according to two fundamental mechanisms of charge transfer (28,238,242): (1) reactions where reactants, intermediates, and products interact negligibly with the electrode surface, and (2) reactions where strong metal-particle interactions take place.

5.2 Simple Electron Exchange Reactions

Redox reactions without specific adsorption of reacting ionic species belong to this group. Equation 40 shows that if the chemical potentials of ions in the activated complex are metal independent, the activation energy must be metal independent (28) provided that the comparison is made at the equilibrium potential or at the same overvoltage $\eta = (\Delta_S^M \phi - \Delta_S^M \phi_e)$. Experiments (28,29) show that this is indeed the case for Fe^{3+}/Fe^{2+} redox couple on a number of metals. Constancy in experimental activation energy is even observed with metals such as Ta (29) and oxides such as RuO_2 (243) and SnO_2 (244).

Although no effect of the electrode material on the rate constant would be expected, results (28,29) do show that the exchange current for the Fe^{3+}/Fe^{2+} couple is metal dependent. Figure 23

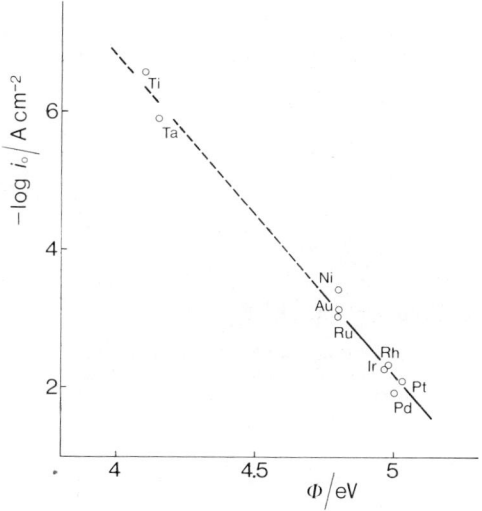

Fig. 23. Change in reaction rate at the equilibrium for the Fe^{3+}/Fe^{2+} redox couple (28,29) with substrate work function.

shows that i_o is a closely linear function of the metal work function. Rationalization of this dependence requires some discussion about factors indirectly affected by a change in the nature of the electrode material. Two possible models have been proposed to account for substrate effects in redox reactions. The two models are not alternatives, but each of them stresses the importance of a different parameter.

One model makes use of double-layer concepts (245) and has been proposed by Frumkin (246). Reactants are assumed to be located to a first approximation at the outer Helmholtz plane. Hence, the actual potential drop operating on charge transfer is the inner layer potential drop $\Delta_2^M \phi$ according to

$$\Delta_S^M \phi = \Delta_2^M \phi + \Delta_S^2 \phi \qquad (41)$$

where $\Delta_S^2 \phi$ is the potential drop across the diffuse layer. Furthermore, if particles are positioned in the outer Helmholtz plane (OHP), their concentrations are governed by the Gouy-Chapman equation (13):

$$_{OHP}c_{M^z} = {_b}c_{M^z} \exp \frac{-ze\Delta_S^2 \phi}{kT} \qquad (42)$$

where

$$\Delta_S^2 \phi \propto (\Delta_S^M \phi - \Delta_S^M \phi_{\sigma=0}) \qquad (43)$$

Equation 43 is a complex equation; however, it reduces to essentially linear relationships at high positive and negative charges (247). $_b c_{M^z}$ is the concentration of M^z in the bulk of the solution. Thus the rate of the reaction is expected to be governed by the structure of the double layer, and since the potential of zero charge $E_{\sigma=0}^M$ is linearly related to the work function, linear dependence of reaction rates on ϕ^M are conceivable at constant E^M.

The quantitative proof of this view requires a knowledge of the structure of the double layer for the various metals. This is unfortunately impossible in the case of the data in Fig. 23. The equilibrium potential for the Fe^{3+}/Fe^{2+} system is always more positive than $E_{\sigma=0}$ for the metals investigated. The reaction should thus occur on positively charged surfaces. However, owing to specific adsorption of the anion of the supporting electrolyte, some reversal of charge is to be expected (247) if extension to other metals of what occurs with Hg (212) is made. Thus the effective electrode charge could be increasingly negative as ϕ^M decreases. Because the particles are positively charged, a negative effective charge should represent an accelerating factor and i_o should increase as ϕ^M decreases, which is quite at variance with experimental findings. Reasons for this may be a decreasing rather than an increasing specific adsorption of anions because of competition with water. This would not be unreasonable in view of preceding discussions. An alternative possibility (29) is the existence of Fe^{2+} and Fe^{3+} as negative complex ions in H_2SO_4, which cannot be ruled out a priori. However, the location of Au would be difficult to understand in light of the former hypothesis since adsorption of water is thought to be weak on this metal. Thus the results in Fig. 23 cannot be accounted for quantitatively for lack of double-layer data, even though it is thought that they may indeed represent double-layer effects. In fact, consideration (29) of the activation entropy and preexponential factor does not seem to be able to account for the magnitude of the change in i_o.

An interesting point is that the extension of the plot in Fig. 23 to low-ϕ^M metals surprisingly reveals that Ti and Ta apparently conform to the same pattern, although they are known to form oxide films in the air. If Fig. 23 really represents a double-layer effect, the location of Ti and Ta would require that thin, probably monomolecular layers of oxide on these metals do not affect their work functions. Some evidence for this may be found for Ta (248) and also for Al (249); further experimental evidence is needed to support this view. If this is the case, then electrons would be able to tunnel through the oxide layer with a negligible change in the transmission coefficient.

A different view exists on the effect of the electrode material on simple electron exchange reactions. According to this view, the nature of the metal is directly implied in the kinetic equation through the transmission coefficient (250 - 252). In concentrated solutions as regards the supporting electrolyte, $\Delta_S^2\phi$ changes from metal to metal are assumed to be negligible and variations in i_o are thought (253) to be governed by the density of electronic states at the Fermi level (2), $D(\varepsilon_F)$, explicitly introduced into the transmission coefficient considered as metal independent in classical kinetic formulations. Thus electrons would be considered as chemical reactants (254) and their concentration would appear explicitly in the kinetic equation (250,251, 254 - 256). If this is the case, a linear relationship with unity slope is to be expected between log i and $D(\varepsilon_F)$. The results in Fig. 24 showing a linear dependence between log i_o for the Fe^{3+}/Fe^{2+} redox couple (28,29) and density of states (2) for metals of the Pt group and Au could apparently be taken as a support to the above view. Ta, Ti, and Ni fall far from the line in a way that could indicate an actual, much lower value for $D(\varepsilon_F)$ because of the presence of surface oxide layers (257).

The results in Fig. 24 are, however, thought not to be definite proof of the idea that $D(\varepsilon_F)$ for metals may play a role in kinetics. Other, quantitatively more manageable, results seem to assign a marginale role to the density of the states. First, the rate of change in i_o with the nature of the substrate appears to depend markedly on the nature of the supporting electrolyte

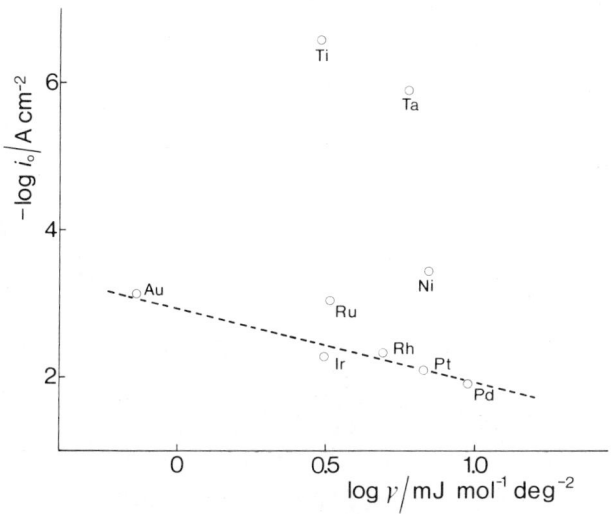

Fig. 24. Reaction rate at the equilibrium for the Fe^{3+}/Fe^{2+} redox couple (28,29) plotted as a function of density of states (2) for the substrate. γ is actually the electronic specific heat related to $D(\varepsilon_F)$.

(258,259) and on the charge of particles. This would become inconceivable if $D(\varepsilon_F)$ were a determining factor. On the contrary, on an assumption of double-layer effects, rearrangement (245) of Eq. 39 shows that

$$\vec{i}_{obs} = \vec{i}_i \exp\left(\frac{(z-\alpha)e\Delta_S^2\phi}{kT}\right) \quad (44)$$

where $\vec{i}_i$ is the metal independent reaction rate. Equation 44 shows that the charge of particles is expected to have a strong effect. This has led Capon and Parsons (260) to minimize charge effects by using the benzoquinone anion-radical couple for which $z = 0$. With respect to the Fe^{3+}/Fe^{2+} system, the rate of change of the reaction rate with the nature of the metal because of diffuse layer effects should thus be reduced by a factor of 5. If the structure of the interface were comparable to that governing the results in Fig. 23, the change of log i upon a change of ϕ^M by 0.5 eV should amount only to 0.5 instead of 2.5. Unfortunately, the structure of the double layer is not known for the nonaqueous system used by Capon and Parsons, but it may be reasonable to expect a smaller variation in log i since smaller capacities and charges are likely to be involved. Thus the observed metal independence of the rate constant (260), as shown in Fig. 25, may not be surprising. In the same figure, the dependence of the reaction rate on the work function for the $Fe(C_2O_4)_3^{3-}/Fe(C_2O_4)_3^{4-}$ system (259) is also shown. A linear correlation can again be observed. If $D(\varepsilon_F)$ played a determining role, then the reaction rate would be expected to be the same for Au and Ag, and that of Au would be lower than that for Pt by a factor of 10.

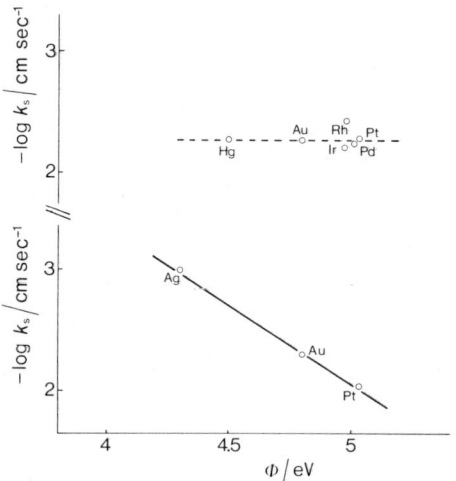

Fig. 25. Rate constant as a function of substrate work function for the benzoquinone anion/radical couple (260) (broken line) and for the $Fe(C_2O_4)_3^{3-}/Fe(C_2O_4)_3^{4-}$ system (259) (solid line).

Strong evidence of a determining double-layer effect in redox reactions is apparently given by results for the reduction of $Fe(CN)_6^{3-}$ and $S_2O_8^{2-}$ on sp metals (261 - 263) under conditions in which allowance for the structure of the double layer can rigorously be made. With reference to Fig. 12, if the reduction reaction takes place at potentials where the charge is linearly related to the potential, then the same rate should be observable on different metals at potentials linearly related to the work functions. Alternatively, in the same region, log i on various metals at the same thermodynamic potential should be linearly related to the work function. If the reaction occurs in a range of charge where σ^M is not linearly related to the potential, then the same rate could be observed at potentials linearly related to the work functions only for metals within the same group in Fig. 16.

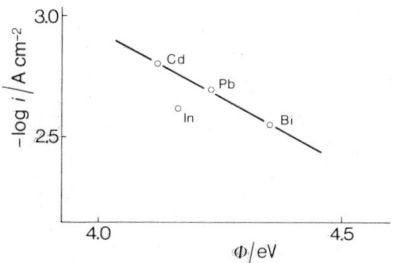

Figure 26 shows a linear dependence of log i on Φ^M for Cd, Pb, and Bi (262,263). The electrochemical work function of In would predict an activity between Cd and Pb, but in fact the activity is seen to increase from In to Pb to Cd in accordance with a greater hydrophilicity for In. Reversal in the order with respect to that expected from the position of the potential of zero charge suggests that In should probably be located in Fig. 16 in a different group from that of Pb and Cd. This is in agreement with the hydrophilicity estimation from C^i, ΔH_f, and peak potential for

organic desorption, but it is definitely in contrast with the
electrochemical work function from charge-potential curves. An
unavoidable conclusion is that at a given negative rational potential the charge on In must be higher than expected from σ/E
curves. Thus experimental charge-potential curves apparently fail
somewhere for this metal. It is to be noted that a correction of
10% of the charge to allow for the higher value of C on In than
on Hg at the far cathodic end has actually been adopted (264).
If the value for C^i places In correctly, it is not understandable
why the σ/E curve should be incorrect. The apparent inconsistency
of these results seems to require some more detailed explanation.

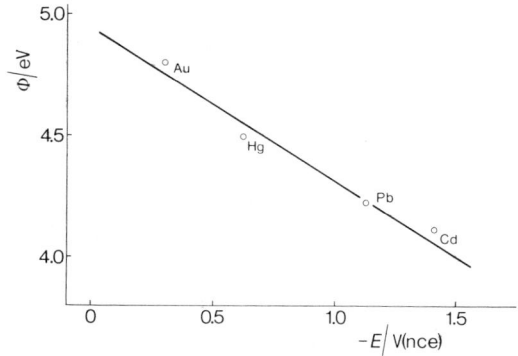

Fig. 27. Potential for constant rate of 3×10^{-3} N $K_3Fe(CN)_6$
reduction (261) as a function of electrode work function. Hg is
amalgamated Cu.

Figures 27 and 28 show linear relationships between E and
Φ^M at a constant rate of $Fe(CN)_6^{3-}$ and $S_2O_8^{2-}$ reduction on various
metals (261). Pb, Cd, Hg (amalgamated Cu), and Au can be seen
to fit the same straight line as expected from Fig. 16. Figure
28 shows that Ag apparently does not obey the same relationship.
Whether this is due to a different state of the surface or indi-

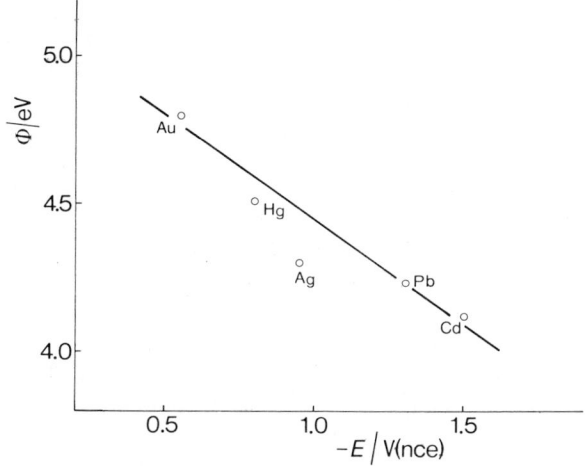

Fig. 28. Potential for constant rate of 10^{-3} N $K_2S_2O_8$ reduction (262) as a function of electrode work function. Hg is amalgamated Cu.

cates a higher hydrophilicity requires further study. After correction for double-layer effects, reaction rates for the above systems have been found (261,262) to become substantially metal independent. No effect of the density of states can thus be detected even in the case of Bi known for its low value of $D(\varepsilon_F)$. This seems to support the view that the concentration of electrons may not be operative in electrochemical kinetics. It is also possible that, like any other surface property, the density of states on the surface differs from the value in the bulk. No quantitative support can be given to this hypothesis, but for the above results to be explainable, $D(\varepsilon_F)$ would be required to exhibit the same surface value on all the metals investigated.

Finally, we dwell further on the relevance of water orientation to linearity of log i versus ϕ^M plots. Figure 12 suggests that linearity can be observed only in the region where σ^M changes linearly with E. At a potential where the orientation of water is incomplete, no linear dependence of log i on ϕ^M could be expected. A linear dependence can be obtained only after allowance for the potential drop due to the adsorbed water layer. Thus comparable conditions in the double layer can be found at the same $[E-E_{\sigma=0} - g_{(M)}^{H_2O}(\text{dipole})]$ value and not simply at the same $(E-E_{\sigma=0})$ value. A consequence of this is that Au in Fig. 23 can be found on the same line with transition metals because at strongly positive charges, as implied in the value of the reversible potential for the Fe^{3+}/Fe^{2+} redox couple, the orientation of water is comparable on all metals.

5.3 Reactions with Strong Metal-Particle Interactions

Although most of the electrochemical reactions belong to this group, only a few have been extensively studied on different metals. Among these, the reaction of hydrogen evolution lends itself especially well to attempting some rationalization of the role of the solid substrate, in view of both the relative simplicity of the system and the number of results available.

The hydrogen evolution reaction has been discussed in the literature (20,21,70,150,238,265-272) and correlations of the reaction rate with the metal work function have been presented repeatedly (19-26). Although some definite trends and separations into groups have been envisaged, no relationships capable of allowing a quantitative approach to the problem have been agreed on by the various authors.

Figure 29 shows a plot of log i_o as a function of work function. Critical selection of i_o values is presented elsewhere (24). Metals can be seen to split into two groups, transition and sp

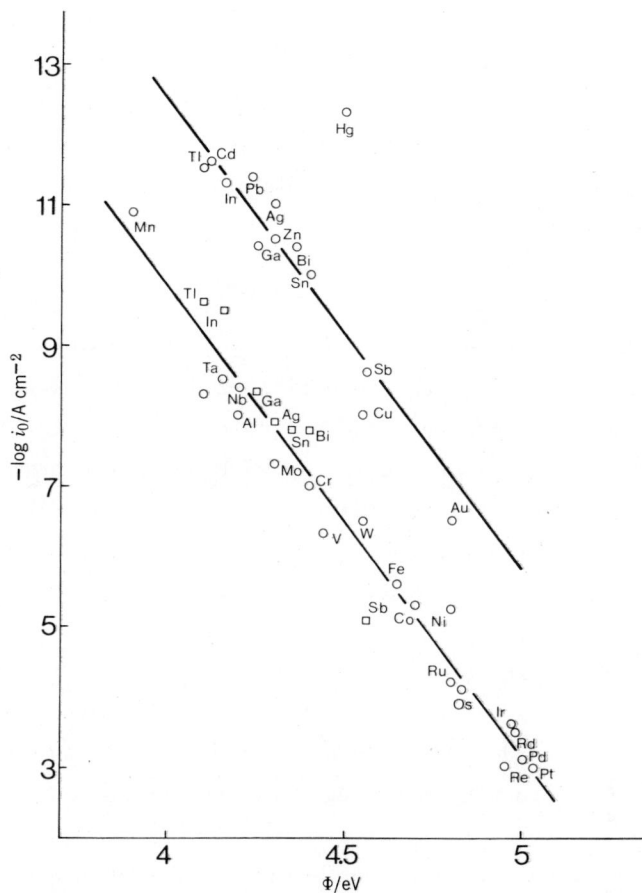

Fig. 29. Dependence of the equilibrium rate for hydrogen evolution on the substrate work function. Open squares, exchange currents on positively charged surfaces of sp metals.

metals. Within each separate group a closely linear relationship is obeyed. There are results (272-277) suggesting that i_o values for sp metals may be different on positively charged surfaces from their values on negatively charged surfaces. Although it is not clear (278,279) whether this is a definite feature, for metals for which it has been observed, i_o values for positive surfaces are seen to fall definitely in the d-metal group. This charge effect has been attributed (273-276) to specific adsorption of anions, which would enhance the reaction rate through a double-layer effect. A tentative, alternative explanation is proposed later in the framework of the theory on the role of the solvent in the double-layer.

It is possible to rationalize on a reasonable basis the way the work function may be implied in the rate equation. A direct double-layer effect of the type discussed for redox reactions, although possibly present, can be shown to be negligible. Calculations based on the change in $\Delta_S^2\phi$ with potential from capacity data (247) show that on negatively charged surfaces the double-layer effect can account only for few percent of the observed change in $\log i_o$. On positively charged surfaces the double-layer effect should be opposite. However, because of the presence of specific adsorption of anions, reversal of charge is possible, and the double-layer effect may be in the same direction as on negatively charged surfaces and about the same magnitude. Thus double-layer effects can be neglected to a first approximation in what follows.

The effect of the electrode material on the rate of the hydrogen evolution is commonly rationalized in terms of different metal-hydrogen bond strength (268,269). Simple considerations on the basis of parabolic energy curves show that the rate for the discharge of the proton can be written (268,272) as

$$\vec{i} = \kappa \frac{ekT}{h} c_{H^+} (1-\Theta) \exp\left(\frac{-\Delta \vec{G}^{\neq}_{chem} - \alpha\mu_H^M - \alpha e \Delta_S^M\phi}{kT} \right) \qquad (45)$$

where $\Delta \vec{G}^{\neq}_{chem}$ is the metal-independent activation energy and μ_H^M is the gain in free energy upon interaction of one hydrogen atom with the metal surface. It has been shown in Figs. 8 and 9 that the adsorption energy of hydrogen on metals apparently is linearly related to the work function. Thus, equating for the moment adsorption energy to adsorption free energy, Eq. 45 may be rewritten in the form

$$\vec{i} = \kappa \frac{ekT}{h} c_{H^+} (1-\Theta) \exp\left(\frac{-\Delta \vec{G}^{\neq}_{chem} - \alpha r \phi^M - \alpha e \Delta_S^M \phi}{kT}\right) \quad (46)$$

where r is the slope of the linear μ_H^M versus ϕ^M relationship.

Equation 46 shows that, although the work function is not explicitly involved in the rate equation (88,89,238), it may be implicitly contained in the activation energy term. Some comparison between calculated and experimental log i_o versus ϕ^M slopes may be attempted but problems may be encountered in doing such a comparison. On the one hand, since μ_H^M is a free energy term, possible entropy effects (280,281) should be kept in mind. On the other hand, desorption of the solvent should not be without any effect since the discharged proton will occupy a definite site on the surface. If reference is first made to the gas-phase situation (72), on sp metals $\mu_H^M > 0$ and $\Theta_H \simeq 0$. Thus, from Eq. 46, since the slope of the dependence of μ_H^M on ϕ^M for sp metals (Fig. 9) is 26 kcal/(mol)(eV), the expected slope for the log i_o versus ϕ^M relationship, assuming $\alpha = 0.5$, should be about 9.5 eV^{-1}. The observed slope is about 7 eV^{-1}. Entropy effects would reduce the theoretic estimate, but solvent displacement effects would increase it since the adsorption of water increases as ϕ^M decreases. More complicated situations may be expected if entropy effects are interconnected with the presence of the solvent. In addition, more complex preexponential factor effects (282-284) cannot be ruled out. Further study is necessary to include these effects quantitatively in the analysis of the log i_o versus ϕ^M plot.

Extension of the preceding simple approach to d metals is not immediate. Rearrangement of Eq. 45 to take into account the appropriate isotherm for hydrogen adsorption (268) [on the basis of μ_H^M values for d metals in the gas phase (72), Θ_H is no longer expected to be negligible] shows that log i_o is now predicted to decrease linearly with μ_H^M. Thus, on the basis of Eq. 45, simultaneous consideration of all metals predicts (268,269) a volcano-shaped curve for the change of log i_o with μ_H^M whose maximum is placed at $\mu_H^M = 0$. sp metals are located on the ascending branch, while d metals appear on the descending branch. This seems indeed

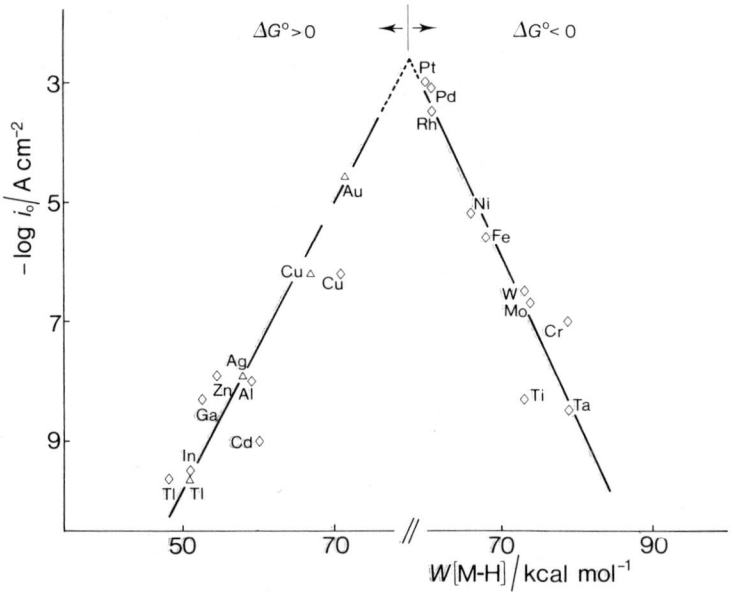

Fig. 30. Equilibrium rate for hydrogen evolution as a function of M-H bond strength derived from adsorption heats in the gas phase (64) for transition metals, and from hydride formation heats (73) for sp metals. Open triangles, spectroscopic dissociation energies.

to occur, since Fig. 29 shows that parallel straight lines are obtained for d and sp metals. However, this fact cannot be accounted for if experimental values of μ_H^M in the gas phase are considered. The volcano-shaped curve would be asymmetric in this case, as Fig. 30 shows. Furthermore, the slope of the straight line in Fig. 8 is 11 kcal/(mol)(eV), which predicts for d metals a slope of about 4 eV^{-1} for the line in Fig. 27. The observed slope is about 7 eV^{-1}. Entropy effects cannot account for this; they would tend to decrease the calculated slope even more. Solvent replacement effects would act in the same direction. Thus the presence of the solvent is more likely to be the determining factor by operating a change in the slope of the linear dependence of μ_H^M on ϕ^M. This effect seems to be dramatic with d metals.

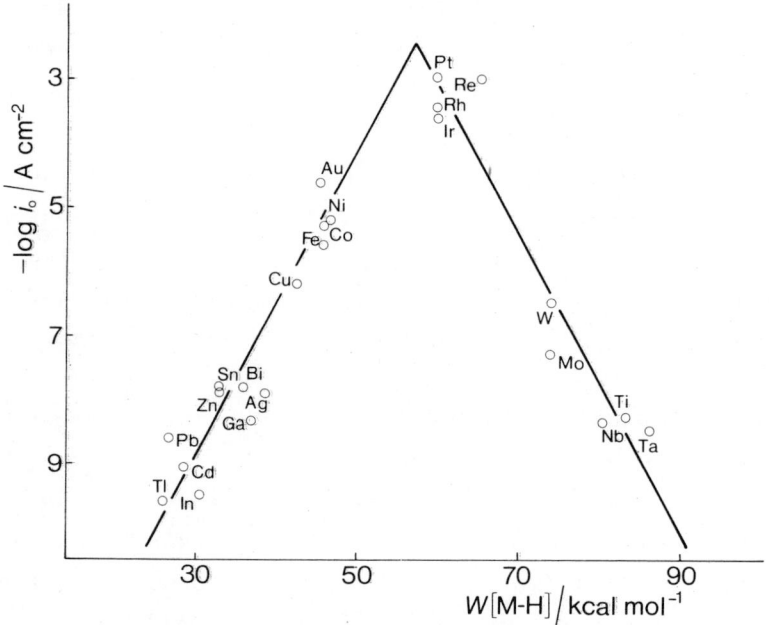

Fig. 31. Equilibrium rate for hydrogen evolution as a function of M-H bond strength calculated (150) from overvoltage data.

Operative μ_H^M values have been obtained by Krishtalik (150) directly from overvoltage data. If these values are used, Fig. 31 shows that a symmetric volcano-shaped curve is obtained as expected from the parallelism of the straight lines in Fig. 29. If Krishtalik's values are enthalpy terms as implied in the procedure adopted for calculation, the same slope for sp metals in Figs. 30 and 31 supports the view that formation heats for bulk compounds parallel those for surface compounds. Solvent effects should thus be minor in this case.

A point deserving attention is that the volcano-shaped curve can be obtained only if i_o values for sp metals with positively charged surfaces are used. Thus for Cu and Au, hypothetical values for positively charged surfaces have to be interpolated in Fig. 29. This means that results must be homogeneous as regards the structure of the double layer. This point is discussed further later. No reasonable plot could be obtained with i_o values for negatively charged surfaces.

The entire discussion given so far has been based on the assumption that d metals in solution adsorb hydrogen as readily as they do from the gas phase. When $\mu_H^M < 0$, $\Theta_H > 0$ is expected on all d metals and especially $\Theta \to 1$ on strongly adsorbing metals such as Ta and Ti. However, although on metals of the group of Pt adsorbed hydrogen can be readily evidenced experimentally (147, 285-287), evidence for adsorbed hydrogen in solution cannot be as easily given in the case of other metals. Even in the case of Ni, adsorbed hydrogen (288,289) is suggested to amount to only a few percent coverage at high overvoltages (290). Absorption of H is known (150,291-296) to occur in the case of Ta, Nb, and other metals, but no underpotential deposition is observed as would be expected. Furthermore, the operative values (150) for μ_H^M place Ni, Fe, and Co on the ascending branch of the volcano curve (Fig. 31), that is, together with metals not adsorbing hydrogen.

Starting from the interesting suggestion made by Barclay (151), it is possible to think that troubles in the case of d metals may arise as a result of weakening rather than strengthening of the M-H bond as the work function decreases. In this case, adsorption rather than removal of hydrogen is to be postulated as the rate determining step also for these metals. There seems to be experimental evidence (150) for this with Fe, Ni, and Co. Krishtalik's calculations, however, suggest strengthening in the M-H bond as the work function decreases for all other d metals, but these calculations start from the assumption that W adsorbs hydrogen as strongly in solution as in the gas phase. Opposite results would probably be obtained for the operative μ_H^M values if calculations started from the opposite model using a different mechanism.

Barclay has suggested that the hydridic nature of the surface M-H bond is responsible for the change in trend in μ_H^M values for d metals. A hypothesis that follows from the picture of the double-layer made here is that the energy for desorption of water molecules may be the cause for the reversal in slope of the μ_H^M versus ϕ^M relationship. The two views are probably similar, although the physical models behind them are not identical.

Some evidence for hydrogen-solvent competition may be found. Adsorption of organic substances decreases (297) instead of increasing from Pt to Ru. Hydrogen adsorption starts at more cathodic potentials on Os than on Pt (149), and on Rh than on Pt (148). The hydrogen adsorption heat measured in solution is less for Rh than for Pt at low coverages (148). Furthermore, the maximum coverage with adsorbed hydrogen is 0.7 on Ru (292) compared to 1 on Pt.

Some conflicting evidence is also available. The experimental hydrogen adsorption enthalpy on Ta in solution is still negative (293). Hydrogen adsorbed on W from the gas phase does not seem to be displaced by H_2O (298). The order of adsorption strength in solution shows some reversal as the coverage increases (148). However, absorption of hydrogen may complicate measurements on low-ϕ^M metals such as Ta and W. In this case experimental μ_H^M values may refer to the absorption process.

It is not clear, because of the lack of definite experimental evidence, whether the possible weakening in μ_H^M with d metals actually corresponds to the true situation. The strongest evidence in favor is probably the inability of d metals other than metals of the Pt group to dissociate hydrogen in solution. Furthermore, the experimental slope for the μ_H^M versus ϕ^M relationship is certainly inadequate to explain Fig. 29. A solvent effect should in any case be present. That this may be dramatic for d metals and not for sp metals may be derived from the fact that, besides a much stronger solvent adsorption on d metals, the increases in adsorption strength for H and O follow the same direction on d metals, but they are opposite on sp metals as a function of the work function, so that in the latter case no reversal of order may be expected in solution with respect to the gas phase. The interesting point is that if this effect is real, there must be a complicated compensation effect, once more emphasizing possible simple interconnections among various quantities. In fact, Fig. 29 clearly suggests that the resulting rate of change in μ_H^M with ϕ^M in solution is possible the same for sp and d metals. Further study is necessary to gain a decisive insight into the role of the various factors determining the rate of hydrogen evolution and to account quantitatively for the points discussed here.

Two alternative interpretations may be given for the splitting of metals into two groups as seen in Fig. 29. It may be simply a result of sp metals and d metals having different μ_H^M values for the same work function. In this case, the distance along the i_o

axis between the two straight lines would be a measure of the difference between μ_H^M on sp and d metals possessing the same work function. However, this approach is unable to account for sp metals with positive surfaces falling in the group of transition metals.

The alternative view is that the splitting may be connected with different orientation of water molecules. On negative surfaces, the surface potential due to water dipoles is at a minimum, whereas on positively charged surfaces it attains a maximum value. For the same reaction rate the difference in work function between the two groups is 0.4 eV. This is the same amount as that arising from the difference in water orientation between the two extreme positions. It has been also shown in the case of redox reactions that linear correlations with the work function can be obtained only with comparable orientations of water molecules. In the case of hydrogen evolution, if the shift of i_o for sp metals with the sign of the charge is a real effect, the fact that sp and d metals are grouped when the surface of sp metals is positively charged is an indication that energy terms should not be operative. The orientation of water may be comparable only if on d metals solvent molecules do not reorient and the surface potential contribution is always that pertaining to positive surfaces, that is, with the oxygen atom toward the metal. This is possible only if metal-water interactions are such that molecules are subjected to strongly directional bonds. This conclusion is consistent with the picture of the double layer in Section 4.4.

6 CONCLUSIONS

The essential promising aspect of this approach is the satisfactory consistency of the resulting picture in the various fields.

This suggests that the ground of the model is able to unify the treatment of the phenomenological aspect of electrochemistry, and it enables a transition to be made consistently from the situation in the gas phase to that in solution. The scale of reactivity of metals seems to be closely respected under quite different circumstances, and some account can be made for behaviors that apparently deviate. The work function turns out to be the best identity card to identify and classify metals. Metals generally split into two groups, sp and d metals. This happens to be a quite general rule in solution as well as in the gas phase. However, in solution, at comparable potentials, the main factor governing the splitting of metals is possibly the orientation of water, which is a parameter closely related to metal-solvent interactions. If the behavior of the solvent is comparable, the work function is able to lump all metals in a compact group.

Although the present research work is continuously open to improvement, the general trends envisaged are not seriously questionable. It may be that the work function values for some metals may require some further adjustment, but it is never expected to be able to upset the general picture. Conversely, it is hoped that some work function values derived from electrochemical measurements could be confirmed by experiments in a vacuum, thus corroborating the validity of the present approach.

Some particular points should stimulate experimental investigations. The nature of the approach has made it possible to predict the behavior of metals in cases where the picture has shown some inconsistency. Thus, a number of results for In from both the double-layer field and the kinetics field point consistently to a metal-water interaction strength higher than that expected from the value of the electrochemical work function. Present evidence suggests that the value of ϕ^{In} is probably too low for this metal as derived from charge-potential curves, and reasons for this should be searched by capacitance measurements. Similarly, the affinity for oxygen as introduced here places Sn and Sb in positions requiring more experimental support.

This approach was able (16) to predict a semiquantitative hydrophilicity scale before the relevant experimental data were available for most of the metals. Thus recent results (112), that call for further study to be finalized quantitatively seem to confirm that Zn is located definitely beside Ga. The exact position probably requires minor adjustments.

Au, Ag, and Cu are gathered with sp metals. This view (24,299) is not universally shared (22,26,30,300). For the hydrogen evolution reaction, there is some trend to include these metals in the group of d metals. If some d character in valence electrons may be understandable for the elements, there is no reason, in principle, to expect the same to occur with bulk metals. Certainly, the situation seems clearer with Cu and Ag than with Au. It should be recalled that in all the conclusions derived here for this metal the value of the electrochemical work function has a strong weight. If its derivation from charge-potential curves concealed any trap, the situation of Au should be reassessed. Inclusion of Au in the d group would imply a strong orientation of water. A strong affinity for oxyen is not expected. Nevertheless, experimental data for adsorption of water from the gas phase are suspiciously controversial (301,302). Electrochemical data (167) point to a relative weak adsorption of water, but adsorption of oxygen on different single crystal faces (303) does not appear to be straightforwardly interpretable in the above context. A possibility for Au is that its behavior may change from sp to d metal on passing from negative to positive charges. The position of the Fermi level close to the top of the d band (126) may make reasonable

the hypothesis that at anodic potentials some d levels may remain
unfilled. Results in catalysis (304) suggest that isolated atoms
of Au on a surface may be deprived from electrons and behave
catalytically as a d metal. The surprising increase in activity
of a Pt or Pd electrode with Au coverages below the monolayer
(305) are probably attributable to the same effect. For these
reasons it is believed that the situation for gold is not yet
settled.

The problems outlined above are just some of the problems
arising from the present approach, which in turn is a proof of
its effectiveness. The situation is probably not completely
settled in some general aspects, also. The role of the density of
electronic states does not appear to be emphasized by results in
this work, but this point requires more study to be fully assessed.
It plays certainly a decisive role in electron transfer reactions
at semiconductors (254 - 256).

The present approach has enabled us to rationalize the present
evidence from a general point of view, but the position of each
metal should probably be tested on the basis of ever-increasing
interdisciplinary knowledge. Certainly, electrochemistry suffers
for some definite limitation in the use of experimental techniques for direct observation of the surface. Nevertheless, it is
necessary to resort in an appropriate way to techniques from the
field of surface science. This is going to be a definite trend
(306 - 308). Experiments in the gas phase are able to show single
atoms leaving a surface (309), and electrochemical measurements
should be supplemented by Auger spectroscopy, LEED, and work function measurements before and after the electrochemical measure
to establish a rigorous cross-examination of the state of the
surface during the investigation. Although it is believed that
a substantial time will be required for these measurements, where

applicable, to become a routine in electrochemical laboratories, this is felt to be a necessary trend. At that time much less speculation will be necessary in an attempt to understand the behavior of metals in solution.

7 LIST OF SYMBOLS

ϕ^α	electron work function of phase α
$\phi^\alpha_{(\beta)}$	electron work function of phase α at the interface with phase β
$\Delta^\alpha_\beta \phi$	difference in work function between phases α and β
V	potential energy inside a phase
ε_F	kinetic energy of electrons at the Fermi level
χ^α	surface potential of phase α
$\delta\chi^\alpha_{(\beta)}$	modification in χ^α as a result of contact with phase β
$g^\alpha_\beta(\text{dip})$	electric potential drop between phases α and β in contact related to dipolar layers
$g^\alpha_{(\beta)}(\text{dip})$	surface potential of phase α at the interface with phase β
$\Delta^\alpha_\beta g^\gamma(\text{dip})$	difference in surface potential of phase γ when in contact with α and β, respectively
$g^\alpha_\beta(\text{ion})$	electric potential drop between phases α and β in contact related to free charges
σ^α	charge density on phase α
$\Delta^\alpha_\beta \phi$	total electric potential drop between phases α and β

$\Delta_2^M \phi$	electric potential drop across the inner layer
$\Delta_S^2 \phi$	electric potential drop across the diffuse layer
$\Delta \phi_e$	equilibrium value of $\Delta \phi$
$\Delta \phi_{\sigma=0}$	$\Delta \phi$ at the point of zero charge
E	cell potential
E^M	potential of electrode M as measured on a conventional scale
$_{abs}E^M$	absolute value of the potential E^M
$E^M_{\sigma=0}$	potential of zero charge of electrode M
$_r E^M$	reduced absolute potential of electrode M
$E(M^+/M)$	electrode potential associated with the system in brackets
E^o, $\Delta \phi^o$	standard quantities
$\Delta_M^{M'} E_d$	value of underpotential for deposition of M' on M
$\Delta_\beta^\alpha E$	difference in electrode potential between α and β
μ_i^α	chemical potential of species i in phase α
x_e^i	electronegativity of element i
x_s^i	electronegativity of solid i
ΔH_{ads}	enthalpy of adsorption
ΔH_f	enthalpy of formation
$W(M-M)$	M-M bond energy
I^M	ionization potential of M
θ	surface coverage with adsorbate
C^i	differential capacity of the inner layer

$\Delta G^{\neq}$ free energy of activation

i_o exchange current

η overvoltage

κ transmission coefficient

c_i concentration of species i in solution

$D(\varepsilon_F)$ density of states at the Fermi level

8 REFERENCES

1. J.C. Rivière, in M. Green (Ed.), Solid State Surface Science, Vol. 1, Chap. 4, Dekker, New York, 1969.
2. C. Kittel, Introduction to Solid State Physics, 4th ed., Wiley, New York, 1971.
3. C. Herring and M.H. Nichols, Rev. Mod. Phys. 21, 185 (1949).
4. G.A. Haas and R.E. Thomas, in E. Passaglia (Ed.), Measurement of Physical Properties, Vol. 6, Part 1, Chap. 2, Interscience, New York, 1972.
5. L. Pauling, The Nature of the Chemical Bond, University Press, Ithaca, N.Y., 1960.
6. J. Tafel, Z. Phys. Chem. 50, 641 (1905).
7. P.J. Boddy, in H. Reiss (Ed.), Progress in Solid State Chemistry, Vol. 4, Chap. 4, Pergamon, Oxford, 1967.
8. S. Srinivasan, H. Wroblowa, and J. O'M. Bockris, Advan. Catal. 17, 351 (1967).
9. A. Frumkin and A. Gorodetzkaya, Z. Phys. Chem., 136, 215 (1928).
10. J. O'M. Bockris, J. Chem. Ed. 48, 352 (1971).
11. J. O'M. Bockris, Trans. Faraday Soc. 43, 417 (1947).

12. A.N. Frumkin, Svensk Kem. Tidskr. 77, 300 (1965).
13. R. Parsons, in J. O'M. Bockris (Ed.), Modern Aspects of Electrochemistry, Vol. 1, Chap. 3, Butterworths, London, 1954.
14. R. Vasenin, Zh. Fiz. Khim. 27, 878 (1953); 28, 1672 (1954).
15. S.D. Argade and E. Gileadi, in E. Gileadi (Ed.), Electrosorption, Chap. 5, Plenum, New York, 1967.
16. S. Trasatti, J. Electroanal. Chem. 33, 351 (1971).
17. L.I. Antropov, Ukr. Khim. Zh. 29, 555 (1963).
18. A. Frumkin, B. Damaskin, I. Bagotskaya, and N. Grigoryev, Electrochim. Acta 19, 75 (1974).
19. D.B. Matthews, Ph.D. Thesis, Univ. of Pennsylvania, 1965.
20. J.O'M. Bockris, in J.O'M. Bockris (Ed.), Modern Aspects of Electrochemistry, Vol. 1, Chap. 4, Butterworths, London, 1954.
21. B.E. Conway and J. O'M. Bockris, J. Chem. Phys. 26, 532 (1957).
22. A.T. Petrenko, Zh. Fiz. Khim. 39, 2097 (1965).
23. H. Kita, J. Electrochem. Soc. 113, 1095 (1966).
24. S. Trasatti, J. Electroanal. Chem. 39, 163 (1972).
25. A.T. Kuhn, C.J. Mortimer, G.C. Bond, and J. Lindley, J. Electroanal. Chem. 34, 1 (1972).
26. H. Kita and T. Kurisu, J. Res. Inst. Catal., Hokkaido Univ. 21, 200 (1973).
27. J.O'M. Bockris, A. Damjanovic, and R.J. Mannan, J. Electroanal. Chem. 18, 349 (1968).
28. J.O'M. Bockris, R.J. Mannan, and A. Damjanovic, J. Chem. Phys. 48, 1898 (1968).
29. D. Galizzioli and S. Trasatti, J. Electroanal. Chem. 44, 367 (1973).
30. O. Johnson, J. Res. Inst. Catal., Hokkaido Univ. 19, 152 (1972); 20, 125 (1972).

31. O. Johnson, J. Catal., 28, 503 (1973).
32. S. Trasatti, J.C.S. Faraday I 68, 229 (1972).
33. O. Johnson, J. Chem. Ed. 47, 431 (1970).
34. S. Trasatti, Surface Sci. 32, 735 (1972).
35. G. Paasch, H. Eschring, and W. John, Phys. Stat. Sol. (b) 51, 283 (1972).
36. N.D. Lang, Solid State Commun. 7, 1047 (1969).
37. J.R. Smith, Phys. Rev. 181, 522 (1969).
38. N.D. Lang and W. Kohn, Phys. Rev. B 3, 1215 (1971).
39. R.A. Oriani and C.A. Johnson, in J.O'M. Bockris and B.E. Conway (Eds.), Modern Aspects of Electrochemistry, Vol. 5, Chap. 2, Butterworths, London, 1969.
40. C. Kittel, Elementary Solid State Physics, Wiley, New York, 1962, p. 138.
41. D.F. Gibbons, in R.W. Cahn (Ed.), Physical Metalurgy, North-Holland, Amsterdam, 1970, p. 80.
42. V. Heine and C.H. Hodges, J. Phys. C 5, 225 (1972).
43. P.K. Rawlings and H. Reiss, Surface Sci. 36, 580 (1973).
44. J.O'M. Bockris and S.D. Argade, J. Chem. Phys. 49, 5133 (1968).
45. E. Lange and K. Mischenko, Z. Phys. Chem. 149, 1 (1930).
46. E. Lange and H. Göhr, Thermodynamische Elektrochemie, Hüttig Verlag, Heidelberg, 1962.
47. F. Seitz, The Modern Theory of Solids, McGraw-Hill, New York, 1940, p. 395.
48. A.O.E. Animalu and V. Heine, Phil. Mag. 12, 1249 (1965).
49. T. Schneider, Phys. Stat. Sol. 32, 323 (1965).
50. R. Smoluchowski, Phys. Rev. 60, 661 (1941).
51. R. Bouwman and W.M.H. Sachtler, Surface Sci. 24, 350 (1971).
52. B.E. Nieuwenhuys, R. Bouwman, and W.M.H. Sachtler, Thin Solid Films 21, 51 (1974).

53. R. Suhrmann and E. Wedler, Z. Angew. Phys. 14, 70 (1962).
54. G. Wedler, C. Wölfing, and P. Wissmann, Surface Sci. 24, 302 (1971).
55. J.C. Rivière, Proc. Phys. Soc. London 70, 676 (1965).
56. A. Kashetov and N. Gorbatyi, Fiz. Tverd. Tela 10, 2135 (1968).
57. O.K. Kultashev and A.P. Makarov, Fiz. Metall. Metalloved 30, 924 (1970).
58. R. Bouwman and W.M.H. Sachtler, Surface Sci. 24, 140 (1971).
59. S. Trasatti, Chim. Ind. Milan 53, 559 (1971).
60. R.S. Mulliken, J. Chem. Phys. 2, 782 (1934).
61. D.P. Stevenson, J. Chem. Phys. 23, 203 (1955).
62. H.O. Prichard and H.A. Skinner, Chem. Rev. 55, 745 (1955).
63. W. Gordy and W.J.O. Thomas, J. Chem. Phys. 24, 439 (1956).
64. S. Cerny and V. Ponec, Catal. Rev. 2, 249 (1968).
65. G.C. Bond, R.I.C. Rev. 3, 1 (1970).
66. G.C. Bond, Catalysis by Metals, Academic Press, London, 1962.
67. M. Boudart, Chem. Eng. Progr. 57, 33 (1961).
68. A.A. Balandin, Advan. Catal. 10, 120 (1958).
69. W.M.H. Sachtler and J. Fahrenfort, Proc. 2nd. Int. Cong. Catal., Vol. 1, Technip., Paris, 1961, p. 831.
70. P. Ruetschi and P. Delahay, J. Chem. Phys. 23, 195 (1955).
71. D.D. Eley, Discus. Faraday Soc. 8, 34 (1950).
72. G. Ehrlich, in W.M.H. Sachtler, G.C.A. Schuit, and P. Zwietering (Eds.), Proc. 3rd. Int. Congr. Catal., Vol. 1, North-Holland, Amsterdam, 1965, p. 113.
73. Natl. Bur. Stand. U.S. Circ., 1952.
74. B. Feuerbacher and B. Fitton, Phys. Rev. Lett. 30, 923 (1973).
75. A.N. Frumkin, Z.A. Iofa, and M.A. Gerovich, Zh. Fiz. Khim. 30, 1455 (1956).
76. J.E.B. Randles, Advan. Electrochem. Electrochem. Eng. 3, 1 (1963).

77. J.E.B. Randles, Trans. Faraday Soc. 52, 1573 (1956).
78. A.N. Frumkin, Electrochim. Acta 2, 351 (1960).
79. G.A. Kenney and D.C. Walker, Electroanal. Chem. 5, 1 (1971).
80. B.E. Conway, in J.O'M. Bockris and B.E. Conway (Eds.), Modern Aspects of Electrochemistry, Vol. 7, Chap. 2, Butterworths, London, 1972.
81. E.J. Hart and M. Anbar, The Hydrated Electron, Wiley-Interscience, New York, 1970.
82. A.M. Brodsky and Yu.V. Pleskov, in S.G. Davison (Ed.), Progress in Surface Science, Vol. 2, Part 1, Pergamon, Oxford, 1972.
83. A. Henglein, Ber. Bunsenges. Phys. Chem. 78, 1078 (1974).
84. A.M. Brodskii and A.V. Tsarevskii, Elektrokhimiya 9, 1671 (1973).
85. R.R. Dogonadze, L.I. Krishtalik, and Yu.V. Pleskov, Elektrokhimiya 10, 507 (1974).
86. K. Fueki, D.-F. Feng, L. Kevan, and R.E. Christoffersen, J. Phys. Chem. 75, 2297 (1971).
87. Z.A. Rotemberg, Elektrokhimiya 8, 1198 (1972).
88. A.N. Frumkin, J. Electroanal. Chem. 9, 173 (1965).
89. A.N. Frumkin, Elektrokhimiya 1, 394 (1965).
90. M. Heyrovsky, Croat. Chem. Acta 45, 247 (1973).
91. Z.A. Rotemberg, V.J. Lakomov, and Yu. V. Pleskov, Elektrokhimiya 9, 152 (1973).
92. S. Trasatti, J. Electroanal. Chem. 52, 313 (1974).
93. S. Trasatti, J.C.S. Faraday I 70, 1752 (1974).
94. Z.A. Rotemberg, Yu.A. Prishchepa, and Yu.V. Pleskov, J. Electroanal. Chem. 56, 345 (1974).
95. B.V. Ershler, Usp. Khim. 21, 237 (1952).
96. B. Jakuszewski, Bull. Soc. Sci. Lett. Lodz., Cl. III 8, 1 (1957).

97. J.O'M. Bockris, Energy Convers. 10, 41 (1970).
98. B.B. Damaskin and A.N. Frumkin, personal communication.
99. R. Parsons, Handbook of Electrochemical Constants, Butterworths, London, 1959.
100. S. Trasatti, J. Electroanal. Chem. 54, 437 (1974).
101. A. Frumkin, N. Polianovskaya, N. Grigoryev, and I. Bagotskaya, Electrochim. Acta 10, 793 (1965).
102. V.A. Kiryanov, V.S. Krylov, and N.B. Grigoryev, Elektrokhimiya 4, 408 (1968).
103. D.C. Grahame, J. Amer. Chem. Soc. 76, 4819 (1954).
104. I.A. Bagotskaya, A.M. Morozov, and N.B. Grigoryev, Elektrochim. Acta 13, 873 (1968).
105. A. Frumkin, I. Bagotskaya, and N. Grigoryev, Ext. Abstrs. Japan-USSR Seminar of Electrochemistry, Tokyo, 1974 (A.N. Frumkin, personal communication).
106. V.Ya. Bartenev, E.S. Sevastyanov, and D.I. Leikis, Elektrokhimiya 4, 745 (1968).
107. K.V. Rybalka and D.I. Leikis, Elektrokhimiya 3, 383 (1967).
108. R.Ya. Pullerits, U.V. Palm, and V.E. Past, Elektrokhimiya 5, 886 (1969).
109. N. Van Huong and J. Clavilier, C.R. Acad. Sci. Paris 272, 1404 (1971).
110. S. Trasatti, J. Electroanal. Chem. 54, 19 (1974).
111. N.B. Grigoryev, V.P. Kuprin, and Yu.M. Loshkarev, Elektrokhimiya 9, 1842 (1973).
112. V.V. Batrakov, B.B. Damaskin, and Yu.P. Ipatov, Elektrokhimiya 10, 144 (1974).
113. G. Valette and A. Hamelin, J. Electroanal. Chem. 45, 301 (1973).
114. K. Müller, Z. Phys. Chem. Leipzig 243, 239 (1970).

115. A.J. Sargood, C.W. Jowett, and B.J. Hopkins, Surface Sci. 22, 343 (1970).
116. C.W. Jowett, P.J. Dobson, and B.J. Hopkins, Surface Sci. 17, 474 (1969).
117. L. Formaro and S. Trasatti, Electrochim. Acta 12, 1457 (1967).
118. D.R. Flinn, M. Rosen, and S. Schuldiner, Collect. Czech. Chem. Commun. 36, 454 (1971).
119. J. Amosse, R. Durend, B. Nguyen, and M.J. Barbier, C.R. Acad. Sci. Paris 274, 1720 (1972).
120. S. Schuldiner and M. Rosen, J. Electrochem. Soc. 118, 1138 (1971).
121. R. Durand, C.R. Acad. Sci. Paris 278, 821 (1974).
122. D.M. Zehner, T.S. Noggle, and L.H. Jenkins, Surface Sci. 41, 601 (1974).
123. P.E. Højlund Nielsen, Surface Sci. 36, 778 (1973).
124. P.J. Dobson and P.N.J. Dennis, Surface Sci. 36, 781 (1973).
125. R.W. Joyner, Surface Sci. 39, 450 (1973).
126. T.N. Rhodin, P.W. Palmberg, and E.W. Plummer, in G.A. Somorjai (Ed.), The Structure and Chemistry of Surfaces, Wiley, New York, 1969, contribution 22.
127. E. Kamp, Dissertation, Hannover, 1966.
128. A. Cetronio and J.P. Jones, Surface Sci. 44, 109 (1974).
129. I. Fried and H. Barak, J. Electroanal. Chem. 30, 279 (1971).
130. D.M. Kolb, M. Przasnyski, and H. Gerischer, J. Electroanal. Chem. 54, 25 (1974).
131. R.L. Gerlach and T.N. Rhodin, Surface Sci. 19, 403 (1970).
132. D.L. Fehrs and R.E. Stickney, Surface Sci. 24, 309 (1971).
133. J.M. Chen and C.A. Papageorgopoulos, Surface Sci. 26, 499 (1971).
134. J.R. Anderson and N. Thompson, Surface Sci. 26, 397 (1971).

135. A.G. Fedorus, A.G. Naumovets, and Yu.S. Vedula, Phys. Stat. Sol. (a) 13, 445 (1972).
136. H. Gerischer, D.M. Kolb, and M. Przasnyski, Surface Sci. 43, 662 (1974).
137. A.K. Vijh, Surface Sci. 46, 282 (1974).
138. R. Adzic, E. Yeager, and B.D. Cahan, J. Electrochem. Soc. 121, 474 (1974).
139. Z. Sidorski, Acta Phys. Pol. A42, 437 (1972).
140. W.J. Lorenz, H.D. Hermann, N. Wüthrich, and F. Hilbert, J. Electrochem. Soc. 121, 1167 (1974).
141. J.P. Jones, Surface Sci. 32, 29 (1972).
142. A. Abon, J. Chim. Phys. 67, 1275 (1970).
143. C.E. Carroll and J.W. May, Surface Sci. 29, 60 (1972).
144. K. Besocke and H. Wagner, Phys. Rev. B 8, 4597 (1973).
145. M. Heyrovsky, S. Vavricka, and R. Heyrovska, J. Electroanal. Chem. 46, 391 (1973).
146. M. Heyrovsky and R. Heyrovska, J. Electroanal. Chem. 52, 141 (1974).
147. A. Capon and R. Parsons, J. Electroanal. Chem. 39, 275 (1972).
148. T.M. Grishina, L.I. Logacheva, and G.D. Vovchenko, Elektrokhimiya 9, 1247 (1973).
149. A.A. Sutyagina, I.N. Golyanitskaya, and G.D. Vovchenko, Elektrokhimiya 8, 908 (1972).
150. L.I. Krishtalik, Advan. Electrochem. Electrochem. Eng. 7, 283 (1970).
151. D.J. Barclay, J. Electroanal. Chem. 44, 47 (1973).
152. A.J. Appleby, Surface Sci. 27, 225 (1971).
153. R.S. Perkins and T.N. Andersen, in J.O'M. Bockris and B.E. Conway (Eds.), Modern Aspects of Electrochemistry, Vol. 5, Chap. 5, Butterworths, London, 1969.

154. A. Frumkin, B. Damaskin, N. Grigoryev, and I. Bagotskaya, Electrochim. Acta 19, 69 (1974).
155. D.I. Leikis, K.V. Rybalka, E.S. Sevastyanov, and A.N. Frumkin, J. Electroanal. Chem. 46, 161 (1973).
156. V.Ya. Bartenev, E.S. Sevastyanov, and I.D. Leikis, Elektrokhimiya 6, 1197 (1970).
157. A.N. Frumkin, Proc. 3rd Symposium on Double Layer and Adsorption on Solid Electrodes, Tartu, 1972, p. 5.
158. T.N. Andersen, J.L. Anderson, and H. Eyring, J. Phys. Chem. 73, 3562 (1969).
159. T.N. Andersen, J.L. Anderson, D.D. Bodé, and H. Eyring, J. Res. Inst. Catal., Hokkaido Univ. 16, 449 (1968).
160. A.N. Frumkin and O.A. Petrii, Electrochim. Acta 15, 391 (1970).
161. O.A. Petrii and T.Ya. Kolotyrkina, Elektrokhimiya 9, 254 (1973).
162. B.E. Nieuwenhuys and W.M.H. Sachtler, Surface Sci. 45, 513 (1974).
163. I.M. Novoselskii, N.I. Maksimyuk, and L.Ya. Egorov, Elektrokhimiya 6, 521 (1970).
164. L.Ya. Egorov and I.M. Novoselskii, Elektrokhimiya 6, 521 (1970).
165. S. Trasatti, J. Electroanal. Chem. 53, 335 (1974).
166. S. Trasatti, J. Electroanal. Chem. 28, 257 (1970).
167. J.W. Schultze, Electrochim. Acta 17, 451 (1972).
168. C. Kemball, Proc. Roy. Soc. London A190, 117 (1947).
169. J.W. Schultze, Ber. Bunsenges. Phys. Chem. 73, 483 (1969).
170. E.V. Osipova, N.A. Shurmovskaya, and R.Kh. Burshtein, Elektrokhimiya 5, 1139 (1969).
171. R. Suhrmann, J.M. Heras, L. Viscido de Heras, and G. Wedler, Ber. Bunsenges. Phys. Chem. 72, 511 (1968).

172. R. Suhrmann, J.M. Heras, L. Viscido de Heras, and G. Wedler, Ber. Bunsenges. Phys. Chem. 68, 511 (1965).
173. E.E. Huber and C.T. Kirk, Surface Sci. 5, 447 (1966).
174. P.J. Page, D.L. Trimm, and P.M. Williams, J.C.S. Faraday I, 70, 1769 (1974).
175. F.C. Tompkins, in E.A. Flood (Ed.), The Solid-Gas Interface, Vol. 2, Chap. 25, Dekker, New York, 1967.
176. J. Patigny, Y. Barbaux, and J.-P.A. Beaufils, in F. Ricca (Ed.), Adsorption - Desorption Phenomena, Academic, London 1972, p. 49.
177. J.C.P. Mignolet, in W.E. Garner (Ed.), Chemisorption, Butterworths, London, 1957, p. 118.
178. J.C.P. Mignolet, Discus. Faraday Soc. 8, 105 (1950).
179. D.K. Klemperer and J.C. Snaith, Surface Sci. 28, 209 (1971).
180. B.E. Nieuwenhuys, O.G. Van Aardenne, and W.M.H. Sachtler, Chem. Phys. 5, 418 (1974).
181. J. Müller, Surface Sci. 45, 314 (1974).
182. D.F. Klemperer and J.C. Snaith, Surface Sci. 45, 314 (1974).
183. A.G. Knapp and M.H.B. Stiddard, J.C.S. Faraday I 68, 2139 (1972).
184. M.A. Chesters, M. Hussain, and J. Pritchard, Surface Sci. 35, 161 (1973).
185. M.A. Chesters and J. Pritchard, Surface Sci. 28, 460 (1971).
186. R.R. Ford and J. Pritchard, Trans. Faraday Soc. 67, 216 (1971).
187. G.V. Samsonov, The Oxide Handbook, Plenum, New York 1973.
188. A. Frumkin, N. Polianovskaya, I. Bagotskaya, and N. Grigoryev, J. Electroanal. Chem. 33, 319 (1971).
189. B.B. Damaskin, O.A. Petrii, and V.V. Batrakov, Adsorption of Organic Compounds at Electrodes, Plenum, New York, 1971.
190. A.N. Frumkin, N.B. Grigoryev, and I.A. Bagotskaya, Elektrokhimiya, 2, 329 (1966).

191. J. Clavilier and N. Van Huong, J. Electroanal. Chem. 41, 193 (1973).
192. K. Palts, R. Pullerits, and V. Past, Uch. Zap. Tartu. Gos. Univ. 235, 64 (1969).
193. V.A. Panin, K.V. Rybalka, and D.I. Leikis, Elektrokhimiya 8, 383 (1967).
194. N.B. Grigoryev, I.A. Gedvillo, and N.G. Bardina, Elektrokhimiya 8, 409 (1972).
195. V. Bartenev, E. Sevastyanov, and D. Leikis, Elektrokhimiya 6, 1868 (1970).
196. M. Khaga and V. Past, Elektrokhimiya 5, 618 (1969).
197. T.N. Andersen and J.O'M. Bockris, Electrochim. Acta 9, 347 (1964).
198. D.C. Grahame, J. Amer. Chem. Soc. 80, 4201 (1958).
199. V.A. Panin and K.V. Rybalka, Elektrokhimiya 8, 1202 (1972).
200. K.V. Rybalka, Elektrokhimiya 8, 400 (1972).
201. K.A. Kolk, M.A. Salve, and U.V. Palm, Elektrokhimiya 8, 1533 (1972).
202. A.M. Morozov, N.B. Grigoryev, and I.A. Bagotskaya, Elektrokhimiya 2, 1235 (1966).
203. J. Clavilier and N. Van Huong, C.R. Acad. Sci. Paris 270, 982 (1970).
204. N.B. Grigoryev and D.N. Machavariani, Elektrokhimiya 5, 87 (1969).
205. L.E. Rybalka, B.B. Damaskin, and D.I. Leikis, Elektrokhimiya 9, 414 (1973).
206. A.R. Alumaa and U.V. Palm, Elektrokhimiya 9, 396 (1973).
207. N.B. Grigoryev and I.A. Bagotskaya, Elektrokhimiya 2, 1449 (1966).
208. V.I. Lukyanicheva, E.M. Strochkiva, V.S. Bagotskii, and L.L. Knots, Elektrokhimiya 7, 267 (1971).

209. N.A. Epshtein, B.I. Podlovchenko, O.A. Petrii, and V.A. Safonov, Elektrokhimiya 10, 561 (1974).
210. V.I. Lukyanycheva, E.M. Strochkova, and V.S. Bagotskii, Elektrokhimiya 6, 701 (1970).
211. A. Hamelin and J. Lecoeur, Collect. Czech. Chem. Commun. 36, 714 (1971).
212. D.C. Grahame, Chem. Rev. 47, 411 (1947).
213. R. Parsons, Proc. Roy. Soc. London 261A, 79 (1961).
214. J.R. Macdonald, J. Chem. Phys. 22, 1857 (1954).
215. N.F. Mott and R.J. Watts-Tobin, Electrochim. Acta 4, 79 (1961).
216. B.B. Damaskin and A.N. Frumkin, Electrochim. Acta 19, 173 (1974).
217. S. Trasatti, J. Chim. Phys. 72, 561 (1975).
218. J.R. Macdonald and C.A. Barlow, J. Chem. Phys. 73, 3577 (1969).
219. K. Müller, J. Res. Inst. Catal., Hokkaido Univ. 14, 224 (1966).
220. K.W. Allen, D.R. Lewis, and K.G.A. Pankhurst, J. Chem. Soc. (A), 3028 (1971).
221. M.C. Phillips, D.A. Cadenhead, R.J. Good, and H.F. King, J. Colloid Interface Sci. 37, 437 (1971).
222. M.E. Nicholas, P.A. Joyner, B.M. Tessem, and M.D. Olson, J. Chem. Phys. 65, 1373 (1961).
223. J.O'M. Bockris, M.A. Devanathan, and K. Müller, Proc. Roy. Soc. London 274A, 55 (1963).
224. R.W. Rice and G.L. Haller, Proc. Int. Congr. Catal., 5th, 1972, North-Holland, Amsterdam, 1973, p. 317.
225. B.B. Damaskin and A.N. Frumkin, J. Electroanal. Chem. 34, 191 (1972).
226. B.E. Conway and H.P. Dhar, Croat. Chem. Acta 45, 109 (1973).

227. J.A. Harrison, J.E.B. Randles, and D.J. Schiffrin, J. Electroanal. Chem. 48, 359 (1973).
228. G.J. Hills and S. Hsieh, J. Electroanal. Chem. 54, 289 (1974).
229. R. Parsons and P.C. Symons, Trans. Faraday Soc. 64, 1077 (1968).
230. A. De Battisti and S. Trasatti, J. Electroanal. Chem. 59, 137 (1975).
231. E. Schwartz, B.B. Damaskin, and A.N. Frumkin, Zh. Fiz. Khim. 36, 2419 (1962).
232. S. Levine, G.M. Bell, and A.L. Smith, J. Phys. Chem. 73, 3534 (1969).
233. J.R. Macdonald and C.A. Barlow, J. Chem. Phys. 36, 3062 (1962).
234. R.J. Watts-Tobin, Phil. Mag. 6, 133 (1961).
235. D.C. Grahame, J. Amer. Chem. Soc. 79, 2093 (1957).
236. R. Parsons, Rev. Pure Appl. Chem. 18, 91 (1968).
237. B.B. Damaskin, Elektrokhimiya 1, 1258 (1965).
238. R. Parsons, Surface Sci. 2, 418 (1964).
239. A.N. Frumkin, quoted in Ref. 265.
240. J.N. Butler and A.C. Makrides, Trans. Faraday Soc. 60, 1656 (1964).
241. M. Temkin and A. Frumkin, Zh. Fiz. Khim. 29, 1513 (1955).
242. R.A. Marcus, in E. Yeager (Ed.), Transactions of the Symposium on Electrode Processes, Wiley, New York, 1961, p. 239.
243. D. Galizzioli, F. Tantardini, and S. Trasatti, J. Appl. Electrochem. 5, 203 (1975).
244. R. Memming and F. Moellers, Ber. Bunsenges. Phys. Chem. 76, 457 (1972).
245. R. Parsons, Advan. Electrochem. Electrochem. Eng. 1, 1 (1961).

246. A.N. Frumkin, Z. Phys. Chem. 164, 121 (1933).
247. D.C. Grahame and B.A. Soderberg, J. Chem. Phys. 22, 449 (1954).
248. T. Krivachy and J. Kluegel, Surface Sci. 15, 358 (1969).
249. M.W. Roberts and B.R. Wells, Surface Sci. 15, 325 (1969).
250. V.G. Levich, Advan. Electrochem. Electrochem. Eng. 4, 249 (1966).
251. R.R. Dogonadze and Yu.A. Chizmadzhev, Dokl. Akad. Nauk SSSR 145, 849 (1962).
252. J.M. Hale, in N.S. Hush (Ed.), Reactions of Molecules at Electrodes, Wiley-Interscience, New York, 1971, p. 229.
253. M.V. Vojnović and D.B. Sepa, J. Chem. Phys. 51, 5344 (1969).
254. H. Gerischer, Advan. Electrochem. Electrochem. Eng. 1, 139 (1961).
255. H. Gerischer, Rec. Chem. Progr. 23, 135 (1962).
256. H. Gerischer, Z. Phys. Chem. Frankfurt 26, 223 (1966).
257. A.C. Makrides, J. Electrochem. Soc. 111, 392 (1964).
258. D.H. Angell and T. Dickinson, J. Electroanal. Chem. 35, 55 (1972).
259. J.E.B. Randles and K.W. Somerton, Trans. Faraday Soc. 48, 937 (1952).
260. A. Capon and R. Parsons, J. Electroanal. Chem. 46, 215 (1973).
261. N.V. Nikolaeva-Fedorovich, B.N. Rybakov, and K.A. Radyushkina, Elektrokhimiya 3, 1086 (1967).
262. A.N. Frumkin, N.V. Fedorovich, and S.I. Kulakovskaya, Elektrokhimiya 10, 330 (1974).
263. A.N. Frumkin, S.I. Kulanovskaya, and N.V. Fedorovich, Elektrokhimiya 10, 837 (1974).
264. I.A. Bagotskaya, S.A. Fateev, N.B. Grigoryev, and A.N. Frumkin, Elektrokhimiya 9, 1676 (1973).

265. J. Horiuti and M. Polanyi, Acta Physicochim. URSS 2, 505 (1935).
266. J.A. Butler, Proc. Roy. Soc. London A157, 423 (1936).
267. R. Parsons and J.O'M. Bockris, Trans. Faraday Soc. 47, 914 (1951).
268. R. Parsons, Trans. Faraday Soc. 54, 1053 (1958).
269. H. Gerischer, Bull. Soc. Chim. Belg. 67, 506 (1958).
270. J.G.N. Thomas, Trans. Faraday Soc. 57, 1603 (1961).
271. R. Parsons, Surface Sci. 18, 28 (1969).
272. A.N. Frumkin, Advan. Electrochem. Electrochem. Eng. 1, 65 (1961); 3, 287 (1963).
273. A.B. Kilimnik and A.L. Rotinyan, Elektrokhimiya 5, 1235 (1969).
274. A.L. Rotinyan, A.B. Kilimnik, and E.D. Levin, Proc. 2nd Symposium on Double Layer and Adsorption on Solid Electrodes, Tartu, 1970, p. 321.
275. A.B. Kilimnik and A.L. Rotinyan, Elektrokhimiya 6, 330 (1970).
276. A.L. Rotinyan and E.D. Levin, Elektrokhimiya 6, 328 (1970).
277. O.L. Kabanova and A.N. Doronin, Elektrokhimiya 6, 222 (1970).
278. T.T. Tenno and U.V. Palm, Elektrokhimiya 8, 1381 (1972).
279. U. Palm and T. Tenno, J. Electroanal. Chem. 42, 457 (1973).
280. R. Parsons, Trans. Faraday Soc. 59, 1340 (1960).
281. A.J. Appleby, Catal. Rev. 4, 221 (1970).
282. E.N. Potapova, L.I. Krishtalik, and I.A. Bagotskaya, Elektrokhimiya 10, 49 (1974).
283. L.I. Krishtalik and V.M. Tsionsky, J. Electroanal. Chem. 31, 363 (1971).
284. T. Biegler and R. Woods, J. Electroanal. Chem. 20, 347 (1969).
285. F.G. Will and C.A. Knorr, Z. Elektrochem. 64, 258, 270 (1960).

286. D.A.J. Rand and R. Woods, J. Electroanal. Chem. 36, 57 (1972).
287. M.W. Breiter, Electrochemical Processes in Fuel Cells, Chap. 6, Springer-Verlag, New York, 1969.
288. G.T. Burstein and G.A. Wright, J. Electroanal. Chem. 50, 399 (1974).
289. J. Sobkowski and J. Uminski, Rocz. Chem. 44, 1135 (1970).
290. G. Vesman, K. Kabin, and V. Past, Uch. Zap. Tartu. Gos. Univ. 193, 65 (1966).
291. N.M. Kozhevnikova and A.L. Rotinyan, Elektrokhimiya 1, 664 (1965).
292. E.K. Tuseeva, A.M. Skundin, and V.S. Bagotskii, Elektrokhimiya 9, 1541 (1973).
293. I.V. Kudryashov, A.P. Popkov, and N.G. Kopalin, Elektrokhimiya 10, 1468 (1973).
294. J.O'M. Bockris, M.A. Genshaw, and M. Fullenwider, Electrochim. Acta 15, 47 (1970).
295. W. Beck, J.O'M. Bockris, J. McBrennen, and L. Nanis, Proc. Roy. Soc. London A290, 220 (1966).
296. I.I. Phillips, P. Poole, and L.L. Shreir, Corrosion Sci. 12, 855 (1972).
297. V.S. Bagotskii, Yu.B. Vasilyev, O.A. Khazova, and S.S. Sedova, Electrochim. Acta 17, 913 (1971).
298. L.I. Krishtalik and B.B. Kuz'menko, Elektrokhimiya 9, 664 (1973).
299. S. Trasatti, J. Electrochem. Soc. 118, 1961 (1971).
300. A.K. Vijh, J. Electrochem. Soc. 118, 1963 (1971).
301. R.P. Bajpai, H. Kita, and K. Azuman, Phys. Stat. Sol. (a) 16, K125 (1973).
302. R.L. Wells and T. Fort, Jr., Surface Sci. 32, 554 (1972).

303. D. Dickertmann, J.W. Schultze, and K.J. Vetter, J. Electroanal. Chem. 55, 429 (1974).
304. G.C. Bond and P.A. Sermon, Gold Bull. 6, 102 (1973).
305. N. Furuya and S. Motoo, Denki Kagaku 41, 364 (1973).
306. Extended Abstracts of the Int. Symposium on Characterization of Adsorbed Species in Catalytic Reactions, Ottawa, 1974.
307. K.S. Kim, N. Winograd, and R.E. Davis, J. Amer. Chem. Soc. 93, 6296 (1971).
308. C.C. Schubert, C.C. Page, and B. Ralph, Electrochim. Acta 18, 33 (1973).
309. K. Ishimoto, H.M. Pak, T. Nishida, and M. Doyama, Surface Sci. 41, 102 (1974).
310. R.Ya. Kamilova and E.P. Sytaya, Dokl. Akad. Nauk Uzb. SSR 27, 20 (1970).
311. R.W. Strayer, W. Mackie, and L.W. Swanson, Surface Sci. 34, 225 (1973).
312. N.G. Imangulova and E.P. Sytaya, Izv. Akad. Nauk Uzb. SSR, Ser. Fiz.-Mat. Nauk 17, 30 (1973).
313. P.O. Gartland, S. Berge, and B.J. Slagsvold, Phys. Rev. Lett. 28, 738 (1972).
314. R. Parsons, J. Electroanal. Chem. 59, 229 (1975).
315. A.N. Frumkin and B.B. Damaskin, Dokl. Akad. Nauk SSSR 221, 395 (1975).
316. A. Frumkin and B. Damaskin, J. Electroanal. Chem. 66, 155 (1975).
317. S. Trasatti, J. Electroanal. Chem. 66, 150 (1975).

Ionically Conducting Solid-State Membranes
ROBERT A. HUGGINS
Center for Materials Research, Stanford University, Stanford, California

1	Introduction	324
2	Unusual Characteristics of Fast Ionic Conductors	328
3	The Beta Alumina Family	336
	3.1 Introduction	336
	3.2 Structures of the phases in the beta alumina family	336
	3.3 Phase equilibria	339
	3.4 Ionic and electronic transport	341
	3.5 Structure of the bridging layer	351
4	Analogues of the Beta Aluminas	355
5	Other Materials that Exhibit Fast Ionic Conduction	356
	5.1 Introduction	356
	5.2 Cation conductors	356

Work in this area within the Solid State Electrochemistry Group at Stanford has been supported by The Environmental Protection Agency, The National Science Foundation, the Office of Naval Research, and the Advanced Research Projects Agency.

	5.2.1 Materials with cubic structures	356
	5.2.2 Materials with hexagonal structures	361
	5.2.3 Materials with unidirectional tunnel structures	361
	5.2.4 Materials with layer structures	362
5.3	Anion conductors	363
	5.3.1 Materials with cubic structures	363
	5.3.2 Materials with layer structures	365
6	Analogous Behavior of Interstitial Species in Metals	366
7	Fast Diffusion in Semiconductors	370
8	Comments on Theoretic Models Related to Fast Ionic Conduction in Solids	372
9	The Application of New Measurement Techniques to Problems in this Area	375
	References	377

1 INTRODUCTION

It has been known for more than a hundred years that a number of solid materials conduct electric charge by the translation of ionic species, and the physical processes responsible for this behavior have been actively studied for a number of decades. Indeed, this phenomenon was employed in a practical device, the Nernst Glower, near the turn of the century (1-4), although the mechanism of its operation was not understood until much later (5). This application involved sending direct current through a sample of doped ZrO_2 with resistive heating, causing it to become a very high-temperature light source. The frequency distribution of the

Introduction

light emitted was much more favorably balanced toward the visible spectrum than the carbon fiber sources previously used.

A material of this same family, ZrO_2 doped with 15 wt % Y_2O_3, was used by Baur and Preis in 1937 (6) as an oxide ion-conducting solid membrane in a fuel cell operated at 1000°C.

There also has been a long but sporadic history of the use of solid-state membranes as electrolytes in cells for making thermodynamic measurements (7-14). Among the materials used as solid electrolytes in these early studies were glass, porcelain, and various halides.

Interest in the practical utilization of solid ionically conducting membranes has gone through two stages of acceleration in recent years. The first was initiated by the important work by Carl Wagner in 1953 (15) in which he used silver iodide as a solid electrolyte in a simple solid-state cell configuration to measure a number of properties of another solid, silver sulfide. This paper also introduced the concept of coulometric titration using a solid electrolyte, and showed the extreme precision with which changes in composition can be controlled by this technique.

Just a few years later, a pair of papers by Kiukkola and Wagner (16,17) expanded this horizon considerably by pointing out that solid electrolytes could be used for a variety of possible applications. This paper also indicated a number of ionically conducting solids that could be used for this purpose. In addition to a group of cationically conducting binary salts, it was shown that the cubic oxides with the fluorite structure, such as calcia-doped ZrO_2, could be of special interest as oxide ion conductors in the range 750 to 1000°C. Another important contribution was the very thorough review of the important relevant principles presented by Wagner at the 1956 meeting of the CITCE (18).

Subsequently, workers in a number of laboratories have pursued these ideas, as well as extending them. Reviews of much of the work during the following decade on the use of ionically conducting solid membranes for making both thermodynamic and kinetic measurements have been presented in several places (19-22).

The second major stimulus to broader consideration of the use of solid electrolytes arose as a result of the important pioneering work at the Ford Motor Company (23,24) on cation-conducting materials of the "beta alumina" family. The introduction of this group of materials led to the recognition that a much wider variety of ionically conducting solids might be of practical importance than had previously been recognized. Of special interest was the conclusion that materials should exist which have very large values of ionic conductivity at relatively low temperatures, and in which a wide variety of charge-carrying species could exist, including several alkali metal ions, as well as a considerable number of other monovalent cations.

Work by the Ford group (25) also showed that by the use of such a solid membrane, one could design batteries with liquid electrodes and a solid electrolyte, just the inverse of traditional battery designs. It has subsequently been found that such an approach can lead to some very distinct advantages in certain cases. This work on the beta alumina family, the most prominent member of which is commonly known as sodium beta alumina, and has the nominal formula $NaAl_{11}O_{17}$, has led to a great deal of interest in the development of the sodium/sulfur rechargeable (secondary) battery system. There are currently over a dozen commercial development programs relating to this type of cell in at least five different countries.

In addition to leading the way to possible exciting applications in battery systems, work on the beta aluminas pointed out

Introduction

the feasibility of the development of solid membranes that transport electrical charge by the motion of a wide variety of cations at relatively low temperatures. Previously, the only recognized cationically conducting solid electrolytes usable at moderate temperatures contained silver or copper. It was pointed out several years ago (26) that as a result of the discovery of the beta alumina family, with both good ionic and poor electronic (27) conductivity over a very wide temperature range, and the expectation of finding other related materials, solid membranes should attain a much wider utilization for a variety of both scientific and technologic purposes.

Work had already begun to appear in which sodium beta alumina was used as an electrochemical transducer to study thermodynamic properties of the sodium-lead (28), sodium-sulfur (29), and sodium-mercury (30) systems. It was also shown (26) that the potassium member of the beta alumina family could be used to measure differences in potassium activity between two solid phases.

A further stimulus that occurred at about the same time as the discovery of the beta alumina family and contributed to the second stage of accelerated interest in ionically conducting solid membranes was the recognition of the high values of ionic conductivity in a group of alkali silver iodides. These materials, whose nominal formula is MAg_4I_5, were discovered independently by Bradley and Greene in England (31,32), and Owens and Argue in the United States (33,34). Of special importance was the observation of very high values of cationic conductivity at and below ambient temperatures. The primary member of this family, $RbAg_4I_5$, has the greatest ionic conductivity at room temperature of any presently known solid. These observations very quickly led to the development of new types of all-solid batteries, capacitors, and so on, based on the use of these materials as silver ion-conducting solid electrolytes. This work was thoroughly discussed in the review by Owens (35).

Because of the substantial number of scientific questions that have been raised in connection with the unusual properties of these solid ionic conductors, as well as the wealth of possible technologic applications, this area is currently attracting a large measure of attention. One indication of this is the number of recent symposia or meetings dealing with such matters (36).

The following sections of this chapter discuss the special characteristics of materials in which ions are found to be unusually mobile, as well as some of the presently available experimental data on several structural groups. Particular emphasis is given to the important beta alumina family. Analogous behavior in metallic and semiconducting materials is also mentioned. Finally, some comments are made about theoretic models relating to fast ionic conduction, and the application of new measurement techniques.

2 UNUSUAL CHARACTERISTICS OF FAST IONIC CONDUCTORS

Measurements of ionic conductivity and diffusion on a large number of materials, predominantly the halides, over a period of many years led to the establishment of the presently standard theoretic approach to the point defect structure of ionic solids, as well as to models for mass and charge transport in such materials. Experimental results have shown that the transport behavior is dependent upon the disorder type, which indicates the intrinsic defect pair of lowest free energy. It is now evident that materials with Frenkel disorder, and thus interstitial species, typically have much greater values of ionic conductivity than those with Schottky disorder, where the predominant defects are vacancies.

Quite early, it became recognized that a number of these simple ionic compounds undergo phase transformations at elevated temperatures, with the ionic conductivity going through a corresponding abrupt transition. Much higher values were sometimes found in the high-temperature phase, with a corresponding reduction in the temperature dependence. The ionic transport properties of these unusual high-temperature phases, most of which were silver or copper compounds, were investigated by a number of workers, and some attention was given as early as the 1920s (37-40) to the reasons for this unusual type of behavior, which is clearly different from that of the more "normal" Schottky- and Frenkel-disorder materials.

A great wealth of experimental information has become available since that time, and it is now known that this type of "high-temperature" behavior, characterized by unusually rapid ionic motion, can persist to ambient temperatures and below in some materials, and that it is not necessarily related to the existence of a first-order phase transformation. Instead, there are fundamental differences in the transport behavior of this third type of materials from the Schottky- and Frenkel-disorder types. It is also recognized that there is nothing magic about silver and copper compounds; fast ionic conduction can involve a wide variety of ions, depending on the structure.

The difference between these three classes of ionic conductors can best be seen from the temperature dependence of their ionic conductivities. Figure 1 indicates the ranges typically observed.

It is generally found that in all three cases the variation in the conductivity σ with temperature can be represented by an expression of the Arrhenius type,

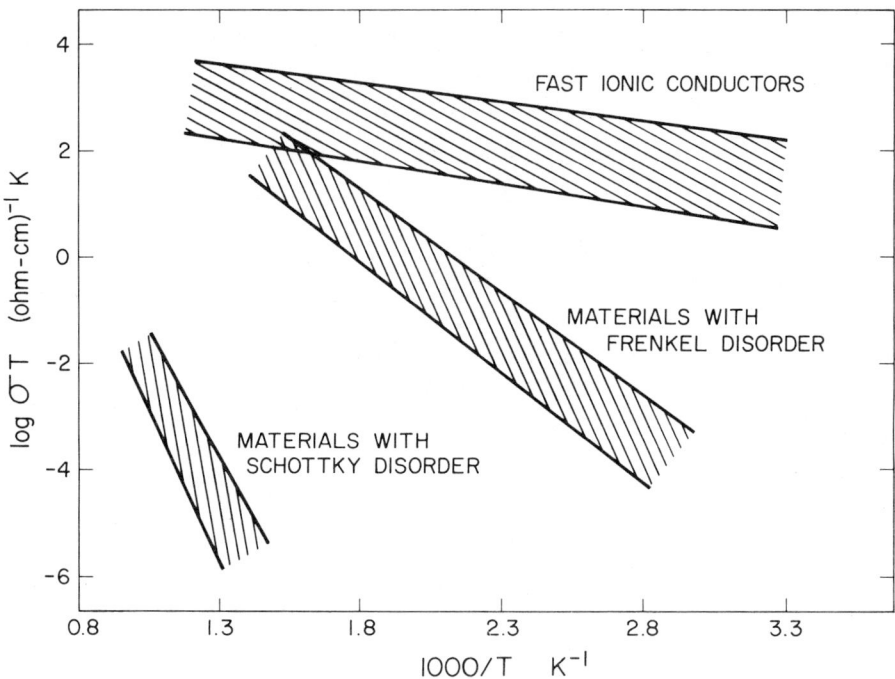

Fig. 1. Bands showing the temperature dependence of the ionic conductivity of three general classes of materials.

$$\sigma = \left(\frac{\sigma_o}{T}\right) \exp\left(\frac{-\Delta H}{RT}\right)$$

The data in Fig. 1 indicate that there are three principal features that differentiate the fast ionic conductor group from the others. The most obvious difference is in the absolute magnitude of the ionic conductivity, which can be unusually high over a wide range of temperature. It has been known for a long time that some materials have such high values, for it was shown by Tubandt and Lorenz in 1914 (41) that the ionic conductivity of AgI is actually more than 20% greater in the solid near the melting point than it is in the liquid melt. In addition to the unusual magnitude of the ionic conductivity, the temperature dependence, represented by the activation enthalpy ΔH, is considerably lower in the fast ionic conductors. The value of the pre-exponential constant σ_o, related to the intercept when $(1/T)$ approaches zero, is also considerably lower. These characteristics are obviously interrelated, rather than independent. Together, they clearly indicate that the process of ionic transport in the fast ionic conductors group is quite different from that in the other two. As will be seen later, the unusually small value of the pre-exponential factor provides especially important information about the mechanistic differences between ionic transport in these materials. If one can use a simple hopping model, it implies either an especially low jump attempt frequency, or a very small (or negative) value of the entropy of migration. It is not consistent with the contention that the major difference between fast ionic conductors and the other groups lies in the presence of an unusually large concentration of defects or potentially mobile ions.

It should be pointed out that there is actually a fourth group of materials, typified by a number of halides with the fluorite (CaF_2) structure, which have anti-Frenkel disorder and relatively high activation enthalpies at lower temperatures, but which go through a gradual transition to fast ionic type of conduction (lower ΔH and σ_o) as they approach their melting points.

Another macroscopic indication of the difference between the fast ionic conductor group and the Schottky- and Frenkel-disorder groups lies in their thermal properties. As early as 1935 Ketelaar (42) showed that the specific heat in the β (low temperature) phase of Ag_2HgI_4 near the β-α transition undergoes an abnormal rise, and that the enthalpy change at the transformation is unusually low. This was interpreted as a disordering of the ions on a dilutely populated set of essentially identical sites. The ionic conductivity also rises appreciably in this temperature range. X-ray diffraction experiments (43) also indicated that the high-temperature α phase of this material (which behaves as a fast ionic conductor) has an "averaged" or disordered arrangement of ions upon the cation sublattice, with three ions distributed at random upon four lattice sites.

A similar conclusion concerning disordered occupancy of a dilutely populated set of cationic sites had also been arrived at by Strock (44,45) for the case of the highly conductive high-temperature α phase of AgI. Both Strock and Ketelaar (46,47) recognized the connection between this microscopic structural feature and the unusually high ionic conductivity of these materials, although Strock (45), who introduced the concept of a molten sublattice, contended that there was a significant difference between the two cases. Incidentally, only four fast purely ionic conductors (α-AgI, α-CuI, α- and β-CuBr, and α-Ag_2HgI_4) were known in 1938 (47).

The existence of unusually large values of specific heat in cation-disordered phases was emphasized by Wiedersich and Geller (48), who pointed out that there are several cases in which the heat capacity at constant pressure decreases with increasing temperature over extensive ranges. Such measurements were reported by Johnston et al. (49) for the fast ionic conductor $RbAg_4I_5$. Thermal effects within the α-AgI phase, possibly resulting from changes in the order parameter, have also been reported (50-52).

The existence of small entropy changes on melting in fast ionic conductors, as well as large entropy changes associated with first-order phase transformations in cases in which one phase (generally the one stable at the higher temperature) is a fast ionic conductor, and the other is not, has been pointed out by O'Keeffe (53-55) and van Gool (56). O'Keeffe has also emphasized thermal evidence for order-disorder reactions in a number of anionic conductors, mentioned earlier as a fourth group in which there is a transition from behavior typical of the Frenkel-disorder group to properties characteristic of the fast ionic conductor class. From the magnitudes of the thermodynamic data, he suggests that the mobile ions in materials that are fast ionic conductors should be thought of as being in a liquid state. As mentioned previously, this molten sublattice model was proposed earlier for α-AgI by Strock (44) on the basis of X-ray diffraction results.

It should be pointed out that thermal effects have been observed in a number of cases, for example, Ag_2HgI_4 (42) and Cu_2HgI_4 and CuI (57), which indicate a high degree of disorder in phases at temperatures below first-order phase transformations. Disordered sublattices are obviously not restricted to high-temperature phases.

Following the early X-ray diffraction work by Strock (44,45) and Ketelaar (43), which showed a large measure of disorder in materials then recognized as fast ionic conductors, other investigators have used more refined techniques to study the details of this phenomenon from a structural point of view.

Some years ago, attention was given to the structure of the copper sublattice in "high chalcocite," which has the nominal formula Cu_2S. This phase is a mixed conductor, since both ionic and electronic species transport electrical charge, and transforms to an ordered "low chalcocite" upon cooling at about 105°C. It was first concluded (58,51) that the cation sublattice was completely disordered in the higher-temperature phase. Careful work by Buerger and Wuensch (60) showed that routine distribution of atoms over the equipoints would not lead to a solution. Refinement of the data showed that the electron density related to the copper sublattice was appreciably smeared out, extending between the three different types of possible copper sites derived from the geometry of the hexagonal close-packed sulfur lattice.

They concluded that, even if anisotropic temperature factors were used, placement of discrete atoms at the sites of twofold, threefold, and fourfold symmetry would not be a satisfactory description of the structure, for the data clearly indicated a continuous copper distribution. The electron density projection that they derived is shown in Fig. 2.

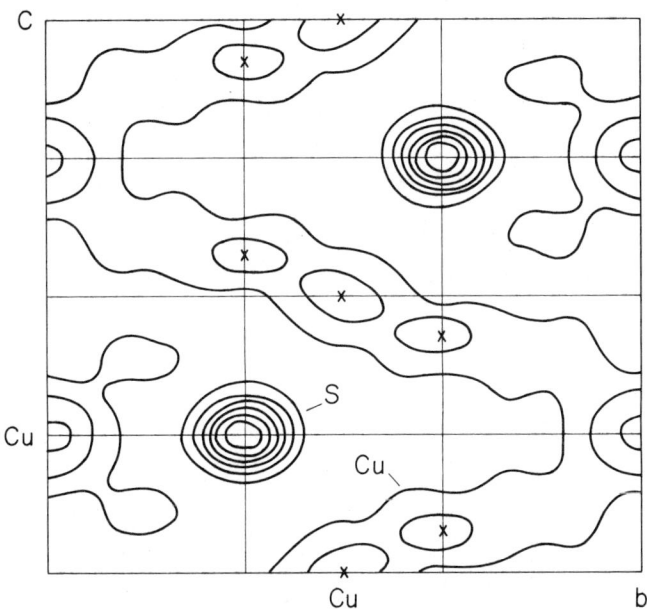

Fig. 2. Electron density section $\rho(o\ x\ y)$ through the orthohexagonal cell for high chalcocite, from Buerger and Wuensch (60).

Work on powdered samples of CuI and CuBr by Miyake and his collaborators (61, 62) showed an anomalous increase in the intensities of certain reflections with temperature. This was interpreted either as a greater Debye temperature factor for cations than for anions, or in terms of an array of four possible minimum-energy positions displaced in the <111> directions from the center of symmetry of the normally assumed copper sites. The occupancy probability then varies with the temperature. Such an explanation was found to fit the entropy data reasonably well. Hoshino (63) also reported similar conclusions from a study of polycrystalline Ag_2HgI_4, and remarked that another possible interpretation might involve asymmetric anharmonic thermal vibration of the cations, with greater amplitudes in the <111> directions.

Single-crystal X-ray diffraction experiments have been recently reported on the high-temperature α phase of Ag_2HgI_4 by Kasper and Browall (63). They found that one could fit the data adequately by the Ketelaar (43) model, but this required the use of large thermal parameters. However, a better fit was obtained by assuming Hoshino's (63) static displacement model. A comparable fit can also be obtained by analysis in terms of anharmonic vibrational motion, as has been found to be present in other crystals (e.g., CaF_2) containing ions at or near sites of $\bar{4}\bar{3}$ m symmetry (64-66) in which tetrahedral displacement effects are indicated. The theory of anharmonic effects, and its application to zincblende structures was presented by Cooper et al. (67), and Kasper and Browall found that their data fit the necessary criteria quite well.

Thus it appears that in at least some fast ionic conductors it is quite reasonable to think in terms of the mobile cations having unusually large vibrational amplitudes in preferred directions, with a considerable degree of anharmonic character. More is said about this later, in connection with a discussion of recent theoretic work on potential profiles in several fast ionic conducting structures (68-70). Some data on the apparent structural location of the mobile ions in the beta alumina structure are also discussed subsequently.

3 THE BETA ALUMINA FAMILY

3.1 Introduction

As mentioned earlier, the discovery of fast ionic conduction in members of the beta alumina family of oxides has evoked a large amount of interest, and a great deal of effort is currently being spent on development programs aimed at its use in advanced battery systems. Some of the available information about this group of materials is reviewed in the next few sections.

3.2 Structures of the Phases in the Beta Alumina Family

Although one finds the lable "beta alumina" widely used in the literature, there actually are several different phases and structures within the beta alumina family, and a wide variety of compositions can occur among these various phases. Present usage gives the labels β, β'', β''', β'''' to the major phases within this group. When the beta alumina phase was first identified by Rankin and Merwin (71,72) in 1916, it was thought to be a metastable form of corundum (α-alumina) with the same composition but a different crystal structure. Later Browmiller and Bogue (73) found that the beta-alumina phase crystallized as a stable structure from samples containing 3.5% Na_2O and 96.5% Al_2O_3. Morey (74) and Brownmiller (75) observed an analogous K_2O- containing beta-alumina phase. Ridgway et al. (76) and DePablo-Galan and Foster

(77) reviewed and extended this work on the formation and identification of the beta aluminas.

The major features of the crystallographic structure of the beta alumina phase, whose nominal formula is $Na_2O \cdot 11\ Al_2O_3$, have been determined by X-ray diffraction experiments in the 1930s by Bragg et al. (78), Beevers and Brohult (79), and Beevers and Ross (80). Subsequently it has become recognized from the work of Yamaguchi (81-83), Yamaguchi and Suzuki (84), Thery and Briancon (85,86), and Scholder and Mansmann (81-88), that a second phase can exist at higher concentrations of some of the alkali metal oxides, M_2O. This phase is generally called β'', and has a nominal formula $Na_2O \cdot 5.33\ Al_2O_3$. Early Japanese workers (81-84) had also claimed that another phase of intermediate composition exists in this system, which they labeled β'. Subsequently, however, the existence of the β' phase has been questioned, and it is generally assumed that their experiments must have involved samples that were a stoichiometric variant of the β phase. Further structural work has been done by a number of authors (89-97).

The two additional phases labeled β''' and β'''' have been reported by Weber and Venero (98) and Bettman and Terner (99) in ternary oxide systems.

As an aside, it should be mentioned that Ampian (100) pointed out that the ASTM cards (10-414 and 1-1300) apparently contained errors in the d spacing for the 002 plane. The proper value should be about 11.3Å. The data listed in the book by Bogue (101) also do not include the low angle lines, probably because the data he reported were obtained using X-ray diffraction equipment and techniques that are now considered obsolete.

The structures of these various beta alumina phases can be described most simply as being composed of blocks, four oxygen layers thick, of close-packed γ-aluminum oxide with a local structure similar to that of spinel, separated by relatively open "bridging" layers containing only oxygen plus the monovalent cation. This arrangement is shown schematically in Fig. 3.

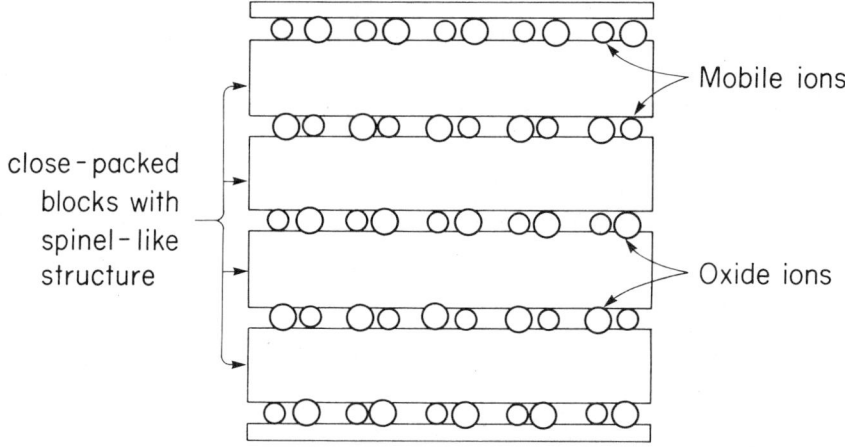

Fig. 3. Schematic representation of the structure of beta alumina showing close-packed blocks with a spinel-like structure separated by bridging layers sparsely populated by M^+ and oxide ions.

The cation arrangement within the spinel-like blocks in the β phase is different from that found in normal γ-alumina, for in the ideal stoichiometric β phase all of the cation sites of the spinel blocks should be occupied, whereas these sites are only two-thirds occupied in the ideal γ-alumina structure.

There are one-fourth as many oxygen ions in the bridging layers as in a normal close-packed plane. They are arranged so that the sodium ions dilutely populate a hexagonal array of parallel tunnels running between them. It is through these crystallographic tunnels that the sodium ions move so readily.

The β" phase differs from the β phase not just with regard to composition. It also has a slightly different structure, the primary difference being the manner in which the spinel blocks are stacked on each other. In the β structure the bridging layers between the spinel blocks are mirror planes, so that the overall structure has hexagonal symmetry. In the case of sodium β-alumina the lattice parameters are a_o = 5.59 Å and c_o = 22.53 Å. The distance between the centers of the bridging layers is therefore 11.27 Å. The β" structure has the same distance between bridging layers, but the spinel blocks are related to each other by rotation about a three-fold screw axis normal to the bridging layer plane. The symmetry is thus rhombohedral, and the c_o value for the sodium β" phase is thus about 33.9 Å. This is sometimes called the three-block structure, as distinct from the two-block structure of the β phase. The details of the atomic arrangement within the bridging layer are also different for both geometric and crystallographic reasons.

The β"' structure is similar to that of the β phase except that the spinel blocks are six, rather than four, oxygen layers thick, and the bridging layers are thus 15.9 Å apart. The β"" structure is analogous to the β" structure except that the spinel blocks are, again, six oxygen layers thick.

Deviations from these ideal structures are discussed shortly.

3.3 Phase Equilibria

There have been a number of reports (92,102-111) concerning phase equilibrium in the sodium oxide-aluminum oxide system. These have led to several different versions of the binary phase diagram, as well as providing some information about related ternary systems (98, 109). These matters are far from settled, and there is no present agreement upon the details of the proper phase diagrams, even in the binary case.

Nevertheless, a number of conclusions can be drawn. Both the β and β" phases can exist over appreciable composition ranges. The β" phase tends to be sodium poor and the β phase is not stable in the binary system above about 1550°C. However, ternary additions of MgO or Li_2O can increase the decomposition temperature substantially (98). The general features of the high-temperature region of the binary phase diagram are illustrated in Fig. 4, which is taken from DeVries and Roth (103).

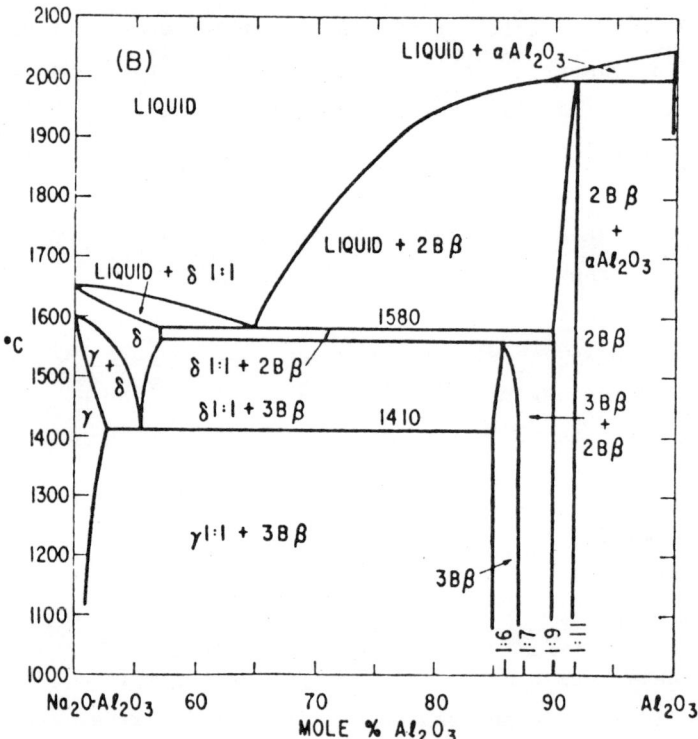

Fig. 4. Version of the $NaAlO_2$-Al_2O_3 phase diagram given by DeVries and Roth (103).

Experiments based upon the use of X-ray diffraction (107,110, 111), and transmission electron microscopy and diffraction methods

(112) have shown evidence for coherent intergrowth of β and β" under some conditions. There has been some controversy (113), however, about the generality of these observations.

Recent experiments involving the low-temperature decomposition of salt-infiltrated polymers (114,115) have shown that an additional phase, called λ-$Na_2O \cdot x\ Al_2O_3$, is formed over a wide compositional range. This phase has a structure similar to that of mullite, and appropriate compositions readily transform to the β" structure at temperatures as low as 1000 to 1100°C.

The influence of pressure on the stability of both the β and β" phases was investigated by Roth (116). He found that both phases decompose to form α-alumina (which is about 14% more dense) as well as the other unidentified products at relatively low temperatures. At a pressure of 5 kbar they both decomposed below 750°C. At 30 kbar, the β phase decomposed at 440°, while the β" phase was found to be stable to 625°C. Their experiments went to 60 kbar, and even in that range the β" phase was found to be more stable under pressure.

3.4 Ionic and Electronic Transport

In 1967 Yao and Kummer (23) showed that the mobile ions in the bridging layer of the beta alumina structure could be exchanged with a number of other monovalent cations by immersion in appropriate molten salts. It has also been shown (92) that this can be done by electrochemical pumping.

Experiments using radiotracer techniques for determining the self-diffusion coefficients of several different monovalent (M^+) ions in the beta alumina structure in the temperature range 200 to 400°C were reported by Yao and Kummer (23) and Radzilowski et al. (24).

The dielectric loss method was also used to evaluate ionic motion in polycrystalline β alumina containing various different M^+ ions by Radzilowski et al. (24). In some cases they found quite good agreement between these results, which were obtained from experiments at relatively low temperatures, and the data obtained from the higher-temperature radiotracer experiments, whereas in others there was a considerable discrepancy in the value of the apparent activation energy.

Measurements of the ionic conductivity of polycrystalline samples of members of the beta alumina family using ion-blocking electrodes and alternating current techniques have been reported by a number of authors. However, the results have typically shown a frequency dependence, and comparison between the results of experiments by different investigators has been disappointing. A step forward was the introduction (117) of the use of mixed conductors as ionically reversible electrodes. In these materials the M^+ ion has a high chemical diffusion coefficient, and by using this technique, frequency-independent values of the ionic conductivity can be obtained over a wide range of temperature. Using properly selected tungsten and vanadium bronze phases as electrodes, reliable data have been obtained for the ionic conductivity of single crystals of beta alumina containing sodium (117), potassium, thallium, and lithium (26) from over 800°C to well below ambient temperature. Comparable results were obtained (118) on samples containing silver ions by the use of silver electrodes. Arrhenius plots of these measurements of the conductivity are linear over an unusually wide temperature range. As an example, the data for sodium beta alumina from 820 to below −150°C are shown in Fig. 5.

It is expected that the information acquired by the use of these three methods, radiotracer self-diffusion, dielectric loss,

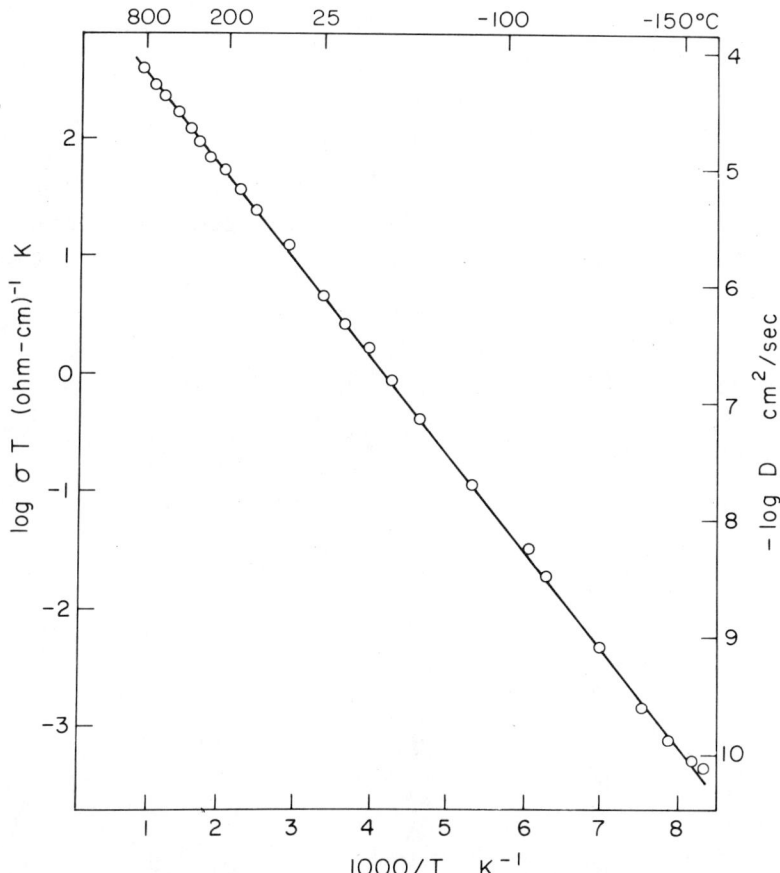

Fig. 5. Temperature dependence of the ionic conductivity of single crystalline sodium beta alumina from Whittingham and Huggins (117).

and ionic conductivity, should be consistent, because they all measure the same phenomenon, ionic transport. That this is actually the case over an extremeley wide temperature range is illustrated in Fig. 6, in which ionic conductivity data were calculated from the radiotracer and dielectric loss results measured by the

Ford group and plotted along with the directly measured single-crystal conductivity values obtained at Stanford for the case of sodium beta alumina. This is easily the greatest range of (1/T) over which either ionic or atomic transport has been measured in any solid. The values extracted from the radiotracer diffusion data by use of the Nernst-Einstein relation are expected to be

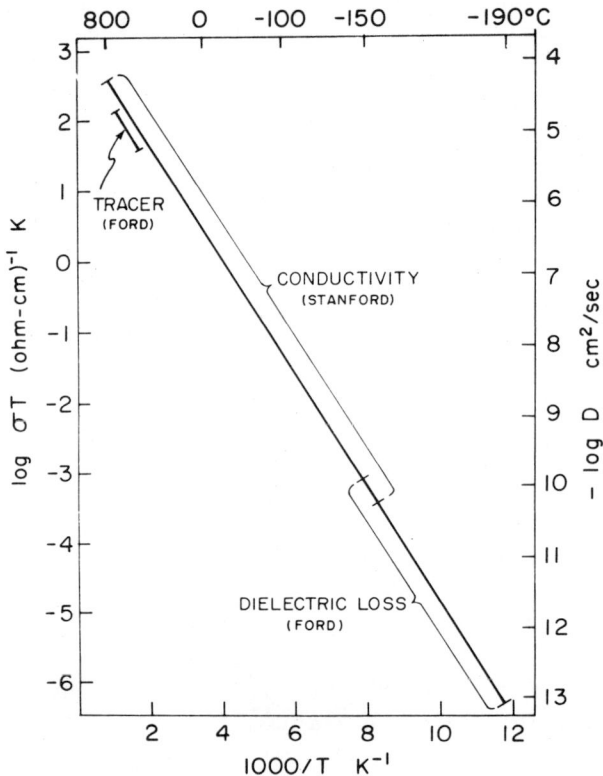

Fig. 6. Temperature dependence of the ionic conductivity and self-diffusion coefficient of sodium beta alumina, as determined by the use of three different techniques, from Whittingham and Huggins (26). Radiotracer data from Yao and Kummer (23), conductivity data from Whittingham and Huggins (117), and dielectric loss data from Radzilowski et al. (24).

somewhat lower than the true conductivity values, as shown in that figure, because of correlation effects. From the magnitude of this difference, the value of the Haven ratio (sometimes called the correlation factor), $H_R = D_T/D_\sigma$, can be calculated. This allows conclusions to be drawn concerning the detailed jump mechanism involved in the transport process.

Similar comparisons of the data for beta alumina containing silver and potassium, thallium, and lithium in the bridging layer have been presented elsehwere (26, 119). In the case of potassium beta alumina, although the slopes of the radiotracer and dielectric loss data are quite difference, their actual magnitudes are reasonably consistent, and they are tied together quite well by the direct conductivity measurements on single crystals.

The ionic conductivity data obtained on single crystals containing these five different M^+ ions by the reversible electrode method are shown in Fig. 7, and the parameters tabulated in Table 1, assuming an Arrhenius relation of the form $\sigma = (\sigma_o/T) \exp(-\Delta H/RT)$. Also shown are the calculated values of the Haven ratio where they are available. In the case of silver, sodium, and thallium beta alumina they were found to be very close to the theoretic value of 0.6 calculated (118) for a noncolinear interstitialcy (indirect interstitial) mechanism in the bridging layer of the beta alumina structure.

The lithium-containing material is evidently somewhat different from the others, since the conductivity data were found to fall upon two connecting straight lines, meeting at about 180°C. The conductivity and radiotracer data essentially coincide in the high-temperature region, giving comparable values of the activation energy. If these data are correct, the Haven ratio is unity in this region, implying that the very small lithium ion is moving primarily by a direct interstitial mechanism in this temperature

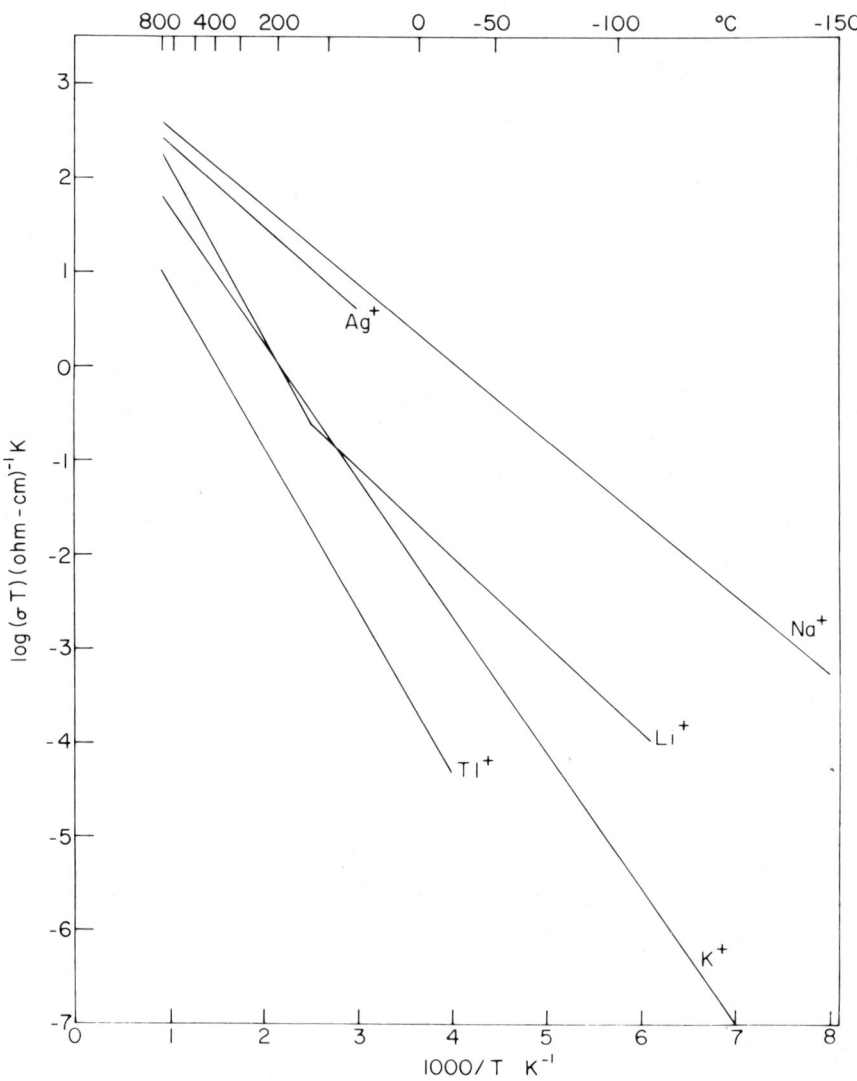

Fig. 7. Variation with temperature of the conductivity of single crystal beta alumina containing sodium, silver, lithium, potassium, or thallium in the bridging layer, from Huggins (122).

Table 1. Ionic Conductivity of Single-Crystal β Alumina Containing Different Mobile Ions

Ion	Temperature range (°C)	σ_o ((ohm-cm)$^{-1}$K)	Activation Enthalpy (kJ/Mole)	Conductivity at 25°C (ohm-cm)$^{-1}$	Correlation factor (D_T/D_σ)
Ag	25 → 800	1.6×10^3	16.6	6.7×10^{-3}	0.61
Na	-150 → 820	2.4×10^3	15.8	1.4×10^{-2}	0.61
K	-70 → 820	1.5×10^3	28.4	6.5×10^{-5}	--
Tl	-20 → 800	6.8×10^2	34.3	2.2×10^{-6}	0.58
Li	180 → 800	9.7×10^3	35.8	--	1.0
Li	-100 → 180	5.4×10^1	18.0	1.3×10^{-4}	--

range, rather than by an interstitialcy mechanism. Because of its lower temperature dependence, another process, perhaps the normal interstitialcy mechanism, evidently dominates the conductivity at lower temperatures. A further complication in lithium-substituted beta alumina is the observation that annealing at temperatures above 800°C causes some structural change -- perhaps diffusion of the lithium into the spinel blocks. This results in an irreversible decrease in the conductivity.

As a result of these transport measurements, it appears that there is an optimum ionic radius for achieving the maximum mobility in the beta alumina structure, with both smaller and larger ions less mobile than those of intermediate size. The importance of this size factor was convincingly confirmed by experiments reported by Radzilowski and Kummer (120) that compared the influence of hydrostatic pressure on the resistivity of beta alumina containing lithium, sodium, and potassium ions. At higher pressures the resistivity of lithium β decreased, whereas it increased for potassium β, and was essentially unchanged in the sodium case, as shown in Fig. 8. A theoretic approach that produces this same type of ionic size dependence is discussed in a later section.

Because of the interest in their potential use as solid electrolytes, it is also important to know the electronic transference number, the fraction of the total conductivity carried by electronic, rather than ionic, species. This has been measured in the case of single-crystal silver beta alumina (118) using the Wagner asymmetric polarization method (121) from 555 to 790°C and over a wide range of oxygen partial pressure. The value of the electronic transference number was found to be about 3×10^{-5} at 750° and 10^{-6} at 560°C. It was also found that below about 750°C changes in the ambient oxygen partial pressure do not appreciably influence the electronic conductivity. Presumably, except for the M^+

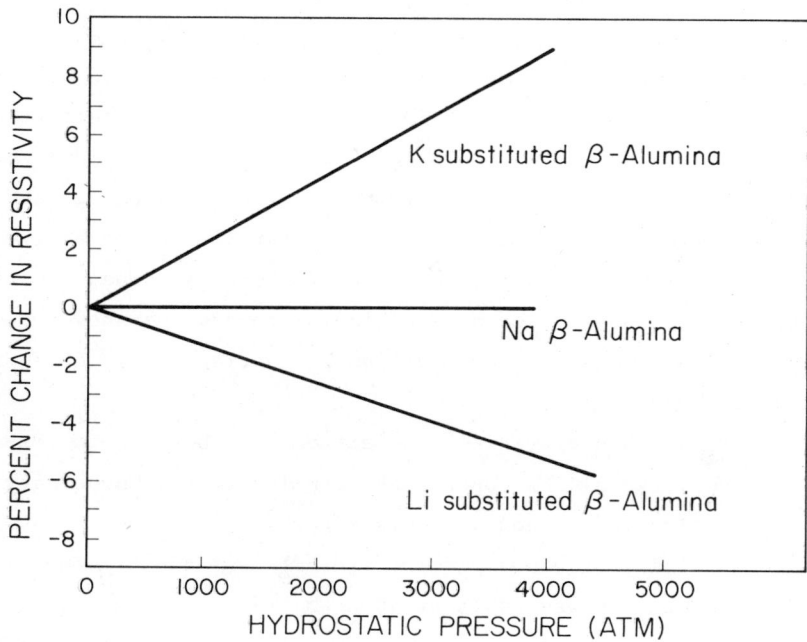

Fig. 8. Influence of hydrostatic pressure on the ionic resistance of beta alumina containing potassium, sodium, and lithium ions, from Radzilowski and Kummer (120).

ions, the structure is effectively "frozen" below that temperature. If one can extrapolate these data to lower temperatures, it is obvious that the beta aluminas should have extremely low values of electronic conductivity at lower temperatures where their ionic conductivity makes them a very attractive solid electrolyte if they do not contain readily reducible species.

An example in which a reducible species was introduced was the work on copper beta alumina (92,122) produced by ion exchange driven by electrochemical pumping as well as the molten salt equilibration method.

The conductivity of samples of copper beta alumina was found to depend strongly upon the oxygen activity, which could be changed by equilibration in different environments at temperatures in the vicinity of 850°C, followed by rapid cooling to lower temperatures. Samples equilibrated with air gave a linear conductivity plot, with an activation energy of approximately 5 kcal/mole. Samples equilibrated under reducing atmospheres and measured at lowered temperatures under nonequilibrium conditions showed much lower values of conductivity, with an activation energy of approximately 25.9 kcal/mole.

The partial conductivities of electrons and holes were determined on polycrystalline copper beta alumina by the Wagner asymmetric polarization method and this material was found to have a hole conductivity of about 5×10^{-3} $(\Omega\text{-cm})^{-1}$ at 850°C. The electronic conductivity was lower by about 5 orders of magnitude. Thus, under these conditions, copper beta alumina has a small concentration of excess holes, whereas the measurements on silver beta alumina (118) indicated a minor amount of electronic conductivity.

It is generally found that the ionic conductivity of the β'' phase is higher than that found for the β materials, although well-defined and reproducible data for different ions in this structure are not yet available. Some early values for sodium β'' samples can be found in the papers by Kummer (109) and Whittingham and Huggins (26).

Beta alumina shapes being produced in several of the technologic programs aimed at the development of sodium/sulfur batteries have microstructures containing both the β and β'' phases, and a great deal of effort is going into the optimization of phase distribution, grain size, and composition, as well as production techniques.

3.5 Structure of the Bridging Layer

The results of these various experiments on ionic transport, as well as observations concerning the apparent wide compositional range over which these phases are stable, have caused attention to be focused on the detailed structure of the bridging layer in beta alumina.

It has been found by several investigators that water intercalates into the bridging layer of at least some beta alumina samples (123, 24, 124, 125, 109). Although it is not clear whether the water is present as neutral H_2O molecules or as H_3O^+ ions, it is obvious that its presence impedes the motion of the other cations present. There have been conflicting reports of the susceptibility of dense bodies of the beta alumina phases to damage by water absorption from the atmosphere. It is generally believed that the β" structure is more susceptible to water pick-up than is the β structure. Lithium- and sodium-containing beta aluminas seem to be more hygroscopic than those containing other ions. However, the major part of this hygroscopic behavior may be the result of the presence of second phases, such as $NaAlO_2$, in the grain boundaries, rather than an intrinsic property of the beta alumina (126).

As in the case of the silver and copper compounds investigated much earlier and discussed in Section 2, the M^+ ions within the bridging layer in the beta aluminas show a large amount of disorder. This was clearly shown by the X-ray diffuse scattering results reported by LeCars et al. (127), who, incidentally, saw signs of some ordering at very low temperatures (77°K). This disorder has also been shown by single-crystal X-ray diffraction studies of sodium beta alumina by Peters et al. (93) and of silver beta alumina by Roth (95). Preliminary neutron diffraction results of Reidinger on sodium beta alumina have also recently been reported by Roth (116).

The results of these studies can be understood by reference to Fig. 9. In all cases, results indicate that the electron density related to the M^+ ions is quite smeared out, characteristic of large-amplitude anharmonic vibration, rather than well-defined sites.

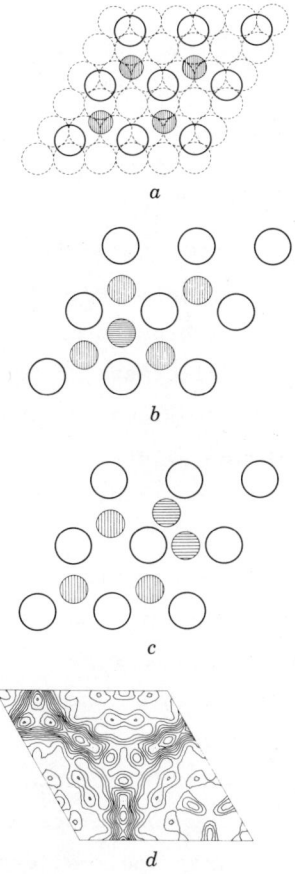

Fig. 9. Structure of the bridging layer in beta alumina. (a) Ideal arrangement, with M^+ ions (cross-hatched) in normal Beevers-Ross sites; oxide ions in bridging layer indicated by heavy circles. (b) Schematic drawing showing presence of extra M^+ ion in alternate Beevers-Ross site. (c) Schematic drawing showing presence of extra M^+ ion accommodated by two ions at in-line sites. (d) Results of neutron diffraction study of sodium ion distribution at 600°C (116).

In the case of silver beta alumina, it was found that the "normal" or "Beevers-Ross" sites, which have trigonal prismatic coordination with oxygens in the adjacent spinel blocks, appear to be about 70% occupied, and the "alternate," or "anti-Beevers-Ross" sites, which lie between pairs of oxygens in the two adjacent spinel blocks, are about 50% occupied.

The distribution in the case of sodium beta alumina is different, in that at low temperatures ($\sim$ 95°K) the sodium ions are localized primarily at the normal sites and at the 6-coordinated positions directly in line with pairs of oxygens in the bridging layers, rather than at the "alternate" sites. As the temperature is raised, there is increased delocalization and also a partial occupation of the "alternate" sites -- a liquid-like distribution. Ionic locations in the bridging layer have also been studied by Feldman et al. (128).

One of the observations made by many workers in the field is that samples of the β phase appear to be quite sodium rich, often containing 15 to 30% excess sodium compared to the nominal formula. Obvious questions are where this excess sodium is in the structure and how its charge is compensated.

The answer to the location of the extra sodium (or other M^+ ions) must surely be that they contribute to the observed electron density distributions, and that their site preference depends on their identity, at least at lower temperatures. In some cases, the "alternate" sites may be preferred, whereas in others the in-line sites are more probable.

The answer to the charge compensation question is more cloudy. It was argued some time ago (118) that the lack of sensitivity of the ionic conductivity to changes in oxygen partial pressure at moderate temperatures means that the excess M^+ ions must be balanced by a small number of aluminum vacancies in the spinel blocks.

On the other hand, recent neutron diffraction results (116) have indicated the presence of excess (interstitial) oxygen ions in the bridging layer, adjacent to interstitial aluminum ions (due to Frenkel disorder) in the spinel blocks.

4 ANALOGUES OF THE BETA ALUMINAS

There are several materials that have structures that are either the same or closely allied to those of the beta alumina family, and should thus have analogous rapid transport of ionic species.

Braun reported that $KFe_{11}O_{17}$ has the same structure as beta alumina, according to Gorter (129), and large values of alkali ion mobility in the hexagonal sodium and potassium ferrites with this formula were found by Hever (130). The situation is actually more complicated, as X-ray and magnetic studies in the potassium ferrite system (131,88,132,133) have indicated the presence of several phases of slightly different composition, as is the case for the beta alumina family. Structural and conductivity work on both polycrystalline and single-crystal samples of the two-block and three-block forms was reported by Roth and Romanczuk (134). They found, as expected, that these materials conduct electric charge by the transport of both ionic and electronic species, the magnitude of the electronic portion being dependent upon the oxygen partial pressure.

There is also a family of gallium compounds that are analogous to the beta aluminas, with gallium, instead of iron, replacing the aluminum in the spinel blocks. This was pointed out by the work of Foster and Stumpf (135). A tentative phase diagram for the $Na_2O-Ga_2O_3$ system was presented by Boilot et al. (136), and Foster and Scardefield (137) recently discussed the preparation of these gallium compounds by growth from molten fluxes. Both structural and conductivity measurements on polycrystalline samples of mixed gallium-aluminum phases were reported by Brinkhoff (138).

5 OTHER MATERIALS THAT EXHIBIT FAST IONIC CONDUCTION

5.1 Introduction

In addition to the beta aluminas, many materials are now known to exhibit high mobility of ionic species. This phenomenon is extremely structure dependent, and fast ionic conductors are all characterized by crystallographic structures which, if modeled by the use of hard spheres of appropriate radii to represent the constituent ions, can be seen to contain pathways of connected space (tunnels) with lateral dimensions comparable to ionic diameters. Because of this geometric feature of their crystal structures, the long-range motion of ions from point to point can occur in such materials without the necessity of significant ionic displacements. Related strain energy effects provide the primary contribution to the activation energy for motion in close-packed metallic systems.

In this section a number of the recognized fast ion conductors are discussed. They are divided into two categories, those in which ionic conductivity is predominantly due to cationic transport, and those that are primarily anion conductors.

5.2 Cation Conductors

5.2.1 MATERIALS WITH CUBIC STRUCTURES As mentioned earlier, the high-temperature (alpha) structure of AgI has been known for some time to have an unusually high value of ionic conductivity. The early results of Tubandt and Lorenz (41) have been confirmed more recently by Kvist and Josefson (139). In this material the iodide ions exist in a body-centered cubic lattice. The silver ions do not have well-defined positions, and the experimental data have sometimes been interpreted in terms of a distribution of the silver ions among the three types of sites within the tunnels that run in the cube edge directions. There are many more possible sites for the silver ions than the number of ions present, and the large Debye-Waller factors obtained from the diffraction data imply unusually large mean displacements (from 0.2 Å at 150°C to 0.5 Å at 450°C) of the atoms from the assumed crystallographic positions. There are several materials with this same crystal structure; they are listed in Table 2.

A group of compounds having the general formula Ag_3SX, where X can be either iodine or bromine, have been studied by several investigators (145-153). The high-temperature alpha form of Ag_3SI also has the same structure, except that half of the iodine ions are randomly replaced by sulfur ions. The silver ions are distributed in the tunnels, and this material also has an unusually high value of silver ion conductivity.

In the case of the beta structure of Ag_3SI, and in Ag_3SBr, the anions are ordered upon the body-centered cubic lattice, forming an anti-perovskite structure. This ordering appears to reduce the mobility of the silver ions within the tunnels somewhat. It has also been shown that beta Ag_3SI and Ag_3SBr exhibit a continuous solid solution range.

A small electronic contribution to the conductivity has been found in Ag_3SI. The data reported for this partial electronic conductivity have not been consistent, the value evidently

Table 2. Phases with Crystal Structures in Which the Anions are Arranged in a Body-Centered Cubic Lattice

Phase	Stability range (°C)	Lattice constant (Å)	Temperature (°C)	Ref.
α – AgI	146 – 555	5.058	146	44, 45
α – Ag$_2$S	179 – 825	4.88	250	140
α – Ag$_2$Se	133 – 880	4.98	250	140
α – CuBr	472 – 490	4.59	485	141, 142
(Na,Li)$_2$SO$_4$ [a]	521 – 600	5.77	556	143
(Ag,Li)$_2$SO$_4$ [a]	417 – 574	5.76	545	144
α – Ag$_3$SI	245 – >400			
β – Ag$_3$SI	20 – 235			
Ag$_3$SBr	0 – 300			

[a] Data are given for equi-molar compositions. These phases are stable over a wide range of cation ratios.

depending on the thermodynamic conditions imposed during sample preparation and measurement.

The high-temperature alpha phase of the compound Ag_2HgI_4, also mentioned earlier, has a structure (43,46,154,63) that is a modification of the zincblende arrangement in which two of the silver ions are replaced by one Hg^{2+} for charge balance, producing a large concentration of empty sites. The geometry of this structure can also be described in terms of the partial occupation by cations of the tetrahedral interstices in a face-centered cubic anion lattice. Motion of the silver ions among the tetrahedral sites is very easy, leading to the high ionic conductivity.

There is also a copper analog, Cu_2HgI_4, and conductivity data have been reported (155) for a solid solution of the silver and copper phases, $(Ag,Cu)HgI_4$. It has also been found (156,141,61) that the α-Ag_2Te, α-CuI, α-Cu_2S, and α-Cu_2Se phases consist of face-centered anion arrangements, with the cations distributed in various ways among the interstitial sites. Lithium sulfate also has a face-centered anion arrangement at high temperatures, and its ionic conductivity was investigated by Kvist and Lunden (157). It has been inferred from conductivity measurements (158) that lithium tungstate probably forms a cubic or psuedo-cubic high-temperature phase. These phases, along with information about the temperatures at which they are stable, are listed in Table 3.

Another group of cubic materials having related structures is based upon the alkali silver iodides with the nominal formula of Mg_4I_5, where M^+ is Rb^+, K^+, or NH_4^+. The crystal structure of the primary member of this group, $RbAg_4I_5$, was found (159,160) to be cubic with four formula units per unit cell. The four rubidium ions are surrounded by distorted iodine octahedra. There are 56 tetrahedral interstitial positions that might be occupied by the

Table 3. Phases with Crystal Structures in Which the Anions are Arranged in a Face-Centered Cubic Lattice

Phase	Stability range (°C)
$\alpha\text{-}Ag_2HgI_4$	50-93
$\alpha\text{-}Cu_2HgI_4$	67-90
$\alpha\text{-}(Ag,Cu)HgI_4$	
$\alpha\text{-}Ag_2Te$	>150
$\alpha\text{-}CuI$	407-600
$\alpha\text{-}Cu_2S$	>91
$\alpha\text{-}Cu_2Se$	>110
$\alpha\text{-}Li_2SO_4$	574-860
$\alpha\text{-}Li_2WO_4$	684-738

16 silver ions present in the structure. The distribution of the silver ions among the three slightly different types of positions were discussed by Wiedersich and Geller (48).

Although the related α-AgI structure is stable only above 146°C, the presence of the large M^+ ions evidently prevents this rather open structure from collapsing until a considerably lower temperature is reached. In the case of KAg_4I_5, the cubic high-temperature structure transforms to one of lower symmetry at -136°C, with an accompanying decrease of a factor of 250 in the ionic conductivity. There is a narrow temperature range (257 to 332°C) in which the analogous potassium copper compound KCu_4I_5 is stable.

A number of related silver ion-conducting materials of the MAg_4I_5 type have also been described in which the M^+ ion is replaced by an organic group (35,161).

There are also a number of materials in which some of the iodine ions are replaced by other large anions. Examples include $Ag_4HgI_2Se_2$, (162), $Ag_7I_4PO_4$, which is stable up to 79°C (163), $Ag_{19}I_{15}P_2O_7$, which is stable to 147°C (162,163), and $Ag_6I_4WO_4$, which melts incongruently at 298°C (164). These are essentially pure silver ion conductors, but have conductivity values that are not quite as high as those found for the MAg_4I_5 phases.

Recent work (165,166) initiated at the Lincoln Laboratory has shown that materials having the cubic $KSbO_3$ structure (167,165) with cations in tunnels running through a skeleton composed of $\overline{SbO_3}$ ions exhibits high values of sodium and silver ion conductivity. In the sodium analog the structure was stabilized by the incorporation of NaF, the fluorine ions evidently going into the structure at the sites where the cation-containing tunnels intersect.

Another group of cubic materials with interesting properties includes the orthosilicates and orthogermanates. Work on Li_4SiO_4 (168,169) has shown it to have the highest value of lithium ion conductivity found in any pure ionic conductor to date, with the exception of lithium beta alumina, which may be metastable. Recent experiments on the germanium analog, Li_4GeO_4 (170), have shown that it is not as good a conductor for lithium ions.

5.2.2 MATERIALS WITH HEXAGONAL STRUCTURES Both Ag_2SO_4 and K_2SO_4 have been found to have hexagonal structures. In the case of K_2SO_4, the hexagonal structure is stable from 584°C to the melting point at 1068°C (171,172).

5.2.3 MATERIALS WITH UNIDIRECTIONAL TUNNEL STRUCTURES Many materials are now known in which the structure is comprised of cation-centered octahedra which share corners or edges. These BX_6 groups are often arranged such that there are parallel open tunnels of ionic size running through the structure between them. In a number of cases these tunnels may be either partially or fully occupied by cations, typically monovalent or divalent species.

It has been found that these tunnel-resident cations are evidently mobile in some cases, and a considerable amount of effort is going into their study. However, data on such materials are currently scattered and contradictory. Part of the problem is surely that many of these materials are mixed, rather than purely ionic, conductors.

Materials in this general class that have been getting attention include those with the tetragonal and hexagonal tungsten bronze structures, the tetragonal hollandites, doped rutile, and several chain-structure vanadium oxides and silicates. The ternary titanium oxide analog of the manganese oxide ramsdellite is also being studied, and gives evidence for high lithium ion mobility (173-175).

It remains to be seen whether any of these materials will become important ionically conducting solid membranes.

5.2.4 MATERIALS WITH LAYER STRUCTURES An increasing number of materials are being investigated that have layer-type crystallographic structures. In some of these it is evident that certain species can have high values of ionic mobility in the layer plane. In cases such as the beta aluminas, these highly mobile species are cations; in other cases they are anions. A considerable amount of attention was given to the beta alumina family in Section 3. Among other materials that have been reported are silver-conducting organic iodide compounds (176) that have relatively rapid ionic motion in two dimensions. Much of the current interest lies in ternary transition metal oxides and chalcogenides and in graphite compounds, both of which are mixed ionic and electronic conductors. These are not discussed here.

Layer-structure anionic conductors are discussed briefly in the next section.

5.3 Anion Conductors

5.3.1 MATERIALS WITH CUBIC STRUCTURES There are a number of materials that have the fluorite (CaF_2) structure in which the ionic conductivity, which occurs by the transport of anions, can reach very high values at elevated temperatures. These fall into two general classes, oxides and fluorides. As mentioned earlier, it has long been known that ZrO_2 is a useful solid electrolyte, the moving species being the oxide ion. Ure (177) studied the ionic conductivity of both pure and doped CaF_2. His results indicated that this material can be employed as a useful fluoride ion-conducting solid electrolyte. Since that time there have been a number of studies of both these and other oxide and fluoride materials, and they have been used as solid electrolytes or electrochemical transducers in an increasing number of applications at high temperatures.

The fluorite structure can be represented as a face-centered cubic arrangement of the cations, with the anions residing in the tetrahedral interstices at low temperatures. Ionic transport takes place by anions moving among these anion positions by a vacancy diffusion mechanism, and it is well known that doping in such a way as to increase the vacancy concentration substantially enhances the ionic conductivity. Interstitial anions can also exist in this structure, and they are mobile as well.

Although the magnitude of the conductivity can be very high in some materials in this group its temperature dependence is very large. As a result, the conductivity does not attain appreciable values until rather high homologous temperatures are reached, and such materials are probably not properly classified as fast ionic conductors at lower temperatures. They are, therefore, discussed only briefly.

The oxide ion conductors that have attracted the greatest interest are those based upon ZrO_2, ThO_2, and CeO_2. These are typically doped with 5 to 15 mole % of another oxide (e.g., CaO or Y_2O_3) to increase the anion vacancy concentration and thereby enhance ionic conductivity. However, ionic transport in these materials is not actually as straightforward as this simple model implies because of the high concentrations of vacancies that result from such large numbers of aliovalent cations. There have been reports of time-dependent changes in the transport properties at temperatures around 800 to 1000°C that may relate to ordering phenomena, phase transformations, or the presence of extended defects, and so on. There have been several quite exhaustive reviews and discussions of the structure and properties of pure and doped ThO_2 and ZrO_2 (178-181) and they are not reiterated here.

Doped CeO_2 has been found (182-185) to have higher values of ionic conductivity than the other two, particularly at lower temperatures. However, the electrolytic domain, the range of oxygen partial pressure within which it is predominantly an ionic conductor, is considerably narrower. For the case of CeO_2 doped with 5 mole % Y_2O_3 the electrolytic domain extends down to $P_{O_2} = 10^{-13}$ at 600°C, but to only 10^{-2} atm at 1000°C (185).

Within the last few years Takahashi and his co-workers have shown (186,187) that oxide solid solutions based upon Bi_2O_3 exhibit rapid transport of oxide ions. Previously, Rao et al.

(188) had shown that Bi_2O_3 itself goes through a phase transformation at 730°C, and that the high-temperature (cubic) phase has a very high ionic conductivity. Takahashi et al. (187) found that they could stabilize the face-centered cubic structure by the addition of 15 to 40 mole % of oxides such as Y_2O_3, Gd_2O_3, WO_3, or Nb_2O_5. These solid solutions have high values of ionic conductivity to quite low temperatures. Information about the possibility of mixed conduction under reducing conditions has not yet been published.

5.3.2 MATERIALS WITH LAYER STRUCTURES There are also a number of anion conductors having layer structures; in some cases they have the characteristics of fast ionic conductors. The primary examples are the materials with the tysonite structure (189) such as LaF_3. In this structure, the lanthanum and one of the fluorines form hexagonal layers similar to those formed by boron and nitrogen in hexagonal boron nitride. The other fluorine atoms lie in planes between these hexagonal layers and are evidently quite mobile.

The tysonite structure is formed by a number of the fluorides of the lighter lanthanide group elements, as well as some of the actinides. Several ternary compounds with the general formula AB_2F_8, where A is an alkaline earth element and B a lanthanide or yttrium, can also have this structure (190).

Nuclear magnetic resonance experiments (191,192) showed that the fluorine ions in the fluorine-only layer are very mobile at low temperatures. At 200 to 250°C, interchange between the fluorines in the two types of positions begins to occur, and thus all the anions contribute to the conductivity.

Ionic conductivity measurements on tysonite structure fluorides (193-196) have been somewhat contradictory, although all indicate very high values extending down to ambient temperatures. Experimental and interpretational problems arose because of an apparent frequency dependence in certain temperatures ranges, aging effects, and the possibility of oxygen contamination.

The trifluorides of yttrium and some of the heavier lanthanides with the hexagonal tysonite structure have been reported to transform to an orthorhombic structure upon cooling at high temperatures (197-199). However, recent evidence (200) indicates that in these materials the tysonite phase is stabilized by the presence of OH^- ions. Crystals grown from the melt in reactive atmospheres designed to minimize OH^- concentrations solidify directly into the orthorhombic structure. Ionic conductivity measurements on some of the tysonite materials have also been reported recently (201), and typical fast ionic conductivity is found at temperatures in which that phase is present.

It was mentioned earlier that solid solutions based upon Bi_2O_3 can have quite high values of oxide ion conductivity. Takahashi et al. (186) reported that a solid solution of 20 to 40 mole % SrO in Bi_2O_3 has a rhombohedral structure, as previously found by Sillen and Aurivillius (202), with good ionic conductivity. Although the crystallographic data were not presented, it seems reasonable to assume that this structure has a layer-like character, with anisotropic transport.

Data for the ionic conductivity of a group of oxide ion-conductors are shown in Fig. 10, and for several fluoride ion conductors in Fig. 11.

6 ANALOGOUS BEHAVIOR OF INTERSTITIAL SPECIES IN METALS

It has long been known that certain elements diffuse with unusual rapidity in some metals. The primary example of this is the

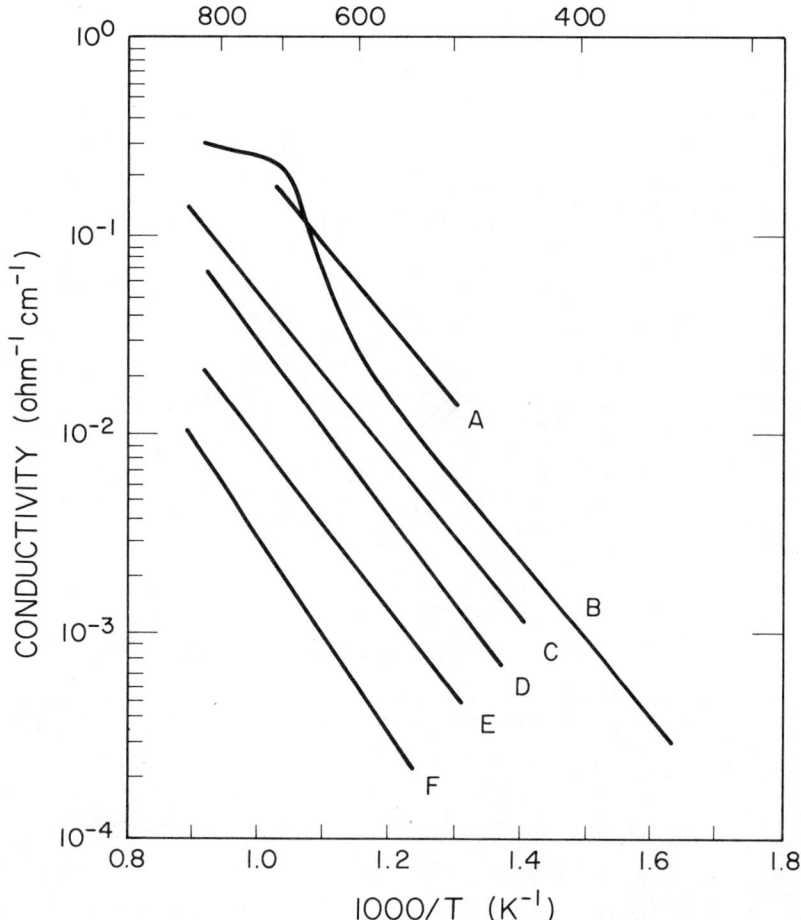

Fig. 10. Ionic conductivity of several oxide-ion conductors: (a) Bi_2O_3-15 M % Nb_2O_5 (187), (b) Bi_2O_3-20 M % SrO (186), (c) CeO_2-5 M % Y_2O_3 (185), (d) ZrO_2-8 M % Yb_2O_3 (229), (e) ZrO_2-9 M % Y_2O_3 (229), (f) $CaTiO_3$-12.5 M % Al_2O_3 (230).

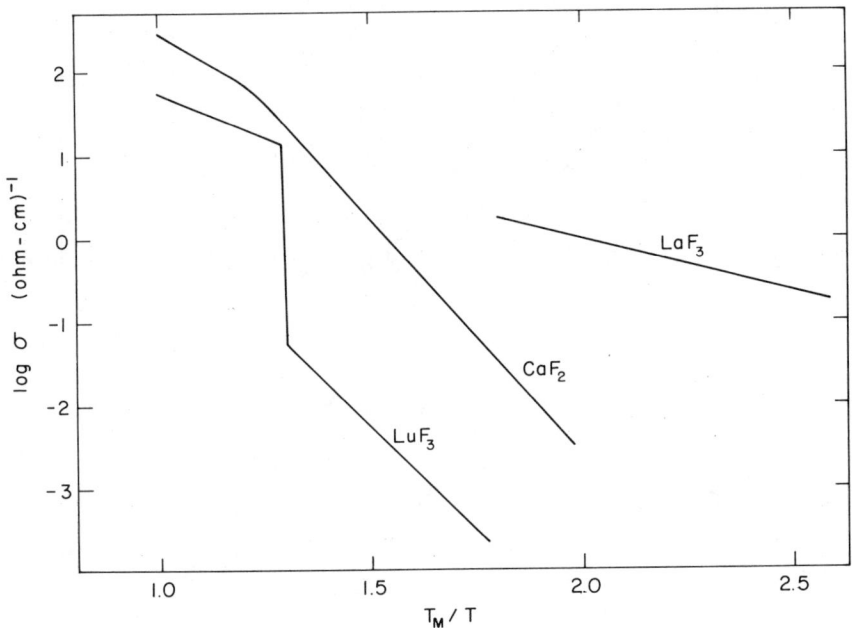

Fig. 11. Temperature dependence of the ionic conductivity for several fluoride ion conductors, from O'Keeffe (201). Note the use of a homologous temperature scale.

transport of hydrogen through palladium, which has led to the technologic use of palladium as a hydrogen-selective membrane or filter. It is used, for example, for the production of unusually high-purity hydrogen.

Another metal within which several elements are known to diffuse rapidly is iron. The transport of carbon in iron alloys over appreciable distances within the microstructure is importantly involved in the mechanisms of phase transformations and the structural rearrangements that occur during the heat treatment of steels. This phenomenon thus has immense technologic importance, and has been studied by a great number of investigators.

Hydrogen moves even more rapidly in steels, and it is now well recognized that its presence is directly related to low-temperature embrittlement problems in many steels, although the exact mechanism whereby this embrittlement occurs, and the details of the role of hydrogen, have not yet been elucidated. This area is also getting a great deal of attention from the research community at the present time.

These matters are mentioned here because it is now apparent that several of the characteristics of the diffusion of hydrogen in metals are quite analogous to the very rapid transport of certain species in ionic crystals. As in the latter case, the crystal structure of the "host" lattice is of primary importance. From available data, one sees that hydrogen diffuses in body-centered cubic metals very rapidly at unusually low temperatures, and the parameters in the Arrhenius equation that describe the experimental data are very similar to those found in fast ionic conductors. The pre-exponential factor D_o and the activation enthalpy ΔH are both unusually low. Available experimental data on hydrogen diffusion in metals were reviewed by Birnbaum and Wert (203) and more recently by Völkl and Alefeld (204).

Although hydrogen also moves rapidly through metals with the face-centered cubic structure, the parameters for the diffusion in such cases are somewhat different from those in body-centered cubic metals, the pre-exponential and the activation enthalpy both being greater in the face-centered cubic case.

Table 4 shows the parameters in the Arrhenius relation $D = D_o \cdot \exp(-\Delta H/RT)$ relating to the diffusion of hydrogen in a group of body-centered metals, as well as in a few important face-centered cubic metals. These data were taken from the compilation by Völkl and Alefeld (204). It can be seen that the diffusion coefficient at room temperature is not greatest in palladium, a face-centered cubic metal. Nonetheless, it is used instead of the body-centered cubic metals as a practical hydrogen-selective membrane because it does not readily oxidize, as do the latter. Such oxide films interfere with the permeation of hydrogen. It should also be pointed out that, although platinum and palladium have comparable diffusion coefficients, the solubility of hydrogen in palladium is much greater, so that the rate of permeation, which is proportional to the product of the solubility and the diffusion coefficient, is considerably greater in palladium than in platinum.

Because of the unusual transport parameters that are observed, Alefeld (205) has argued that hydrogen in body-centered cubic metals should be thought of as having the characteristics of a liquid. This is, of course, analogous to the molten sublattice model of fast ionic conductors.

7 FAST DIFFUSION IN SEMICONDUCTORS

Experimental data indicate that diffusion behavior of solute species in the common group IV semiconductors, and probably also in the III-V compounds, falls into two quite disparate general categories. Solutes are often classified as slow diffusers or fast diffusers. Among the elements generally accepted as fast diffusers in germanium are Li, Cu, Ag, Fe, Co, and Ni. Likewise,

Table 4. Hydrogen Diffusion in Metals

	D_o	ΔH (kJ/Mole)	D at 25°C (cm^2/sec)
BCC metals			
V	2.9×10^{-4}	4.15	5×10^{-5}
Nb	5×10^{-4}	10.2	8×10^{-6}
Ta	4.4×10^{-4}	13.5	2×10^{-6}
Fe	4×10^{-4}	4.5	8×10^{-5}
	7.5×10^{-4}	8.5	1×10^{-5}
FCC metals			
Pt	6×10^{-3}	25.1	3×10^{-7}
Pd	2.9×10^{-3}	22.2	3×10^{-7}
Ni	4.8×10^{-3}	39.6	5×10^{-10}

fast diffusion in silicon is manifested by Li, Na, K, Cu, Ag, Au, Fe, and Ni. This type of behavior was noticed in the early days of the development of semiconductor technology (206,207), as rapid diffusion of some species was causing great practical difficulties. A review of the early work in this area was presented by Letaw (208). It was found that the fast diffusing species was interstitial copper, and it was recognized by van der Maesen and Brenkman (209) that copper atoms may be dissolved on both substitutional and interstitial sites in these diamond structure semiconductors. The interstitial species are very mobile and the total concentration profile involves the interaction of interstitial and substitutional species. Frank and Turnbull (210) showed that the role of vacancies must also be taken into account in this type of transport process, which occurs by what is now commonly known as the dissociative mechanism.

Diffusion data on various species in germanium and silicon were summarized by Seeger and Chik (211). It is generally found that the D_o values for these interstitial diffusers in germanium and silicon are quite low (in the range of 1 to 6×10^{-3} cm^2 sec). The activation enthalpies are also relatively low (from 29 to over 97 kJ/mole). In contrast, the activation enthalpies for diffusion of the group III and V elements in germanium and silicon tend to be between 300 and 400 kJ/mole. The pre-exponential factors are also much higher in those cases.

8 COMMENTS ON THEORETIC MODELS RELATED TO FAST IONIC CONDUCTION IN SOLIDS

Section 2 included a discussion of some of the special characteristics of solids that are fast ionic conductors. It is still an open question whether materials within this general class are fundamentally different from the more classic ionic conductors, or whether they merely exhibit extreme values of "normal" behavior.

It is obvious that this phenomenon is extremely structure sensitive and than an important criterion for its occurrence is the existence of a dilutely populated set of equivalent or nearly equivalent sites on (or near) which the mobile ions might reside. Equally important is the requirement that these positions of possible occupancy be interconnected in such a manner that movement between them is very easy. This means that the potential energy of such species must not vary substantially with position along the interconnecting path. In other words, the potential wells within which the mobile ions sit must be unusually shallow. Such a flat potential profile can lead to the "smeared-out" ionic positions and the large apparent vibrational amplitudes that are observed. Oscillation within such a potential profile should also have an appreciable degree of anharmonic character.

Although there have been attempts to fabricate other types of models, such as those based upon a "free-ion" concept, Haas (212) showed that the resulting formulation could readily be derived from a simple harmonic oscillator hopping type model.

A simple calculation by Rickert (213) showed that experimental data on α-AgI indicate that, if one assumes a simple hopping model and that all of the (mobile) ions move between fixed sites at thermal velocity, nearly every jump attempt must be successful. From a slightly different point of view, O'Keeffe (53) argued that the mobile ions have a very short residence time, and thus spend a good fraction of their time in flight.

It was further shown (119) that the experimental data, especially the very small value of the pre-exponential factor typically found (often about 10^5 lower than "normal"), require that the jump attempt frequency must be unusually low for fast ionic conductors. The entropy of migration may also be much lower than for other materials. However, the dominant influence cannot be an unusually large concentration of the mobile species, as has been suggested by several authors.

This small value of attempt frequency is, of course, consistent with a very shallow potential profile and large vibrational amplitude.

Calculations of the potential profiles within structures that are found to exhibit fast ionic conduction were initiated by Flygare and Huggins (68), who used a modification of the Born-Mayer method to find the potential distribution within the cation-containing tunnels in the α-AgI structure.

This explicit structure-dependent model involves the calculation of the interaction energy between a mobile ion at any arbitrary position within the structure and the ions of the host lattice. Consideration was given to Coulombic, repulsion, and dipolar polarization energies. By using a digital computer and probing a large array of possible positions for mobile ions with various different values of radius, it was possible to determine the minimum-energy path that would be preferably followed for motion from site to site. From this information, the energy profile could then be found along this preferred path.

It was found that the location of the minimum-energy path varies with the ionic size, as does the amplitude of the energy variations along it. For smaller ions, the attractive short-range polarization term is dominant, whereas for larger ions both the path and the energy profile are dominated by the repulsion contribution to the energy. Ions of intermediate size, where these factors tend to balance each other, were found to have the lowest values of activation energy for motion along the site to site path.

This is actually what has been found experimentally in a number of cases.

This model also predicts different site preferences along the tunnel for ions of different radii. As mentioned earlier, this type of observation has been made in the beta aluminas. Another obvious result is the existence of very broad and shallow potential wells, which lead to unusually low vibrational "attempt" frequencies and large amplitudes. All of these features are at least qualitatively consistent with what is found in materials that are fast ionic conductors.

Similar calculations have subsequently been made (69,214) on materials with the more complicated TiO_2 structure and the fluorite structure. Even though this approach involves a number of simplifying assumptions, it appears to be a promising way to evaluate structures that might exhibit fast ionic conduction.

9 THE APPLICATION OF NEW MEASUREMENT TECHNIQUES TO THE STUDY OF IONIC TRANSPORT IN SOLIDS

The traditional ways of evaluating ionic transport in solid membranes have involved macroscopic conductivity measurements and radiotracer diffusion experiments. Because of potential problems with polarization phenomena at interfaces, conductivity data are generally taken under a. c., rather than d. c. conditions.

Structural information has generally been obtained by the use of X-ray, and more recently, neutron diffraction methods.

Nuclear magnetic resonance techniques have also been applied to diffusion processes in solids, and are often quite helpful. In some cases, however, there is some difficulty in clearly separating localized motion or "rattling" from site-to-site long-range atomic or ionic transport.

The last few years have seen the initial application of a group of other techniques to the study of structural features and physical phenomena in these materials. Some of these may well provide new and valuable insight. These include the use of X-ray diffuse scattering (127), particle channeling, as evaluated by measurements of Rutherford back scattering as a function of crystallographic direction (128), optical absorption measurements in the microwave and far infrared regions of the spectrum (215-219), wavelength-modulated optical reflectance and absorption (220), Mössbauer spectroscopy (221), Raman scattering (222), light scattering (223), cold neutron scattering (224), and ultrasonic attenuation (225).

Variants on the traditional ionic conductivity and dielectric loss techniques are also being employed by a number of workers in the field. Particularly useful seems to be the use of complex plane analysis of the frequency dependence of the response of electrode-electrolyte systems to a.c. signals. This approach, whose use in connection with fast ionic conductors was pioneered by Bauerle (226), appears to be of special value in sorting out various electrolyte and electrode phenomena (227,228).

Because of both its wealth of interesting questions and many potential practical applications, this area of science is presently attracting a greatly increased amount of attention. It seems reasonable to expect that progress will be rapid within the next decade.

REFERENCES

1. W. Nernst, German Patent 104872 (1897).
2. W. Nernst, Z. Elektrochem. 6, 41 (1900).
3. W. Nernst and W. Wild, Z. Elektrochem. 7, 373 (1901).
4. H. Reynolds, Ph.D. dissertation, Göttingen (1902).
5. C. Wagner, Naturwissenschaften 31, 265 (1943).
6. E. Baur and H. Preis, Z. Elektrochem. 43, 727 (1937).
7. F. Haber et al., Z. Anorg. Chem. 41, 407 (1904).
8. F. Haber et al., Z. Anorg. Chem. 51, 245 (1906).
9. F. Haber et al., Z. Anorg. Chem. 51, 289 (1906).
10. N. W. Taylor, J. Amer. Chem. Soc. 45, 2865 (1923).
11. H. Reinhold, Z. Elektrochem. 40, 361 *1934).
12. W. D. Treadwell et al., Z. Elektrochem. 22, 415 (1916).
13. W. D. Treadwell et al., Helv. Chim. Acta 19, 1255 (1936).
14. B. A. Rose, G. J. Davis, and H. J. T. Ellingham, Disc. Faraday Soc. 4, 154 (1948).
15. C. Wagner, J. Chem. Phys. 21, 1819 (1953).
16. K. Kiukkola and C. Wagner, J. Electrochem. Soc. 104, 308 (1957).
17. K. Kiukkola and C. Wagner, J. Electrochem. Soc. 104, 379 (1957).
18. C. Wagner, Proc. Int. Comm. Electrochem. Thermo. Kinetics (CITCE) 7, 361 (1957).
19. H. Schmalzried, in Thermodynamics and Atomic Transport in Solids, International Atomic Energy Agency, Vienna, 1965.
20. C. Wagner, Adv. Electrochem. Electrochem. Eng. 4, 1 (1966).
21. D. O. Raleigh, Prog. Solid State Chem. 3, 83 (1967).
22. C. B. Alcock (Ed.), Electromotive Force Measurements in High-Temperature Systems, Institute of Mining and Metallurgy, London, 1968.

23. Y. F. Y. Yao and J. T. Kummer, J. Inorg. Nucl. Chem. 29, 2453 (1967).
24. R. H. Radzilowski, Y. F. Yao, and J. T. Kummer, J. Appl. Phys. 40, 4716 (1969).
25. N. Weber and J. T. Kummer, Proc. Ann. Power Sources Conf. 21, 37 (1967).
26. M. S. Whittingham and R. A. Huggins, in R. S. Roth and S. J. Schneider (Eds.), Solid State Chemistry, Nat. Bur. Standards Spec. Pub. 364, Washington, D.C., 1972, p. 139.
27. M. S. Whittingham and R. A. Huggins, J. Electrochem. Soc. 118, 1 (1971).
28. G. R. Belton and P. T. Morzenti, J. Metals 22, 26 A (1970).
29. N. K. Gupta and R. P. Tischer, J. Electrochem. Soc. 117, 125C (1970).
30. L. Hsueh and D. N. Bennion, J. Electrochem. Soc. 118, 1128 (1971).
31. J. N. Bradley and P. D. Greene, Trans. Faraday Soc. 62, 2069 (1966).
32. J. N. Bradley and P. D. Greene, Trans. Faraday Soc. 63, 424 (1967).
33. B. B. Owens and G. R. Argue, Science 157, 308 (1967).
34. B. B. Owens and G. R. Argue, J. Electrochem. Soc. 117, 898 (1970).
35. B. B. Owens, Adv. Electrochem. Electrochem. Eng. 8, 1 (1971).
36. (a) Meeting on Mass Transport in Non-metallic Solids, Basic Science Section of Brit. Ceram. Soc., London, 1969.
    (b) Symposium on Solid State Chemistry, Nat. Bur. Stand., Washington, D.C., 1971.
    (c) NATO Advanced Study Institute on Fast Ion Transport in Solids, Belgirate, 1972.

(d) Session on Solid Electrolyte Fundamentals and Applications, Electrochem. Soc., Houston, 1972.

(e) 24th Meeting of ISE, Solid Electrolytes, Ion Transport in Insulating Layers, and Their Application, Eindhoven, 1973.

(f) Session on High Temperature Electron and Ion Transport in Solids, Electrochem. Soc., San Francisco, 1974.

(g) 9th Univ. Conf. on Ceram. Science, Cleveland, 1974.

(h) Session on Ceramics for Fuel Cell Applications, Am. Ceram. Soc., Williamsburg, 1974.

(i) Symposium on Superionic Conductors, Solid State Physics Division of Am. Phys. Soc., Denver, 1975.

(j) Session on Ceramic Materials for High Power Density Batteries, Am. Ceram. Soc., Washington, D.C., 1975.

(k) Session on Materials Problems in Battery and Fuel Cell Technology, Electrochem. Soc., Toronto, 1975.

37. A. Smekal, Z. Physik. 26, 707 (1925).
38. A. Smekal, Z. Phys. Chem. B6, 103 (1929).
39. W. Jost, Z. Phys. Chem. B6, 88, 210 (1929).
40. W. Jost, Z. Phys. Chem. B7, 234 (1930).
41. C. Tubandt and E. Lorenz, Z. Phys. Chem. 87, 513 (1914).
42. J. A. A. Ketelaar, Z. Phys. Chem. B30, 53 (1935).
43. J. A. A. Ketelaar, Z. Krist. 87, 436 (1934).
44. L. W. Strock, Z. Phys. Chem. B25, 441 (1934).
45. L. W. Strock, Z. Phys. Chem. B31, 132 (1936).
46. J. A. A. Ketelaar, Z. Phys. Chem. B26, 327 (1934).
47. J. A. A. Ketelaar, Trans Faraday Soc. 34, 874 (1938).
48. H. Wiedersich and S. Geller, in L. Eyring and M. O'Keeffe (Eds.), The Chemistry of Extended Defects in Non-Metallic Solids, North-Holland, Amsterdam, 1970, p. 629.

49. W. V. Johnston, H. Wiedersich, and G. W. Lindberg, J. Chem. Phys. 51, 3739 (1969).
50. C. M. Perrott and N. H. Fletcher, J. Chem. Phys. 48, 2143 (1968).
51. C. M. Perrott and N. H. Fletcher, J. Chem. Phys. 48, 2681 (1968).
52. C. M. Perrott and N. H. Fletcher, J. Chem. Phys. 50, 2770 (1969).
53. M. O'Keeffe, in W. van Gool (Ed.), Fast Ion Transport in Solids, North-Holland, Amsterdam, 1973, p. 233.
54. M. O'Keeffe, J. Electrochem. Soc. 121, 102C (1974).
55. M. O'Keeffe, Bull. Am. Phys. Soc., Series II, 20, 451 (1975).
56. W. van Gool, in H. K. Henisch, R. Roy, and L. E. Cross (Eds.), Phase Transitions, Pergamon, New York, 1973, p. 373.
57. G. W. Herzog and H. Krischner, in J. S. Anderson, M. W. Roberts, and F. S. Stone (Eds.), Proc. 7th Int. Symp. on Reactivity of Solids, Chapman and Hall, London, 1972, p. 140.
58. M. J. Buerger, Anais Acad. Brasil. Cienc. 21, 261 (1949).
59. R. Ueda, J. Phys. Soc. Japan 4, 287 (1949).
60. M. J. Buerger and B. J. Wuensch, Science 141, 276 (1963).
61. S. Miyake, S. Hoshino, and T. Takenaka, J. Phys. Soc. Japan 7, 19 (1952).
62. S. Hoshino, J. Phys. Soc. Japan 7, 560 (1952); J. Phys. Soc. Japan 10, 197 (1955).
63. J. S. Kasper and K. W. Browall, J. Solid State Chem. 13, 49 (1975).
64. B. T. M. Willis, Acta Cryst. 18, 75 (1965).
65. B. Dawson, A. C. Hurley, and V. W. Maslen, Proc. Roy. Soc. A298, 289 (1967).
66. B. T. M. Willis, Acta Cryst. A25, 277 (1969).

67. M. J. Cooper, K. D. Rouse, and H. Fuess, Acta Cryst. A29, 49 (1973).
68. W. F. Flygare and R. A. Huggins, J. Phys. Chem. Solids 34, 1199 (1973).
69. O. B. Ajayi, L. E. Nagel, I. D. Raistrick, and R. A. Huggins, J. Phys. Chem. Solids, 37, 167 (1976).
70. O. B. Ajayi, I. D. Raistrick, and R. A. Huggins, to be published.
71. G. A. Rankin and H. E. Merwin, J. Amer. Chem. Soc. 38, 568 (1916).
72. G. A. Rankin and H. E. Merwin, Z. Anorg. Allgem. Chem. 96, 291 (1916).
73. L. T. Brownmiller and R. H. Bogue, Am. Cer. J. Sci. 5th Ser. 23, 501 (1932).
74. G. W. Morey, Bull. Amer. Ceram. Soc. 13, 79 (1934).
75. L. T. Brownmiller, Amer. J. Sci. 5th Ser. 29, 260 (1935).
76. R. R. Ridgeway, A. A. Klein, and W. J. O'Leary, Trans. Electrochem. Soc. 70, 71 (1936).
77. L. De Pablo-Galan and W. R. Foster, J. Amer. Ceram. Soc. 42, 491 (1959).
78. W. L. Bragg, C. Gottfried, and J. West, Z. Krist. 77, 255 (1931).
79. C. A. Beevers and S. Brohult, Z. Krist. 95, 472 (1936).
80. C. A. Beevers and M. A. S. Ross, Z. Krist. 97, 59 (1937).
81. G. Yamaguchi, Elect. Chem. Soc. Japan 11, 260 (1943).
82. G. Yamaguchi, Preprint, 7th Ann. Meeting of Chem. Soc. Japan, April, 1954, p. 192.
83. G. Yamaguchi, Ph.D. thesis, University of Tokyo (1954).
84. G. Yamaguchi and K. Suzuki, Bull. Chem. Soc. Japan 41, 93 (1968).
85. J. Thery and D. Briancon, Compt. Rend. 254, 2782 (1962).

86. J. Thery and D. Briancon, Rev. Hautes Temp. Refract. 1, 221 (1964).
87. R. Scholder and M. Mansmann, Z. Naturforsch. 156, 681 (1960).
88. R. Scholder and M. Mansmann, Z. Anorg. Allgem. Chem. 321, 246 (1963).
89. J. Felsche, Naturwissenschafterr 54, 612 (1967).
90. F. Felsche, Z. Krist. 127, 94 (1968).
91. M. Bettman and C. R. Peters, J. Phys. Chem. 73, 1774 (1969).
92. M. S. Whittingham, R. W. Helliwell, and R. A. Huggins, U.S. Gov. Res. and Dev. Rept. 69, 158 (1969).
93. C. R. Peters, M. Bettman, J. W. Moore, and M. D. Glick, Acta Cryst. B27, 1826 (1971).
94. M. Harata, Mat. Res. Bull. 6, 461 (1971).
95. W. L. Roth, J. Solid State Chem. 4, 60 (1972).
96. D. J. Dyson and W. Johnson, Trans. J. Brit. Ceram. Soc. 72, 49 (1973).
97. W. L. Roth, W. C. Hamilton, and S. J. La Placa, Am. Cryst. Assoc. Abstracts Ser. 2, 1, 169 (1973).
98. N. Weber and A. F. Venero, Ford Motor Co. Technical Report No. SR 69-102 (1969).
99. M. Bettman and L. L. Terner, Inorg. Chem. 10, 1442 (1971).
100. S. G. Ampian, U.S. Bureau of Mines Report of Investigations, 1964, p. 31.
101. R. H. Bogue, The Chemistry of Portland Cement, 2nd ed., Reinhold, New York, 1955, p. 793.
102. M. Rolin and P. H. Thanh, Rev. Hautes Temp. Refract. 2, 175 (1965).
103. R. C. DeVries and W. L. Roth, J. Amer. Ceram. Soc. 52, 364 (1969).
104. N. Weber and A. F. Venero, Ford Motor Co. Technical Report No. SR 69-86, 1969.

105. J. Fally, C. Lasne, Y. Lazennec, Y. LeCars, and P. Margotin, Electrochem. Soc. Extended Abstracts 72-1, 441 (1972).
106. Y. LeCars, J. Thery and R. Collongues, Compt. Rend. 274, 4, (1972).
107. Y. LeCars, J. Thery, and R. Collongues, Rev. Hautes Temp. Refract. 9, 153 (1972).
108. J. Liebertz, Ber. Dt. Keram. Ges. 49, 288 (1972).
109. J. T. Kummer, Prog. Solid State Chem. 7, 141 (1972).
110. J. Fally, C. Lasne, Y. Lazennec, Y. LeCars, and P. Margolin, J. Electrochem. Soc. 120, 1296 (1973).
111. Y. LeCars, J. Thery, and R. Collongues, presented at 24th Meeting of ISE, Eindhoven, 1973.
112. D. J. M. Bevan, B. Hudson, and P. T. Moseley, Mat. Res. Bull. 9, 1073 (1974).
113. M. Bettman, Mat. Res. Bull. 10, 229 (1975).
114. A. G. Elliot and R. A. Huggins, presented at Am. Ceram. Soc. Meeting, Chicago, 1974.
115. A. G. Elliot and R. A. Huggins, J. Am. Ceram. Soc. 58, 497 (1975).
116. W. L. Roth, General Electric Co. Report No. 74CRD054 (1974).
117. M. S. Whittingham and R. A. Huggins, J. Chem. Phys. 54, 414 (1971).
118. M. S. Whittingham and R. A. Huggins, J. Electrochem. Soc. 118, 1 (1971).
119. R. A. Huggins, in A. S. Nowick and J. J. Burton (Eds.), Diffusion in Solids/Recent Developments, Academic, New York, 1975, p. 445.
120. R. H. Radzilowski and J. T. Kummer, J. Electrochem. Soc. 118, 714 (1971).
121. C. Wagner, Z. Elektrochem. 60, 4 (1956).

122. R. A. Huggins, in A. R. Cooper and A. H. Heuer (Eds.), Mass Transport Phenomena in Ceramics, Plenum, New York, 1975.
123. H. Saalfeld, H. Matthies, and S. K. Datta, Ber. Deutsch. Keram. Ges. 45, 212 (1968).
124. W. L. Roth and S. P. Mitoff, General Electric Co. Report 71-C 277, 1971.
125. D. Kline, H. S. Story, and W. L. Roth, J. Chem. Phys. 57, 5180 (1972).
126. R. S. Gordon, personal communication.
127. Y. LeCars, R. Comes, L. Deschamps, and J. Thery, Acta Cryst. A30, 305 (1974).
128. L. C. Feldman, W. M. Augustyniak, J. P. Remeika, P. J. Silverman, and D. B. McWhan, presented at Meeting of Amer. Phys. Soc., Denver, 1975.
129. E. W. Gorter, Philips Research Reports 9, 363 (1954).
130. K. O. Hever, J. Electrochem. Soc. 115, 826 (1968).
131. W. L. Roth and A. S. Cooper, General Electric Co. Report No. 60-RL-2461 M, 1960.
132. W. L. Roth and F. E. Luborsky, J. Appl. Phys. 35, 966 (1964).
133. C. J. M. Rooymans, C. Langerais and J. A. Schulkes, Solid State Commun. 4, 85 (1965).
134. W. L. Roth and R. J. Romanczuk, J. Electrochem. Soc. 116, 975 (1969).
135. L. M. Foster and H. C. Stumpf, J. Amer. Chem. Soc. 73, 1590 (1951).
136. J. P. Boilot, J. Thery, and R. Collongues, Mat. Res. Bull. 8, 1143 (1973).
137. L. M. Foster and J. E. Scardefield, presented at Meeting of the Electrochem. Soc., Toronto, 1975.
138. H. C. Brinkhoff, presented at 24th Meeting of ISE, Eindhoven, 1973.

139. A. Kvist and A. M. Josefson, Z. Naturforsch. 23a, 625 (1968).
140. P. Rahlfs, Z. Physik. Chem. 31, 157 (1935).
141. J. Krug and L. Sieg, Z. Naturforsch. 7a, 369 (1952).
142. S. Hoshino, J. Phys. Soc. Japan 7, 560 (1952).
143. T. Førland and J. Krogh-Moe, Acta Cryst. 11, 224 (1958).
144. H. A. Øye, Ph.D. thesis, Technical University of Norway, Trondheim, 1963.
145. B. Reuter and K. Hardel, Angew Chem. 72, 138 (1960).
146. B. Reuter and K. Hardel, Naturwissenschaften 48, 161 (1961).
147. B. Reuter and K. Hardel, Z. Anorg. Allg. Chem. 340, 158, 168, (1965).
148. B. Reuter and K. Hardel, Z. Elektrochem. 70, 82 (1966).
149. B. Reuter and K. Hardel, Ber. Bunsenges, Phys. Chem. 70, 82 (1966).
150. T. Takahashi and O. Yamamoto, J. Electrochem. Soc. Japan 32, 174 (1964).
151. T. Takahashi and O. Yamamoto, J. Electrochem. Soc. Japan 33, 191 (1965).
152. T. Takahashi and O. Yamamoto, Electrochim. Acta 11, 779, 911 (1966).
153. B. Reuter, J. Pickardt, and K. Hardel, Z. Physik. Chem. 56, 309 (1967).
154. L. Suchow and G. R. Pond, J. Amer. Chem. Soc. 75, 5242 (1953).
155. L. Heyne, Electrochim. Acta 15, 1251 (1970).
156. P. Rahlfs, Z. Physik. Chem. B31, 157 (1936).
157. A. Kvist and A. Lunden, Z. Naturforsch. 20a, 235 (1965).
158. A. Kvist and A. Lunden, Z. Naturforsch. 21a, 1509 (1966).
159. J. N. Bradley and P. D. Greene, Trans. Faraday Soc. 63, 2516 (1967).
160. S. Geller, Science 157, 310 (1967).

161. S. Geller, in W. van Gool (Ed.), Fast Ion Transport in Solids, North-Holland, Amsterdam, 1973, p. 607.
162. T. Takahashi, O. Yamamoto, and K. Kuwabara, Denki Kagaku 35, 264 (1967).
163. T. Takahashi, S. Ikeda, and O. Yamamoto, J. Electrochem. Soc. 119, 477 (1972).
164. T. Takahashi, S. Ikeda, and O. Yamamoto, J. Electrochem. Soc. 120, 647 (1973).
165. J. B. Goodenough and J. A. Kafalas, J. Solid State Chem. 6, 493 (1973).
166. H. Y-P. Hong, J. A. Kafalas, and J. B. Goodenough, J. Solid State Chem. 9, 345 (1974).
167. P. Spiegelberg, Ark. Kemi 14A, 1 (1940).
168. A. R. West, J. Appl. Electrochem. 3, 327 (1973).
169. I. D. Raistrick, C. Ho, and R. A. Huggins, to be published in J. Electrochem. Soc.
170. B. E. Liebert and R. A. Huggins, Mat. Res. Bull. 11, 533 (1976).
171. A. J. Majumdar and R. Roy, J. Phys. Chem. 69, 1684 (1965).
172. K. Schroeder and A. Kvist, Z. Naturforsch. 24a, 844 (1969).
173. D. H. Whitmore, personal communication.
174. W. A. Spurgeon, H. M. Lee, D. H. Whitmore, and L. B. Walsh, presented at Meeting of Amer. Phys. Soc., Denver, 1975.
175. B. E. Liebert and R. A. Huggins, to be published.
176. P. M. Skarstad, S. Geller, and S. A. Wilber, presented at Meeting of Amer. Phys. Soc., Denver, 1975.
177. R. W. Ure, J. Chem. Phys. 26, 1363 (1957).
178. R. E. Carter and W. L. Roth, in C. B. Alcock (Ed.), Electromotive Force Measurements in High-Temperature Systems, Institute of Mining and Metallurgy, London, 1968, p. 125.
179. T. H. Etsell and S. N. Flengas, Chem. Rev. 70, 339 (1970).

180. A. Kvist, in J. Hladik (Ed.), Physics of Electrolytes, Vol. 1, Academic, New York, 1972, p. 319.
181. B. C. H. Steele, in L. E. J. Roberts (Ed.), Solid State Chemistry, Butterworths, London, 1972, p. 117.
182. L. D. Yushina and S. F. Pal'guev, in M. V. Smirnov (Ed.), Electrochemistry of Molten and Solid Electrolytes, Vol. 2, Consultants Bureau, New York 1963, p. 74.
183. B. C. H. Steele and J. M. Floyd, Proc. Brit. Ceram. Soc. 19, 55 (1971).
184. T. Takahashi, in J. Hladik (Ed.), Physics of Electrolytes, Vol. 2, Academic, London, 1972, p. 989.
185. H. L. Tuller and A. S. Nowick, J. Electrochem. Soc. 122, 255 (1975).
186. T. Takahashi, H. Iwahara, and Y. Nagai, J. Appl. Electrochem. 2, 97 (1972).
187. T. Takahashi, H. Iwahara, T. Esaka, and T. Arao, presented at 24th Meeting of ISE, Eindhoven, 1973.
188. C. N. R. Rao, G. V. Subba Rao, and S. Ramdas, J. Phys. Chem. 73, 672 (1969).
189. R. P. Bauman and S. P. S. Porto, Phys. Rev. 161, 842 (1967).
190. L. S. Garashina and B. P. Sobolev, Soviet Phys. Cryst. 16, 254 (1971).
191. M. Goldman and L. Shen, Phys. Rev. 144, 321 (1966).
192. K. Lee, Solid State Commun. 7, 367 (1969).
193. A. Sher. R. Solomon, K. Lee, and M. W. Muller, Phys. Rev. 144, 593 (1966).
194. W. L. Fielder, NASA Technical Note, D-5505 (1969).
195. L. E. Nagel, Ph.D. dissertation, Arizona State University, 1972.
196. L. E. Nagel and M. O'Keeffe, in W. van Gool (Ed.), Fast Ion Transport in Solids, North-Holland, Amsterdam, 1973, p. 165.

197. A. Zalkin and D. H. Templeton, J. Amer. Chem. Soc. 75, 2453 (1953).
198. R. E. Thoma and G. D. Brunton, Inorg. Chem. 5, 1937 (1966).
199. D. A. Jones and W. A. Shand, J. Cryst. Growth 2, 361 (1968).
200. R. C. Pastor and M. Robinson, Mat. Res. Bull. 9, 569 (1974).
201. M. O'Keeffe, Science 180, 1276 (1973).
202. L. G. Sillen and B. Aurivillius, Z. Krist. 101, 483 (1939).
203. H. K. Birnbaum and C. A. Wert, Ber. Bunsenges. Phys. Chem. 76, 806 (1972).
204. J. Volkl and G. Alefeld, in A. S. Nowick and J. J. Burton (Eds.), Diffusion in Solids/Recent Developments, Academic, New York, 1975, p. 231.
205. G. Alefeld, Ber. Bunsenges. Phys. Chem. 76, 355 (1972).
206. H. C. Theurer and J. H. Scaff, J. Metals 4, 59 (1951).
207. C. S. Fuller, H. S. Theurer, and W. van Roosbroeck, Phys. Rev. 85, 678 (1952).
208. H. Letaw, Jr., J. Phys. Chem. Solids 1, 100 (1956).
209. F. van der Maesen and J. A. Brenkman, Philips Res. Rept. 9, 225 (1954).
210. F. C. Frank and D. Turnbull, Phys. Rev. 104, 617 (1956).
211. A. Seeger and K. P. Chik. Phys. Status Solidi 29, 455 (1968).
212. C. W. Haas, J. Solid State Chem. 7, 155 (1973).
213. H. Rickert, in W. van Gool (Ed.), Fast Ion Transport in Solids, North-Holland, Amsterdam, 1973, p. 3.
214. O. B. Ajayi, Ph.D. dissertation, Stanford University, 1975.
215. K. Funke and A. Jost, Ber. Bunsenges. Phys. Chem. 75, 436 (1971).
216. K. Funke and R. Hackenberg, Ber. Bunsenges. Phys. Chem. 76, 885 (1972).
217. S. J. Allen, Jr. and J. P. Remeika, Phys. Rev. Letters 33, 1478 (1974).

218. U. Strom, P. C. Taylor, and S. G. Bishop, presented at Meeting of Amer. Phys. Soc., Denver, 1975.
219. S. J. Allen, Jr. and J. P. Remeika, presented at Meeting of Amer. Phys. Soc., Denver, 1975.
220. R. S. Bauer and B. A. Huberman, presented at Meeting of Amer. Phys. Soc., Denver, 1975.
221. J. P. Remeika, R. L. Cohen, and K. W. West, presented at Meeting of Amer. Phys. Soc., Denver, 1975.
222. L. L. Chase, C. H. Hao, and G. D. Mahan, to be published.
223. W. H. Flygare, personal communication.
224. K. Funke, J. Kalus and R. E. Lechner, Solid State Commun. 14, 1021 (1974).
225. M. Nagao and T. Kaneda, Phys. Rev. B11, 2711 (1975).
226. J. E. Bauerle, J. Phys. Chem. Solids 30, 2657 (1969).
227. R. A. Huggins, J. Electrochem. Soc. 122, 53C (1975).
228. R. A. Huggins and I. D. Raistrick, presented at NATO Advanced Study Institute, Ajaccio, 1975.
229. R. W. Christy, J. Chem. Phys. 34, 114 (1961).
230. T. Takahashi, H. Iwahara, and T. Ichimura, Denki Kagaku 37, 857 (1969).

The Sulfur Electrode in Nonaqueous Media

RAGNAR P. TISCHER and FRANK A. LUDWIG
Research Staff, Ford Motor Company, Dearborn, Michigan

1	Redox Reactions and EMF Measurements	392
2	Organic Solvent Solutions	409
3	Fused Salt Melts	414
	3.1 Slags containing sulfide	415
	3.2 LiCl-KCl eutectic melts	416
	3.3 Molten sodium polysulfides	425
	3.4 Polysulfides in molten KCNS	456
	3.5 Practical sulfur electrodes	469
4	Sulfur Cations	477
	References	478

The sulfur electrode in recent years has found increasing interest, because the relatively low equivalent weight, the abundant availability, and the low price of sulfur are attractive properties for its use as an active agent in batteries, in particular in secon-

---

Work partially supported by NSF-RANN.

dary systems at a time when there is an urgent need for more efficient and lighter storage batteries.

This chapter concerns itself only with sulfur electrodes using liquid and gaseous sulfur, that is, electrodes that are depolarized by elemental sulfur. Some battery developments have turned to heavy metal sulfide electrodes in despair over the difficulties of making a sulfur electrode practical in certain systems. We do not discuss these heavy metal sulfide electrodes, although in some (e.g., FeS_2) sulfur oxidation and reduction also take place.

Sulfur electrode potential measurements have been used for obtaining thermodynamic data. Some important papers that have contributed to the perfection of this technique are listed, as well as some historically interesting approaches.

The major part of this chapter is devoted to the rather recent attempts at clarifying the kinetics of the sulfur/sulfide electrode reaction in various media and at developing practical battery electrodes. The prime difficulty for the basic work is posed by the variety of polysulfide ions present, which are hard to identify. Practical electrode development has had to contend with the complete lack of conductivity in elemental sulfur. In some approaches using liquid electrolyte, containment of the sulfur is a serious problem, because it cannot be prevented from reaching the opposite electrode.

1 REDOX REACTIONS AND EMF MEASUREMENTS

Tammann and Samson-Himmelstjerna (1), in their study of substitution reactions between a metal and the salt of a different metal, developed an emf series for metals in sulfides without making any electrochemical measurement. The ability of one metal to replace another in its sulfide is related to the free energy of formation of the respective sulfides. Data for many metal sulfides were not available at that time. Experimentally, the following series was established: Mg, Mn, Zn, Al, Cr, Mo, Fe, Cd, Sn, Cu, Ni, Co, Pb, Bi, Ag. Differences in the temperature dependence of the free energies of formation can lead to permutations in this series at higher temperatures. For instance, above 1300°C, Fe becomes more noble than Cu in a sulfide melt.

For the LiCl-KCl eutectic melt, Delarue (2) has defined a pS^{2-} scale. In the expression for the sulfur potential

$$E = E_0^1 - \frac{RT}{2F} \log [S^{2-}]$$

he substitutes

$$pS^{2-} = -\log [S^{2-}]:$$
$$E = E_0^1 + \frac{RT}{2F} pS^{2-}$$

where E_0^1 still depends on the activity or fugacity of sulfur. In this study the oxidizing power of various redox systems was studied. The pS^{2-} in the LiCl-KCl melt obviously depends upon the solubility product of the sulfide supplying the S^{2-} ions, provided the melt is saturated. Since redox systems of sufficiently positive potential oxidize sulfide ions to sulfur, metal sulfides of limited solubility can be brought into solution (as chlorides) by such oxidation depending upon their pS^{2-}. Sufficiently strong oxidants (Cu^{2+}, Fe^{3+}, Au^+, Cl_2) oxidize the sulfur further to sulfur monochloride (S_2Cl_2). These relationships are depicted in an E versus pS^{2-} diagram (Fig. 1). Because Delarue uses potential data from Laitinen and co-workers (3), it can be assumed that his

quantitative results refer to experiments at 450°C. To make this
diagram more useful, it would be desirable to obtain the exact
solubility products and therefrom the corresponding pS^{-2} values
or vice versa.

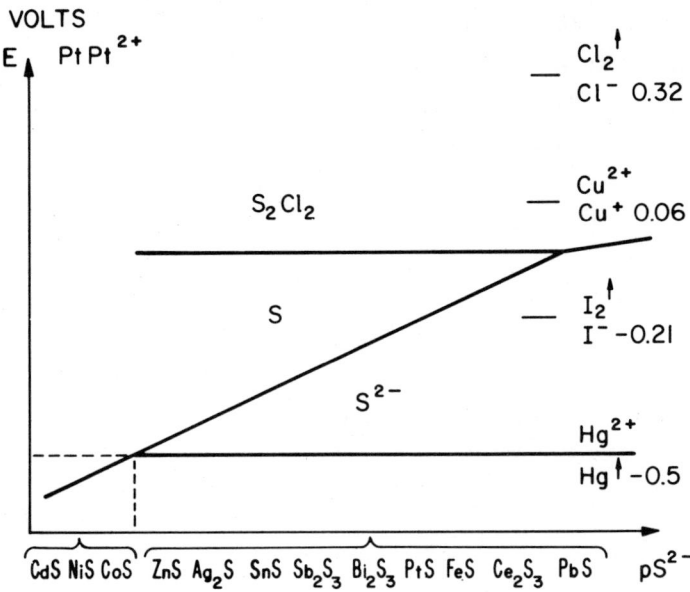

Fig. 1. Potential - pS^{2-} diagram showing areas of stability of
various sulfur species and approximate order of heavy metal sulfides on the pS^{2-} scale (CdS, NiS and CoS are brought into solution by Hg^{2+}), after Delarue (2).

Electromotive force measurements are a very useful tool for
determining thermodynamic data, and they can also be quite helpful
in establishing phase diagrams. This is particularly true at
higher temperatures where exchange currents are high, and potentials are quickly established and generally well reproducible.
However, there are cases of slow reactions hampering measurements,

especially in the solid state at elevated temperatures, so that reversibility cannot be taken for granted. Cell voltages of cells with sulfur electrodes have been measured for many reasons, and it cannot be the task of this review to give a complete listing, but rather to review a variety of applications.

Kiukkola and Wagner (4) have determined the free energy of formation of Ag_2S by measuring the emf of the cell

$$Ag(s)/AgI(s)/Ag_2S(s), S(\ell), C$$

improving on earlier measurements by Reinhold (5). The emf is controlled by the reaction

$$2Ag(s) + S(\ell) = Ag_2S(s)$$

Application of the formula

$$\Delta G^\circ = -2E^\circ F,$$

with E representing emf and F the Faraday constant yields the standard free energy change of the reaction:

T (°C)	E (mV)	ΔG° (kcal)
150	220	-10.15
300	252	-11.62
425	264	-12.18

Application of the Gibbs-Duhem equation permits calculation of the molar free energy of a ternary system from measurements of the partial molar free energy of a single component (6). Calculations of this nature were used by Verduch and Wagner (7) in finding thermodynamic data for double sulfide systems. As an example of the experimental cells involved, Fig. 2 shows the vessel for the cell

$$\text{Pt, } S_x(g)/Ag_2S(s)/AgI(s)/(Ag_2S, Sb_2S_3)(s)/Pt, S_x(g)$$

with the overall cell reaction

$$\tfrac{1}{2} Ag_2S(s) = \tfrac{1}{2} Ag_2S(s) \text{ (in } Ag_2S - Sb_2S_3)$$

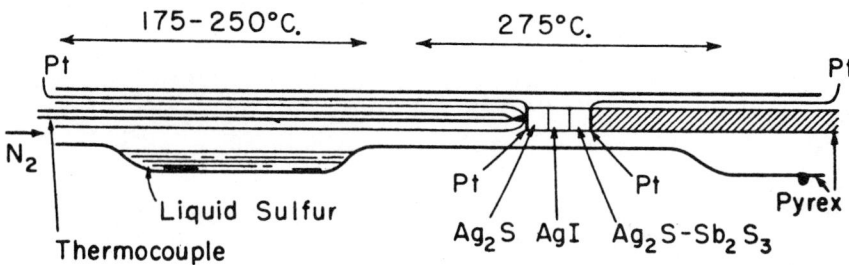

Fig. 2. Vessel for cell with two sulfur electrodes, from Verduch and Wagner (7).

The emf does not depend on sulfur partial vapor pressure, as long as it is the same at both sulfur electrodes. The emf of this cell yields

$$\tfrac{1}{2} \left[\overline{G}_{Ag_2S} - G^o_{Ag_2S} \right] = -EF$$

where $\overline{G}_{Ag_2S}$ is the partial molar free energy of Ag_2S in the system $Ag_2S - Sb_2S_3$. This result could also be obtained by combining emf measurements of the following two cells:

$$Ag(s)/AgI(s)/(Ag_2S, Sb_2S_3)(s)/Pt, S(\ell)$$

$$Ag(s)/AgI(s)/Ag_2S(s)/Pt, S(\ell)$$

An integration over x, the ratio of equivalents of Sb_2S_3 to the total equivalents of Ag_2S and Sb_2S_3, results in (8)

$$\Delta G_{eq}(\xi) = (1 - \xi) \int_0^\xi \frac{\bar{G}_{Ag_2S,S\,(eq)} - G^o_{Ag_2S,S\,(eq)}}{(1 - x)^2} dx$$

the free energy of formation of the double sulfides with composition ξ from the corresponding single sulfides. Results obtained in this fashion at 275°C for

$$Ag_3SbS_3 \quad \Delta G_{eq}(\xi = 0.50) = -710 \text{ cal/eq}$$

and

$$AgSbS_2 \quad \Delta G_{eq}(\xi = 0.75) = -565 \text{ cal/eq}$$

are shown to compare well with those calculated by Verduch and Wagner using the partial free energy of sulfur,

$$\bar{G}_{S_{eq}} = \text{const} + \frac{1}{2} RT \ln q$$

from measurements by Schenck et al. (9) of the ratio $q = p_{H_2S}/p_{H_2}$ in the gas phase in the reduction of the same Ag_2S, Sb_2S_3 mixtures with hydrogen.

In three papers Thompson and Flengas (14) have studied the behavior of a sulfur vapor electrode and its application for determining thermodynamic data for Ag_2S and PbS in chloride melts. They make use of mass spectrometric data for the molecular composition of sulfur vapor given by Berkovitz et al. (15) and the improved data by Detry et al. (16). (The latter authors generated a controlled flow of sulfur vapor in a Knudsen cell by electrolysis of the solid-state cell: Pt, Ag/AgI/Ag$_2$S, Pt. This is another interesting sulfur electrode application.)

Thompson and Flengas (14) used cells of the type

$$Ag/AgCl, \ Ag_2S/S(g), \ C$$

$$C, \ Pb/PbCl_2, \ PbS/S(g), \ C$$

A separation of the electrolyte into two compartments connected by an asbestos frit was necessary to avoid direct reaction of sulfur with the metal electrode.

Thompson and Flengas (14) have calculated a wealth of thermodynamic data for the systems studied, but above all they have developed theory and technique for the electrochemical use of sulfur vapor, an agent of complicated molecular composition. From the literature (15,16) data, formulas and diagrams (Figs. 3 and 4) were developed for the partial pressures of the $S_2,\ldots S_\nu,\ldots,S_8$ species as a function of temperature and total pressure.

The Nernst formula for the silver sulfide cell is

$$E = E^o_{Ag_2S} - \left(\frac{RT}{2F}\right) \ln \frac{a_{Ag_2S}}{P_{S_\nu}^{1/\nu}}$$

where

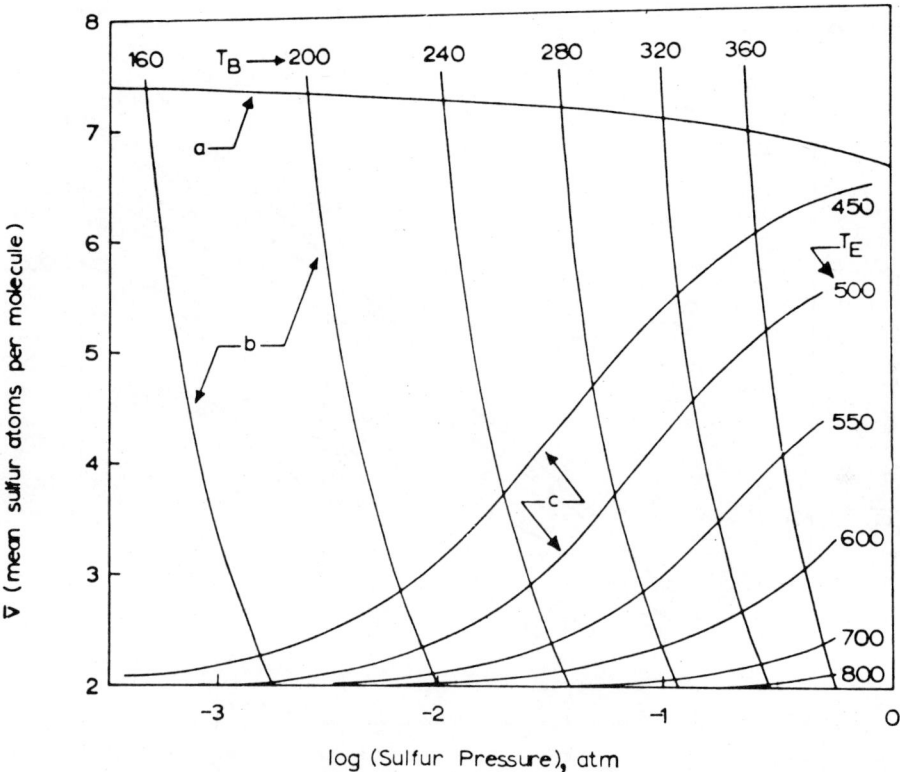

Fig. 3. Sulfur pressures plotted as a function of the number of sulfur atoms per sulfur vapor molecule for the saturated vapor (s) and nonsaturated vapor (c) at various temperatures. Curve (b) is the variation of the mean number of sulfur atoms per molecule with temperature T_E plotted against total sulfur pressure for sulfur vapor in equilibrium with liquid sulfur at temperature T_B, from Thompson and Flengas (14). T_E, electrode temperature; T_B, temperature of sulfur pool for generating vapor.

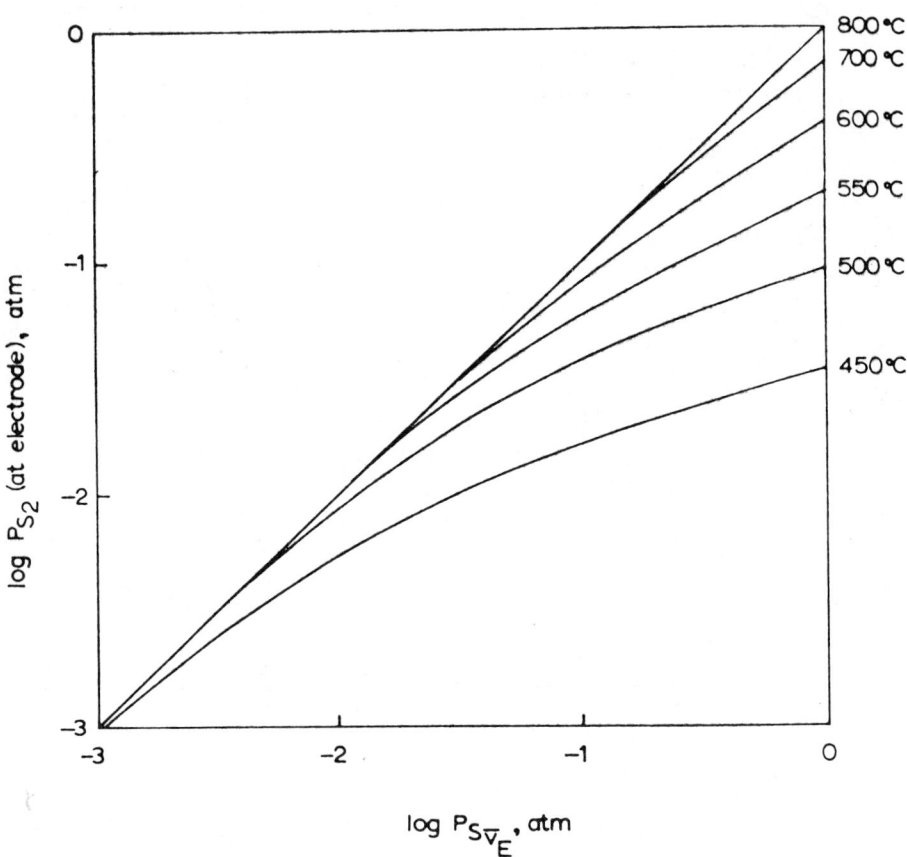

Fig. 4. Variation of the partial pressure of S_2 species at various temperatures versus total sulfur pressure, from Thompson and Flengas (14).

E = cell emf
E^o = standard emf, here chosen for sulfur in a standard state of pure S_2 vapor at 1 atm
a = activity
ν = 1,...,8

for the reaction

$$2Ag(s) + \frac{1}{\nu} S_\nu(g) \rightarrow Ag_2S(s)$$

The choice of ν can be arbitrary as long as the S_ν molecules are in equilibrium at the electrode surface. Results of the emf measurements are plotted against sulfur vapor pressure in Fig. 5. The plot for the total vapor pressure deviates from the straight line at higher pressures, where the molecular composition of the vapor changes, showing that the Nernst formula is not applicable for total pressure. If the data are, however, replotted for the partial pressures of individual S_ν species, then linearity is satisfactory. Figure 6 shows cell emf versus log P_{S_2} at various temperatures. The least-squares slopes give a value of n of 1.91±0.06, indicating a two-electron reaction. Figure 7 shows the temperature dependence of the emf for various total sulfur vapor pressures. In the area under the dashed line the influence of species other than S_2 is negligible.

The resulting data for the standard potential of the Ag_2S and the PbS formation cells were compared with the data by Kiukkola and Wagner (4) obtained at lower temperatures with liquid sulfur. For comparison of these low-temperature data the standard state of sulfur had to be changed from liquid sulfur to S_2 gas at 1 atm pressure. Subsequent extrapolation showed very good agreement of the data.

It is obvious that the tools for the use of the sulfur electrode in emf measurements for obtaining physical and thermodynamic data have been well developed so as to make further investigations almost routine.

Gupta and Tischer (10) used emf measurements to obtain thermodynamic data and to revise the phase diagram of the sodium sulfide-

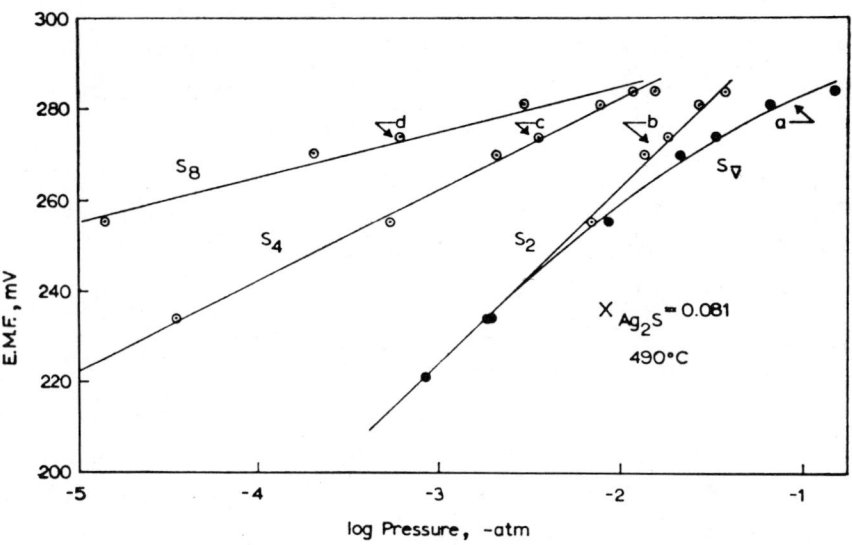

Fig. 5. Variation of cell potential with respect to the logarithm of the partial pressures of S_2, S_4, S_8, and total sulfur vapor pressure, from Thompson and Flengas (14).

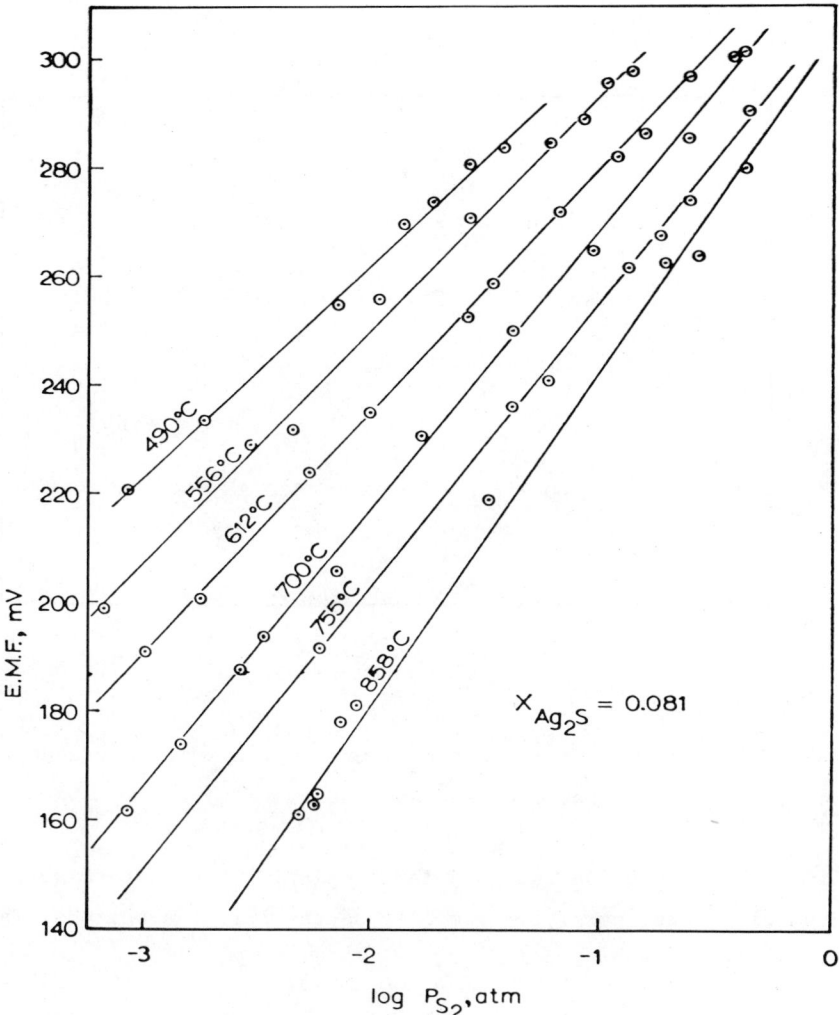

Fig. 6. Cell emf against log P_{S_2} at various electrode temperatures, from Thompson and Flengas (14).

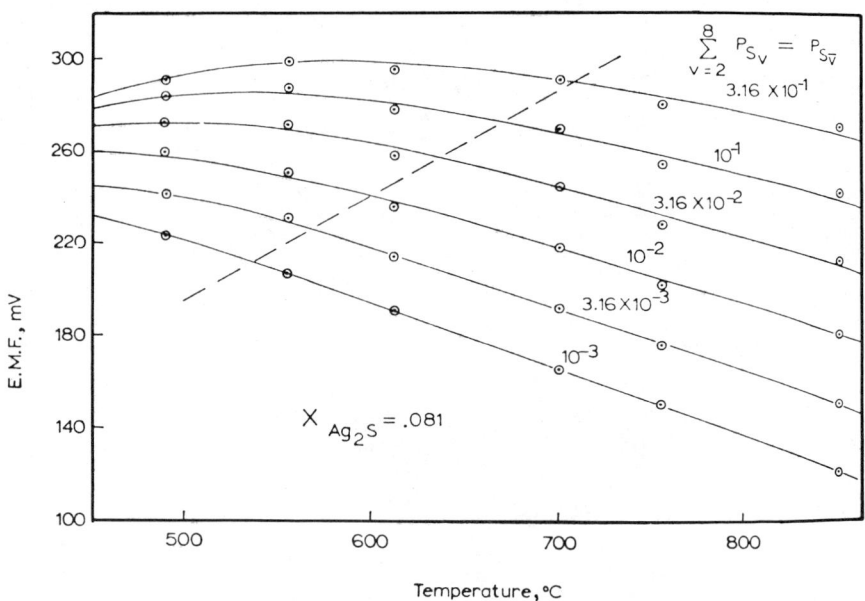

Fig. 7. Effect of total sulfur pressure on the temperature dependence of cell emf, from Thompson and Flengas (14).

sulfur system. Sodium enclosed in a β-alumina tube served as the sodium electrode, pyrolytic graphite as the sulfur electrode. The emf determining reaction is

$$2Na + (x - 1)Na_2S_x = xNa_2S_{x-1} \quad (x = 3,\ldots,5) \tag{1}$$

If free sulfur is present it dissolves in the polysulfide until saturation

$$S + Na_2S_{x-1} = Na_2S_x \quad (x = 3,\ldots,5) \tag{2}$$

for an overall reaction

$$2Na + xS = Na_2S_x \quad (x = 2,\ldots,5) \qquad (3)$$

From the emf for the potential determining reaction, its partial equivalent free energy has been calculated. Its temperature dependence has yielded partial enthalpy and entropy values. By application of the Gibbs-Duhem equation, using the fact that at the point of saturation of the polysulfide with sulfur the free energy of the sulfur dissolution reaction becomes zero, the partial free energy of reaction (Eq. 2) has been determined. The sums of the two partial free energies (Eqs. 1 and 2) gave the molar free energies of formation of the overall reaction of sodium and sulfur (Eq. 3) for various values of x.

Applying emf measurements for the determination of the phase-diagram offered a distinct advantage in this system. Polysulfides have a strong tendency toward supercooling. With emf measurements there is theoretically no limit as to how long one may wait for equilibrium to be established, whereas thermal measurements must be run at a certain minimum rate. For practical reasons, mainly heating curves were used in this investigation. Figure 8 shows the resulting phase diagram with the old data from Pearson and Robinson (11) for comparison. Also shown are data obtained by Oei (12) using DTA heating curves. The diagram compares well with the one developed contemporaneously by Rosén and Tegman (13) with microscope hot-stage observations and thermal analysis.

In Fig. 9, emf is plotted as a function of composition at various temperatures. Unfortunately, it is not possible to draw any conclusions concerning the number of electrons in the potential determining reaction from these emf measurements. Since the individual species actually present in the melt are not known, it is

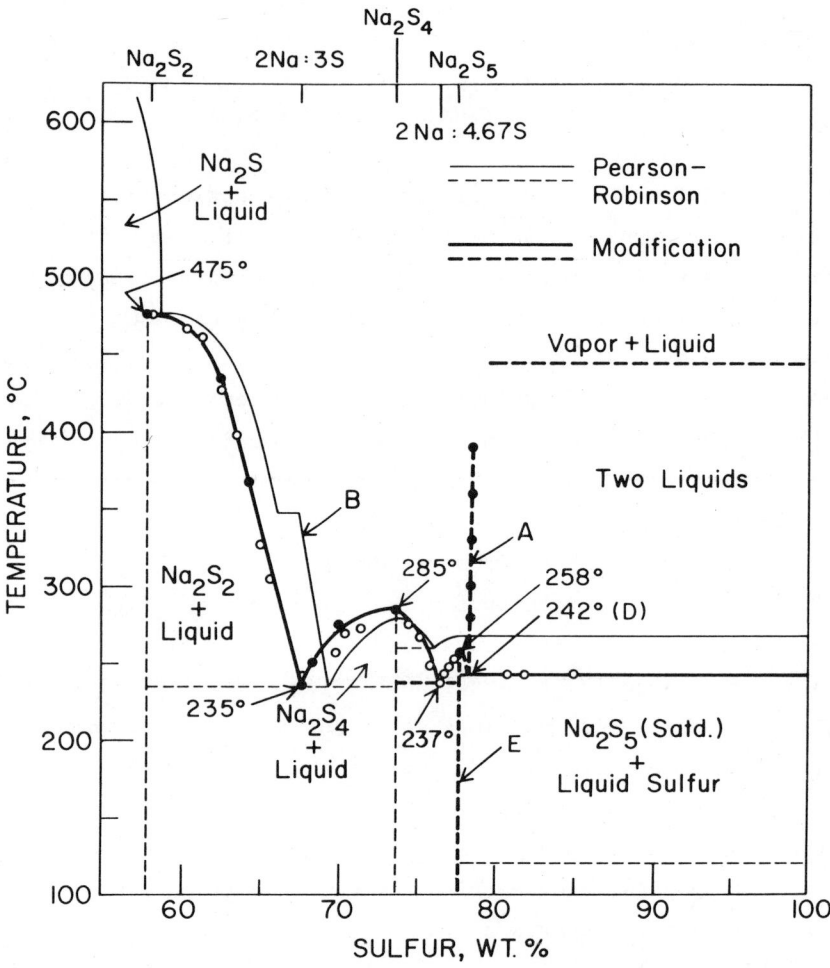

Fig. 8. Na$_2$S$_2$-S phase diagram according to Oei, Gupta, and Tischer (10,12); solid circle, data from emf measurements; open circle, data from DTA measurements.

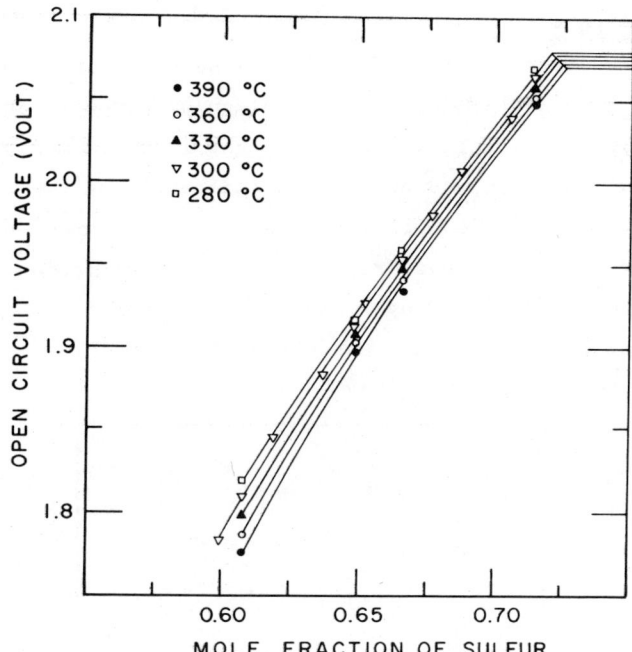

Fig. 9. Emf versus composition at various temperatures, from Gupta and Tischer (10).

impossible to assign activities. Furthermore, even if simplifying assumptions are made as to the species present, the range of concentrations is too narrow to discriminate between different assumptions.

However, it is quite possible to calculate the activity of sulfur as a function of melt composition and compare the results to data independently determined via the vapor pressure. This is precisely what Cleaver and Davies (31) have done. Applying the Gibbs-Duhem equation and using the fact that sulfur activity is equal to one in a sulfur-saturated polysulfide melt, they obtained sulfur activities from emf measurements in a sodium-sulfur cell.

A transpiration method was also used independently to obtain sulfur activities from the rate of weight loss under inert gas flow, taking into account the complicated molecular composition of sulfur vapor (16). Initially, rate of weight loss was measured directly, but later it was determined indirectly by emf measurements for the sulfur-rich polysulfides. In the region where sulfur activity is low and evaporation slow, Ag_2S formation cells similar to those described by Kiukkola and Wagner (4) were used for measuring the sulfur vapor pressure directly. The results are summarized in Fig. 10

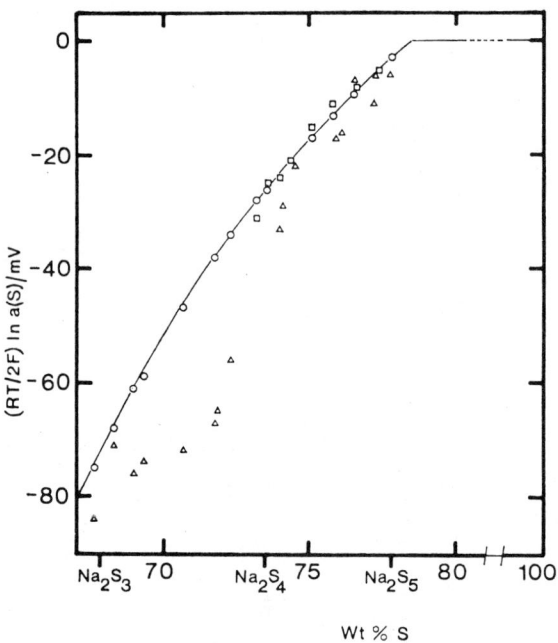

Fig. 10. Variation of sulfur activity with composition for sodium polysulfide melts at 360°C; squares, results from transpiration experiments; triangles, results obtained with Ag_2S cells; circles, values calculated from sodium-sulfur emf measurements, from Cleaver and Davies (31).

The sulfur activity in sodium polysulfide determined by emf measurements is in good agreement with that determined by vapor pressure methods. Sulfur electrode potential differences were also measured in a concentration cell with transference consisting of two compartments, with polysulfide in one and saturated polysulfide in the other, separated by a frit. Cleaver and Davies (31) compare these potential differences with those obtained for the same polysulfides against a sodium electrode.

Their conclusion is that sulfur in the polysulfide melt is present only in the polysulfide ion and not as free sulfur. They arrive at this conclusion via the phenomenologic equations of irreversible thermodynamics. An intuitive physical argument can be presented in support of their rather important conclusion. By basing their argument on the difference in potentials of the cell with transference to the same cell without transference, they are in effect measuring a liquid junction potential. They find this potential to be zero within experimental error. If the sulfur anion were of identical composition in both half cells, and the difference were only in the concentration of salt and free sulfur, then a concentration cell with an appreciable liquid junction potential would exist. If, on the other hand, all the sulfur is incorporated as polysulfide, then the only difference in the two half cells is in the composition of the polysulfide anion. It is reasonable to expect that the mobility of the various polysulfide anions relative to the sodium cation would be very similar, and, therefore, that the liquid junction potential is negligible.

2 ORGANIC SOLVENT SOLUTIONS

The electrochemical reduction of sulfur in DMSO has been thoroughly investigated by Sawyer and co-workers (17) with cyclic voltammetry, coulommetry, and spectrophotometry. These studies are of particular interest because aprotic solvents and, more generally, many organic solvents stabilize higher polysulfide species which are not stable in the pure salt form, in melts, or in aqueous solution. Also, sulfur is sufficiently soluble in some of these solvents to permit study of its oxidation from the elemental state (not via the polysulfide ion) without the usual difficulties posed by a nonconductive solid or liquid electrode reagent.

The cyclic voltammogram of sulfur in DMSO in Fig. 11 clearly shows two pairs of distinct reduction and oxidation peaks plus two minor peaks apparently caused by a protonated sulfide impurity (proved by acid additions). The first reduction peak at −0.60 V is believed to be due to the reaction

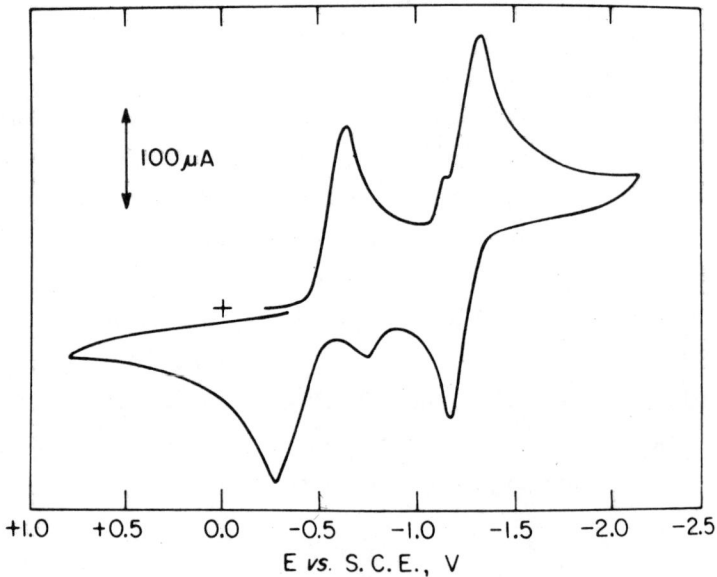

Fig. 11. Cyclic voltammogram of 2.9×10^{-3} F S_8 in DMSO (0.1 F TEAP) at a scan rate of 0.1 V/sec (Au electrode), from Martin, Doub, Roberts and Sawyer (17).

$$S_8 + 2e^- \rightarrow S_8^{2-}$$

which may be followed by a dismutation (18)

$$4S_8^{2-} = 4S_6^{2-} + S_8$$

as evidenced by the appearance of blue color to be discussed further on. The corresponding anodic peak at −0.3 V is due to oxidation of any polysulfide

$$S_x^{2-} \rightarrow \left(\frac{x}{8}\right) S_8 + 2e^- \quad (x = 2,\ldots,8)$$

The second reduction wave at −1.29 V shows further reduction of S_8^{2-},

$$S_8^{2-} + 2e^- \rightarrow S_8^{4-}$$

The anodic peak at −1.17 V for the reverse reaction becomes smaller at lower scan rates because the S_8^{4-} is unstable and dissociates slowly:

$$S_8^{4-} \rightarrow 2S_8^{2-}$$

The potential range accessible in this solvent does not permit further reduction of S_8^{4-} or S_4^{2-}. Although the voltammogram looks at first sight quite similar to those obtained in fused salts and pure polysulfide melts (discussed in Section 3), interpretation aided by additional experiments leads to different mechanisms. The shape of the peaks in DMSO indicates that all reactants are sufficiently soluble so that reactions do not lead to layer formation, as observed in some other media.

The dependence of electrochemical reaction mechanisms of polysulfides on the environment is related to the varied stability of different polysulfide species in different media. In pure polysulfides the stability of a particular polysulfide ion depends on the cation and on temperature; in mixtures and solutions it also depends on the presence of other molecules, especially those forming the solvation shell to the polysulfide ion. To quote well-known examples, K_2S_3 and K_2S_6 are stable compounds, whereas Na_2S_3 is difficult to make and unstable above 100°C (12), and Na_2S_6 exists only in certain solvents. In general, longer chains are stabilized in solution, but dissociation into monobasic sulfide ion radicals is also favored. With increasing temperatures the average chain length decreases and dissociation increases.

All these considerations make the differences of the sulfur electrode behavior in different media quite plausible; however, they do not explain why in DMSO oxidation of S_x^{2-} (2,...x...,6) occurs at practically one and the same potential independent of x, whereas reduction of S_x^{2-} (3,...x...,6) is hindered enough to push the corresponding potential out of reach.

Figure 12 shows the absorption spectra of various polysulfides in DMSO. Note that the sum of the curves for Na_2S_4 and Na_2S_8 is quite different from the curve for Na_2S_6, which indicates that the pronounced peak at 618 nm responsible for the blue color of these polysulfide solutions is connected with the S_6 ion. However, solid Na_2S_6 is orange. Thus it is concluded that in solution S_6^{2-} is dissociated,

$$S_6^{2-} \rightleftarrows 2S_3^-$$

and that S_3^- is causing the absorption at 618 nm and the blue color. This is in agreement with the findings of Bonnaterre and

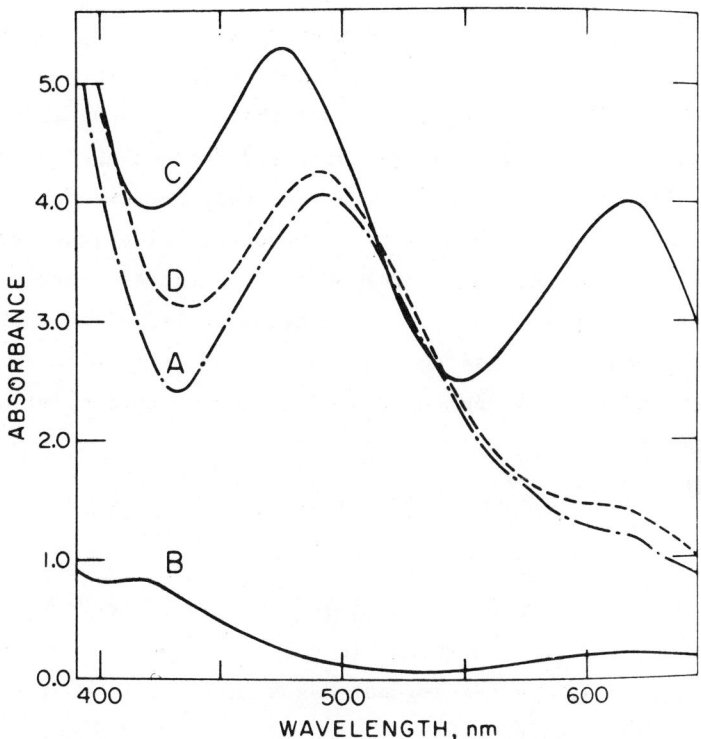

Fig. 12. Absorption spectra of polysulfides in DMSO solutions in 0.1 cm cells. A, 10 mF Na_2S_8; B, 10 mF Na_2S_4; C, 20 mF Na_2S_6 (or 10 mF Na_2S_8 plus 10 mF Na_2S_4); D, sum of curve A plus curve B, from Martin, Doub, Roberts and Sawyer (17).

Cauquis (18) and with Chivers and Drummond (19). Seels and Güttler (20) found that the blue color is strongly temperature dependent in DMF, with the dissociation of dibasic polysulfide ions completely repressed near the freezing point of DMF (-61.0°C), where the solutions become yellow. Giggenbach (21) had earlier assumed that S_2^- is the blue species. The controversy about the origin of the blue color that had been going on for some time has been reasonably settled by these recent papers, as again stated by Chivers (22). There is an extensive review of work in the area of homonuclear sulfur species by Chivers and Drummond (23).

The sulfur electrode has been proposed for use as a cathode in organic solvent primary batteries. However, current densities must be low to achieve practical electrode utilization. For instance, Coleman and Bates (24), using electrodes compressed from sulfur and carbon powders, reached a utilization of 20% with a current density of 2 mA/cm^2 in DMSO at a cutoff voltage of 2.0 V. This figure is calculated with respect to completion of the reaction

$$S + 2_e^- = S^=$$

Even if one considers that a utilization of one electron per sulfur atom might be a more realistic goal that would double the percent utilization, the development has a long way to go toward more acceptable power outputs.

3 FUSED SALT MELTS

The electrode kinetics are best understood in the pure sodium polysulfide melt and in solutions of Na_2S_x in KCNS. We discuss slags and then attempts at clarifying the sulfur electrode kinetics in chloride melts. Afterward, we develop the more consistent picture of the kinetics in pure melts and KCNS solutions, and the similarities and differences of the kinetics in these two media. Finally, we discuss some practical electrodes and an application in a cell where mass transport based on free convection and wetting properties becomes important.

3.1 Slags Containing Sulfide

Sulfur plays an important role in extractive metallurgy. Thus one would expect an interest in the electrochemistry of sulfide ions in slags. However, the experimental difficulties must have deterred researchers, because we have been able to find only one investigation in this field. Toporishchev, Esin, and Kalugin (25) investigated interfacial polarization for transfer of various ions through the boundary cast iron or steel-slag, and calculated diffusion coefficients. In particular, the transfer of sulfur ions from carbon saturated iron into a calcia-alumina-silica slag with 3% CaS was studied. The slopes of an η versus $\log\left(1 - \frac{i}{i_\ell}\right)$ plot (with i_ℓ the limiting current) of polarographic sweeps and of an η versus log t plot from constant current transients clearly indicate a two-electron process. The question is whether to accept the authors interpretation,

$$S(me) + 2e^- \rightarrow S^{2-}(sl)$$

or to assume a more or less fully ionized sulfur already in the iron, so that the interfacial process is just the transfer of S^{2-} ions,

$$S^{2-}(me) \rightarrow S^{2-}(sl)$$

The possibility that the current in this experiment might actually be carried by Fe^{2+} ions, possibly present in the slag, and moving in the opposite direction, was excluded by showing that the calculated diffusion coefficient corresponds to the diffusion coefficient for sulfur in iron, and not to the one for Fe in slag.

Rempel' and Malkova (26) give decomposition potentials for various anions in a NaCl-KCl melt at 700°C. From the current voltage curve for a melt containing 0.01 n sulfide (generated cathodically from a PbS electrode) the authors read a half-wave potential of -0.7 V versus chlorine for the discharge of sulfur. Similar curves taken with S^{2-} in the presence of SO_4^{2-} are less well defined. S^{2-} and SO_4^{2-} ions can hardly be expected to be stable in the presence of each other at this temperature. The results of Rempel' and Malkova seem to bear this out. They report that the addition of sulfur, sulfide, or sulfate with sulfide to magnesium electrolysis baths has been recommended for removing oxygen that is detrimental to the graphite anodes. Here the combination of sulfide and sulfate is obviously used as a convenient means for in situ generation of sulfur that might otherwise be difficult to get into a high-temperature bath in elemental form.

3.2 LiCl-KCl Eutectic Melts

Kennedy and Adamo (28) added a weighed amount of sulfur to an isolated electrode compartment in LiCl-KCl eutectic at 420°C and performed controlled potential electrolysis just negative of the potential of the first cathodic peak shown in Fig. 13, and then continued negative of the potential of the second cathodic peak. They worked on gold electrodes. By isolating the electrode compartment, loss of sulfur vapor was kept to a minimum. Blue color was observed at the electrode surface both in the cyclic sweep and controlled potential electrolysis at the potential of the first cathode wave. The blue color disappeared on electrolysis at the potential of the second cathodic wave. On oxidation at controlled potential of the first anodic wave, blue color reappeared, and on electrolysis at the second anodic wave sulfur condensation occurred in the isolation compartment. The number of coulombs involved in the electrolyses is consistent with

$$S_4 + e^- \rightarrow S_4^-$$
$$S_4^- + 3e^- \rightarrow 2S_2^=$$

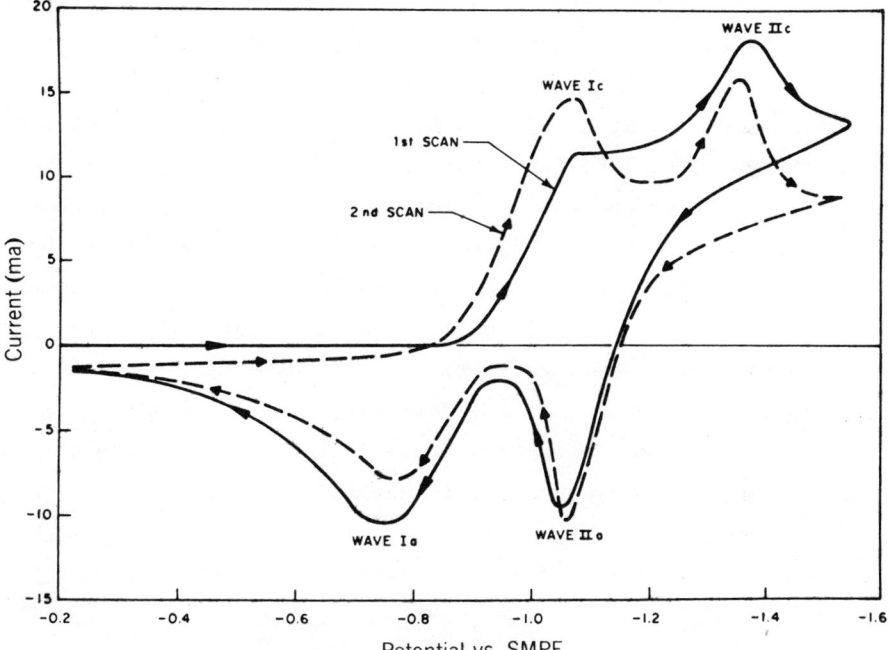

Fig. 13. Cyclic voltammetry of sulfur in LiCl-KCl eutectic at 420°C; sweep rate 200 mV/sec, from Kennedy and Adamo (28).

Kennedy and Adamo point out that the electrode reduction steps need not coincide with the final reduction products of the long-time electrolyses. It is interesting to observe that the features of two cathodic waves and two anodic waves seen in Fig. 13 are similar to those found in the KCNS melts (cf. Fig. 36c) and in aprotic solvents (cf. Fig. 11). In the aprotic solvent DMSO at 25°C, it was shown that the second cathodic wave gave the reduction product $S_4^=$ (17). The second cathodic wave for Na_2S_x (where $1 \leqslant x \leqslant 5$) dissolved in KCNS melts in the range 200° to 250°C was reduction to $S^=$ (68), but the two reduction waves were separated by about 0.8 V with an intervening structure of several additional overlapping reduction waves. Kennedy and Adamo propose that the second reduction wave in LiCl-KCl at 420°C is to $S_2^=$. Cleaver et al. (35) find that a solution of Na_2S_2 in LiCl-KCl eutectic at 420°C gives a cyclic voltammogram (also on a gold electrode) similar to Fig. 13 except that cathodic wave I is a stripping peak and waves I and II are reversible waves. They are separated by about 0.4 V as are I and II in Fig. 13. Kennedy and Adamo found that the cathodic and anodic peak potential separation of wave I was invariant with sweep rate from 40 to 200 mV/sec, but that the peak separation indicated irreversibility. Over a limited sweep rate range, these two seemingly contradictory observations can be reconciled by assuming a catalytic mechanism (42):

$$S_2^- + e^- \rightarrow S_2^=$$
$$S_2^= + S_4 \rightarrow S_2^- + S_4^-$$
$$\overline{S_4 + e^- \rightarrow S_4^-} \quad \text{(net reaction)}$$

Polysulfides, K_2S_x, for $x > 2$, are unstable in LiCl-KCl at 420°C (35). However, both S_2^- and S_3^- exist at these temperatures in LiCl-KCl (30). The S_2^- radical anion is present in higher concentration than S_3^- (32), but S_3^- is responsible for the blue color (30). This S_2^- mechanism could therefore be the primary reaction with a secondary parallel mechanism, which might be

$$2 S_3^- + 4e^- \rightarrow 3S_2^=$$
$$3 S_2^= + 4S_4 \rightarrow 2S_3^- + 4S_4^-$$
$$\overline{}$$
$$S_4 + e^- \rightarrow S_4^- \quad \text{(net reaction)}$$

As the sulfur disappeared at the end of the controlled potential electrolysis of the first cathodic wave, the blue color decreased (28). This is further confirmation for the above mechanism.

Birk and Steunenberg (32) obtained cyclic voltammograms at carbon electrodes by adding 0.065 wt % Li_2S to LiCl-KCl eutectic at 410°C. Their data, summarized in Fig. 14, show that the first peak on cathodic scan decreases and the second one increases at higher sulfide concentrations, at increased temperatures, and at slower scan rates. This is indicative of a coupled slow chemical reaction involving sulfide in which the chemical reaction deprives the first cathodic reaction of a reactant and supplies the second cathodic reaction with a reactant. In addition, they suggest that since the third cathodic peak is unaffected by changes in the relative heights of the first two cathodic peaks, the same product is produced during the first two cathodic waves and is the reactant for the third wave. Thus their proposed reactions are consistent with their observations.

There appears to be a gross inconsistency between the data of Birk and Steunenberg (32) and those of Kennedy and Adamo (28). Although the electrodes used were different, the systems appear sufficiently similar to allow comparisons of the data. Kennedy and Adamo find that their second reduction process is an $S_4^-/S_2^=$ couple, whereas Birk and Steunenberg suggest the couple 0.4 V, negative of the sulfur/polysulfide couple is the $S_2^=/S^=$ couple.

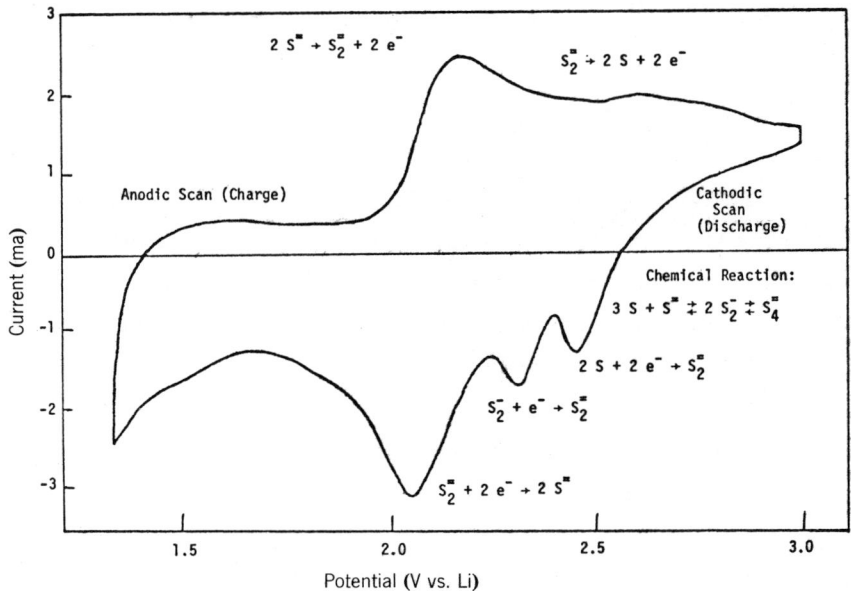

Fig. 14. Cyclic voltammogram of LiCl-KCl 0.006 wt % Li_2S; scan rate 66 mV/sec, from Birk and Steunenberg (32).

The second inconsistency is that Birk and Steunenberg do not have a mechanism that is consistent with the production of blue color even though blue color has been observed in $S^=$ containing LiCl-KCl (29). We suggest that the systems, although similar, are different enough so that the two sets of data are not necessarily in conflict. First, Na_2S is insoluble in the LiCl-KCl eutectic whereas Li_2S is slightly soluble (32,35). Thus, solubilities and stabilities of sulfur species in the molten system depend on the environment (cations, solvation, etc.), and no definitive comparison can be made of the various couples if the systems are originally different.

Unfortunately, the experimenters have not measured the equilibrium potentials for $S/S_2^=$ and $S_2^=/S^=$, which then could be compared with the $E_{1/2}$ values of cyclic voltammograms or chronopotentiograms. It would be helpful to know, for example, the equilibrium potential obtained upon addition of an equimolar mixture of Li_2S and Na_2S_2 to the LiCl-KCl eutectic.

The probable presence of blue color (not mentioned by the authors) due to the reaction products created during cyclic voltammetric experiments by Birk and Steunenberg (32) can be explained by either of the following two reactions involving S_3^-:

$$5S + S^= \rightarrow 2S_3^-$$
$$4S + S_2^= \rightarrow 2S_3^-$$

Bernard, DeHaan, and Van der Poorten (29) in their chronopotentiometric studies started with a LiCl-KCl melt (400 to 420°C) to which CaS had been added. The melt therefore contained only S^{2-} ions. They observed three steps for oxidation and only one for reduction.

At high currents, their $i\tau^{1/2}$ values were constant, whereas at low current they were able to use $i\tau^{1/2}$ values to follow the change with time of the bulk concentrations of the four species in the melt over a time of approximately 2 hr. The first two oxidation steps can be safely assigned to oxidation of S^{2-} and of S_x^- or $S_x^=$. For the third oxidation step the authors give the improbable interpretation

$$S_y^- \rightarrow S^+ + S_{y-1} + 2e^-$$

This seems to be a novelty among two-electron processes proposed. Since sufficient sulfur is produced in the preceding step, a process like

$$S_x \rightarrow S^{2+} + S_{x-1} + 2e^-$$

or

$$S_x + 2Cl^- \rightarrow S_2Cl_2 + S_{x-2} + 2e^-$$

would be more probable. That only one reduction step is observed with this method is probably the result of the long time span between the oxidation and reduction. This time is sufficient to allow the following chemical reaction to go to completion:

$$2S^= + 2S_2^- \rightarrow 3S_2^=$$

S_2Cl_2 escapes or reacts with the sulfide; sulfur also evaporates, leaving only S_2^{2-} behind. The $S_2^=$ is then reduced to $2S^=$ in the one observed reduction step. The evaporation of sulfur was conspicuous by condensation in the cooler parts of the vessel.

Bernard et al. (29) found that sulfur was completely insoluble in a pure, sulfide-free, LiCl-KCl melt. Bodewig and Plambeck (27) found that both sulfur and sulfide must be present in LiCl-KCl to produce a blue color. However, they incorrectly assigned their electrochemical reduction step on graphite and gold electrodes to the reaction $S + 2e^- \rightarrow S^=$, by assuming that sulfide was the reaction product of their coulometric reduction of sulfur and then finding a Nernst slope with n = 2. The concentrations cannot be assumed and then plotted to verify the assumption. They always maintained an excess of liquid sulfur and therefore did not know the number of coulombs per S atom. Their plot is given in Fig. 15. Comparison of their voltammetric data in Fig. 16 with the data of Fig. 13 indicates that they operated at the potential of the first reduction wave, which is most certainly not reduction to $S^=$. Bodewig and Plambeck claimed that earlier work by Delarue (2) was difficult to reproduce because of corrosion of the Pt electrode used by Delarue.

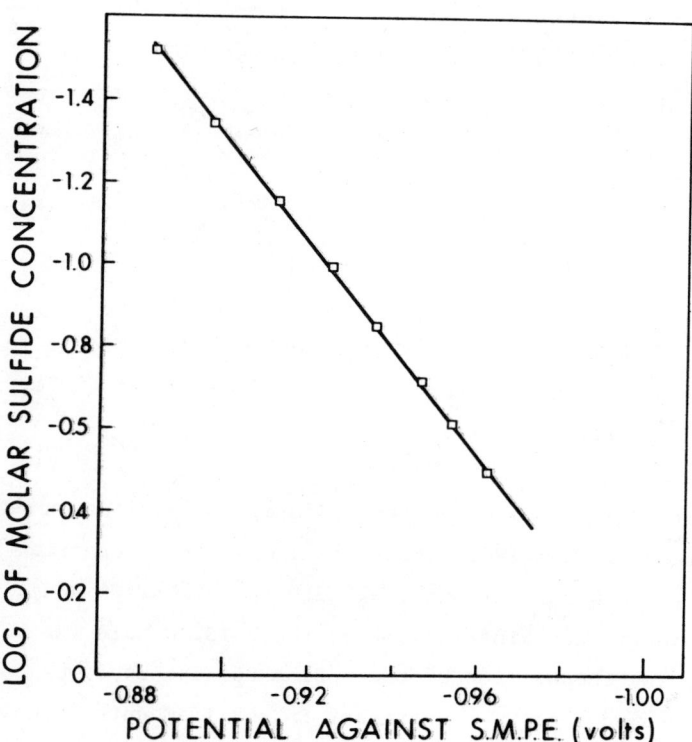

Fig. 15. Emf of sulfide electrode (versus 1 M Pt/Pt^{2+}) as a function of sulfide concentration at 450°C, from Bodewig and Plambeck (27).

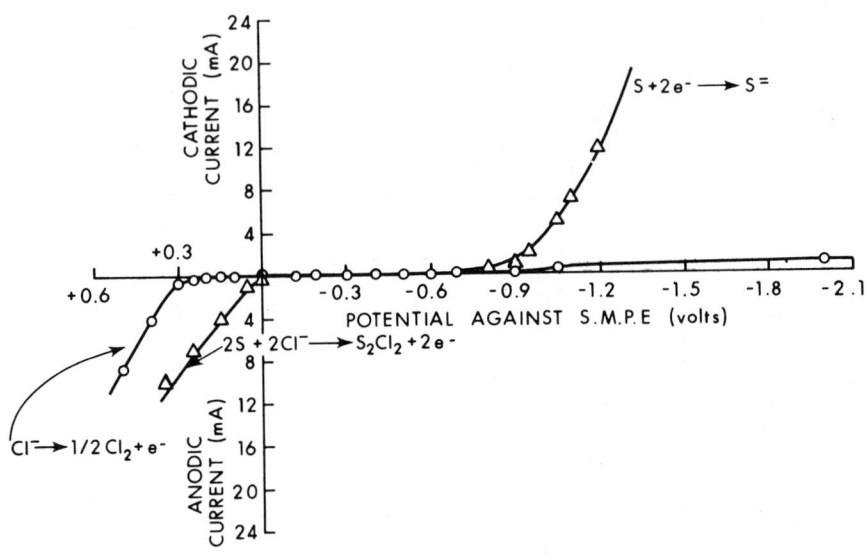

Fig. 16. Voltammetric curves of pure LiCl-KCl eutectic (circles) and of eutectic with sulfur present (triangles) at 420°C, from Bodewig and Plambeck (27).

Weaver (33) has examined the oxidation of sulfide on Au, Pt, and vitreous carbon (VC) electrodes in LiCl-KCl eutectic, 10^{-3} to 10^{-1} molar in Na_2S over the range 360 to 530°C. In addition, some potentiometric and kinetic work on the $S^=/S_2^=$ couple was done; however, the composition of the sodium sulfide used was unknown. Weaver assumed that $S_2^=$ was present in small amounts [however, see (35) concerning the solubility of Na_2S]. Anodic chronopotentiometric data were interpreted on the basis of a blocking sulfur film. The data were similar on Au, Pt, and VC electrodes except that the platinum gave evidence of the additional formation of a platinum sulfide film. Steady-state voltammetric, anodic, and cathodic galvanostatic charging, and open-circuit decay techniques were used to examine the nature of the sulfur film. No definite conclusions were formed, although a mechanism was proposed:

$$S^= \xrightarrow{-2e^-} Sp \text{ (polymer precursor units)} \rightleftarrows S \text{ (polymer film)} \xrightarrow{+S^=} S_2^=$$

The proposed rate controlling step is the chemical reaction of sulfide with the polymer film. Weaver also repeated the above work on Au and Pt electrodes in $NaNO_3$-KNO_3 eutectic at 250°C. Similar results to those in the chloride system were found. Quantitative interpretations were hampered because the sulfide reacts slowly with the melt.

3.3 Molten Sodium Polysulfides

Most investigations of electrode kinetics have been carried out in dilute solutions of the reacting species containing a supporting electrolyte, rather than from a pure melt of a salt of the redox system. However, for the sulfur sulfide system there was strong interest in establishing the electrode kinetics in such a pure melt in conjunction with the development of the sodium-sulfur battery.

Some preliminary discussion of the pure melt as an electrochemical system is needed before an examination of the experimental data. These arguments are a further development of those previously proposed by Ludwig (34). It is shown later in this section that 1.) a Na_2S_4 melt consists primarily of Na^+ and $S_4^=$ ions; 2.) that the entire anion concentration is electroactive; 3.) the $S_2^=$ ion is the product of reduction at the electrode; 4.) there is no chemical reaction before this charge-transfer reaction. It is further shown that the usual electroanalytical equations for diffusion control are obeyed within a few percent and that the experimental errors of the measurements are also within a few percent. How, then, do we perceive of diffusion in a pure binary salt? Furthermore, is it possible to prove that the usual electroanalytical diffusion equations should, indeed, apply to our special case?

With the constraints established so far, diffusion can occur only if the electrode reaction product and the reactant counterdiffuse. For a quantitative study of the nature of the diffusion coefficients of the reactant and product with respect to the electrode as a frame of reference, the methods of the thermodynamics of irreversible processes must be used (36). However, there is adequate justification for the direct application of the usual electroanalytical equations.

Crank (37) and others (38) have shown that interdiffusion in a two-component liquid, in which there is no volume change on mixing, yields $D_O = D_R = D_{OR}$ where O and R are the two components. Furthermore, the volume-fixed frame of reference is identical to the cell-fixed frame of reference. However, in the experiment under consideration, we are eliminating O and creating R in a reaction $O + 2e^- \rightarrow 2R$ ($S_4^= + 2e^- \rightarrow 2S_2^=$) at the electrode. The molar volume of Na_2S_4 is 1.5 times that of Na_2S_2. Therefore, we do not have a constant total volume and the volume-fixed frame (V) is moving with respect to the cell-fixed frame of reference (C). However, Sundheim (36) has pointed out that, under these conditions $D_{OR}^V \simeq D_{OR}^C$, and Dullien and Shemilt (38) have shown that using the diffusion equations without corrections for volume changes introduces small errors only. Laity (39) and Wagner (40) hold that interdiffusion in a mixture of two binary salts with a common ion in the presence of a potential gradient as well as a concentration gradient can still be expressed in terms of $D'_{OR} = D'_O = D'_R$ where D' values are "effective diffusion coefficients" containing averaged mobility terms.

The general theory of diffusion-limited charge-transfer processes in electroanalytical techniques (41,42) assumes Fick's second law to be applicable to both species O and R, the effects of modes of mass transfer other than diffusion, for example, electrical migration, to be negligible, and D_O and D_R to be independent of distance x from the electrode or time t. The polysulfide anions are similar enough so that ideality may be expected. Therefore, assuming ideality in conjunction with the preceding arguments regarding volume changes, D_O and D_R should be nearly independent of x and t. The essence of the counterdiffusion coefficient is that Fick's law applies to both species. It may seem that the requirement above of the independent applicability of Fick's second law to O and R is more general than it is in the pure melt case, where the concentrations of O and R are rigorously interdependent because of electroneutrality requirements. However, in both cases the forcing boundary condition is given by the charge transfer at the electrode where the rates of arrival and removal of diffusing ions are fixed (41,42). Therefore, we can treat the pure melt case as we do the general case with the influence of migration included in the effective diffusion coefficient.

Thus the electroanalytical equations based on the simple Fick diffusion equations should be applicable to an electrochemically reacting pure binary salt. Small errors are expected. A rigorous mathematical treatment is called for in light of the surprisingly close fit of experimental data to the curves predicted by the general theory of diffusion-limited charge transfer.

For an investigation of electrode kinetics, it is quite essential to know the ionic species present in the electrolyte. The DTA and x-ray diffraction results of Oei (43) and, more recently the laser-Raman spectroscopic results of Janz (44) indicate that $S_3^=$ in sodium polysulfide is not stable in the solid above 106°C and therefore it is probably not stable at the higher temperatures of the molten state. Oei and Janz ascertained that above 106°C Na_2S_3 disproportionates into Na_2S_2 and Na_2S_4. The dissociation of polysulfide ions into ion radicals (S_2^- or S_3^-) in solutions in molten salts (22,23,30,45) and aprotic solvents (19,20,22,23) has been well documented. Preliminary e.s.r. measurements in molten sodium polysulfides show only very weak signals (46) between 300 and 400°C. This and the mounting repression of this dissociation with increasing concentration make it most probable that the concentration of ion radicals in the undiluted polysulfide melt is extremely small.

Transference experiments by Cleaver and Davies (31) have shown that there is no free elemental sulfur in the melt, that is, the melt contains Na^+ and polysulfide ions only. Thus the undiluted sodium polysulfide melts in the 300 to 400°C range contain mainly $S_2^=$, $S_4^=$, $S_5^=$, and some $S_6^=$, the relative amount of each depending on the overall composition. This assignment of distinct species to the melt is generally in accord with the results of Rosén and Tegman (47) and Cleaver and Davies (45). Rosén and Tegman (47) relate sulfur vapor pressures to the relative amounts of the various polysulfide anions. However, the existence of the various anion species is postulated and the distribution curves for the amounts of the various species are empirically fitted to the pressure data. They postulate that $S_3^=$ is present in appreciable amounts. Cleaver and Davies (45) determine the number of ions in dilute solution in KCNS by freezing point depression measurements.

The above conclusions concerning the species present in undiluted sodium polysulfide melts are fairly well founded. However, it would be desirable to obtain more direct evidence. The following discussion on sweep voltammetry does lead to some direct evidence in support of the above arguments.

Weber and Kummer (48) demonstrated that stable, steady-state potentials are obtained at graphite electrodes in sodium polysulfide melts. Three separate groups have studied kinetics in the pure sodium polysulfide melts: Selis (49), South et al. (50), and Ludwig et al. (51). Selis worked on platinum and graphite electrodes, whereas the latter two groups worked at vitreous carbon electrodes. The experimental results, however, are quite similar and also do not seem to be influenced by origin or preparation method of the melt. The work of Selis (49) and of the other groups provides clear evidence that the results of investigations with transient methods are independent of the electrode material used (graphite, vitreous carbon, platinum) for reduction of the melt. As is discussed below, the kinetics of oxidation of the melt differs substantially at metal and carbon electrodes.

The salient feature of all voltage sweep diagrams is a clear distinction between the $S/S_x^=$, the $S_x^=/S_2^=$, and the $S_x^=/S^=$ reactions. In addition, the diagrams always show a final rise of current at large polarizations. Cathodically, this rise represents the deposition of sodium (50). The final rise on the anodic side is more difficult to interpret. For lack of any other possibility and since the carbon electrodes do not seem to take part in the reaction, we can safely conclude that it is sulfur that is oxidized in this region to positive sulfur ions.

It is difficult to determine the kinetics of a process without knowing the concentrations of the diffusing species involved in the electrode reaction. Owing to our current inability to measure concentrations of individual ions in the melt, indirect means must be used to obtain the concentrations. In most electroanalytical techniques, the functional relationship between concentration and diffusion coefficient is $CD^{1/2}$. The $CD^{1/2}$ values can be combined with $CD^{2/3}$ values from the rotating disk technique to separate C and D. Ludwig (52) has used this approach to calculate nC and D for the primary reduction reaction (Na_2S_5 or Na_2S_4 to Na_2S_2) in both Na_2S_5 and Na_2S_4 melts. For Na_2S_5 at 350°C, he has combined the rotating disc data of Armstrong et al. (80) with the chronopotentiometric data of South et al. (50), and obtained a value of $D = 6.3 \times 10^{-7}$ cm^2/sec and a value of nC = 18.2 moles/liter. Since the maximum concentration of Na_2S_5 calculated from the molecular weight and density of Na_2S_5 can only be 9.02 moles/liter, it is probable that the entire concentration of $S_5^=$ is reacting and that n = 2.0. The proposition that $S_5^=$ is the reacting species rather than a dissociation product of $S_5^=$ will be supported later in this section.

For Na_2S_5 at 300°C, a value of $D = 2.4 \times 10^{-7}$ cm^2/sec can be calculated from the Armstrong et al. (80) rotating disc data. The activation energy for diffusion from these measurements agrees well with the activation energy for viscosity calculated from the data of Cleaver and Davies (54).

It is advantageous to combine rotating disk with chronopotentiometric data to determine nC and D. The chronopotentiometric equation does not contain a separate proportionality constant, as does the linear sweep equation. At present, for $Na_2S_{4.0}$ the data from both techniques do not exist at the same temperatures. However, the linear sweep and chronopotentiometric data of South et al. (50) at 350°C in Na_2S_4 can be combined to find a value of $n_1^{3/2} \chi(at)/n_2$ where $\chi(at)$ is the current function in the sweep equation:

$$i = nFAC_o^b D_o^{1/2} v^{1/2} \left[\frac{F}{RT}\right]^{1/2} \pi^{1/2} n^{1/2} \chi(at)$$

where $a = nFV/RT$, v is the sweep rate, C_o^b is the bulk concentration of the diffusing species, n_1 is the number of electrons associated with the linear sweep peaks and n_2 is the number of electrons associated with chronopotentiometric and rotating disk experiments. The value of $n_1^{3/2} \chi(at)/n_2$ can then be substituted into rotating disk (53) and linear sweep data (51) for Na_2S_4 at 300°C; a value of $D = 2 \times 10^{-7}$ cm^2/sec is obtained (52). Calculating C from the molecular weight and density of Na_2S_4, $n_2 = 2$ is obtained. The value of $n_1^{3/2} \chi(at)$ is discussed later.

These values of D seem low for fused salts, however the viscosity of 40 cp for Na_2S_4 at 300°C is much larger than the range of 0.5-5.0 cp for most fused salts (54), and the size of the $S_4^=$ ion is larger than that of most fused salt ions (55). The D can be estimated from the equivalent conductance of the melt using the Nernst-Einstein equation. A 10-50% increase of D over that calculated usually required in the application of this equation for molten salts (36); a value of $D = 7-9 \times 10^{-7}$ cm^2/sec is obtained (52). The equivalent ionic conductivity used in the equation was calculated from the equivalent conductance of the melt (56) and the estimated ionic radii of Na^+ and $S_4^=$ (55). The calculated D is in reasonable agreement with the experimentally determined effective D.

Armstrong et al. (80) claim that the presence of a film on the electrode surface precludes the interpretation of linear sweep measurements using the Randles-Sevcik analysis. We will show by means of conductance measurements superimposed on linear sweep experiments, that the film does not form at all during the cathodic peak of the second cycle in $Na_2S_{5.2}$ (see Figure 26); and that in $Na_2S_{4.0}$, most of the film formation occurs immediately after the cathodic diffusion peak. It will be shown in conjunction with Figure 18 that in $Na_2S_{4.0}$, melt plus film resistance at the peak of the linear sweep wave is only slightly higher than the melt resistance determined at the rest potential. Furthermore the "film" apparently precipitates within the vicinity of the electrode rather than on the electrode surface since the conductance data associated with the high currents of the third cycle in Figure 30 are identical to the conductance data shown in Figure 17; i.e., current rises as conductance rapidly falls. Thus we claim that in Na_2S_5, the Randles-Sevcik analysis applies rigorously since film formation does not occur during the formation of the wave peak; also that in Na_2S_4 some precipitation does occur which may alter the diffusion results slightly depending upon the particular electroanalytical technique used. However the precipitation effects are so slight that we believe that quantitative application of electroanalytical theory is justified.

The reduction reactions just discussed are typified by peak A in the cyclic sweep shown in Fig. 17 (51). The sweep begins in the cathodic direction from the rest potential at a vitreous carbon (VC) electrode of 0.00407 cm^2 area. All sweep curves that are discussed here begin from the rest potential. The cases in which diagnostic information is gained by starting a sweep at a potential other than the rest potential or holding a sweep at a certain potential are specified. Because of the very high current densities involved, care was taken to surround the working electrode with the counterelectrode to ensure uniform current distri-

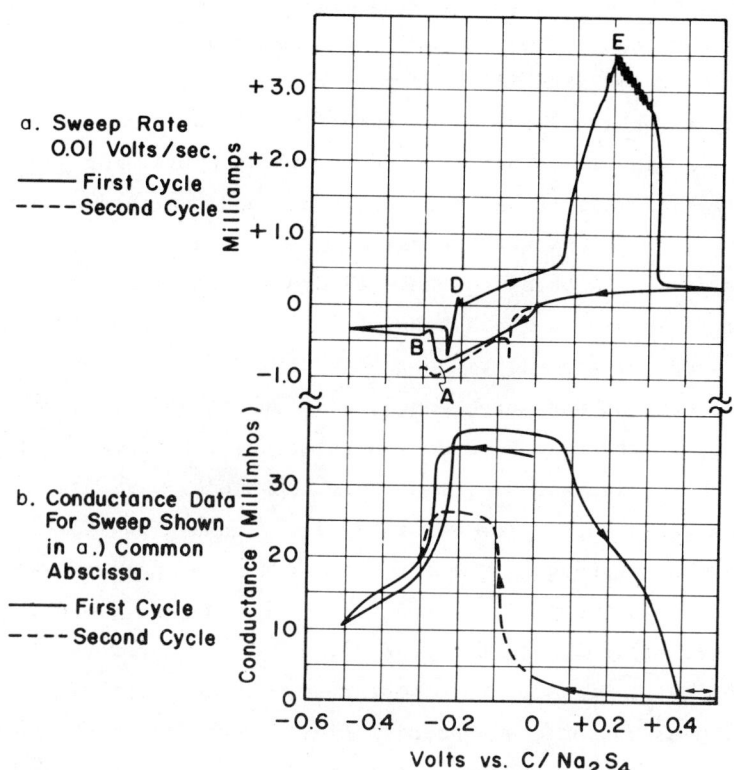

Fig. 17. Cyclic sweep and melt conductance in $Na_2S_{4.0}$ at 300°C. Vitreous carbon electrode area is 0.00407 cm^2. Sweep is iR uncorrected. From Ludwig (34).

bution over the working electrode. Differential conductance measurements were taken during the sweep as shown in Fig. 17 by superimposing a small-amplitude a.c. signal on the triangular sweep, at frequencies between 1592 and 10,000 Hz. The faradaic impedance has dropped to about 1% of the melt resistance at 1592 Hz (52). We return later to these differential melt conductance data; at this time we point out the constancy of the conductance from the

beginning of the wave to the wave peak A. The melt resistance
remains constant during the sweep. This is an important point
because of the desirability of having a truly constant sweep rate
in order to obtain good quantitative data. The increase of
current in Fig. 17 is nearly linear with potential for wave A.
Thus di/dt as well as dE/dt are constant and the iR drop can be
used directly to correct the sweep rate. Because of the high
current densities observed, this iR drop correction must be extremely accurate, since a Luggin capillary cannot be brought
close enough to the electrode to make the correction small. The
counterelectrode was large enough that it did not polarize, and
it was therefore also used as the reference electrode. Both current interruption and a.c. impedance techniques were used to measure the resistance. The values agreed within a few percent.
However, the exchange current for the reduction process was determined by an independent transient technique and found to be
very rapid (57). Since the peak potential for a reversible wave
is independent of sweep rate, observed peak potential was plotted
against peak current, as shown in Fig. 18, and the slope was interpreted as the resistance of the melt, with the intercept being
the true peak potential, E_p. This resistance was in close agreement with the interrupt values and slightly higher than the a.c.
impedance values. Using the value of the resistance from Fig. 18
to obtain a corrected sweep rate, the plot of Fig. 19 was obtained
(51). The linearity of the plot is indicative of simple diffusion
control without any complications from coupled chemical reactions.
The data of South et al. (50), shown in Fig. 20, were taken over
a much more limited sweep rate range and show an upturn at low
sweep rates for Na_2S_4 and Na_2S_5. They interpret this increase in
the current function as indication of a following chemical reaction in which sulfur in the melt reacts with the reduction product,

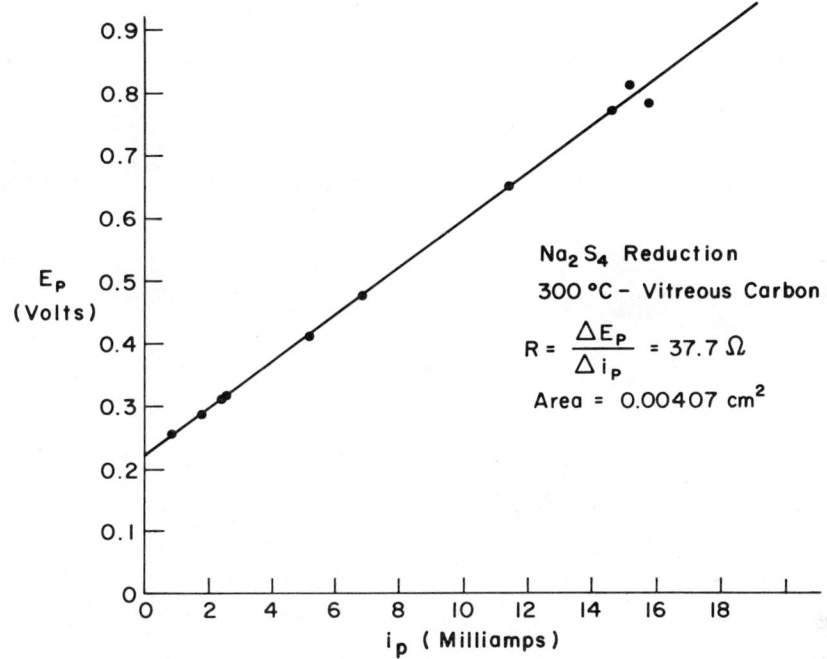

Fig. 18. Peak potential versus peak current for wave A, from Ludwig (52).

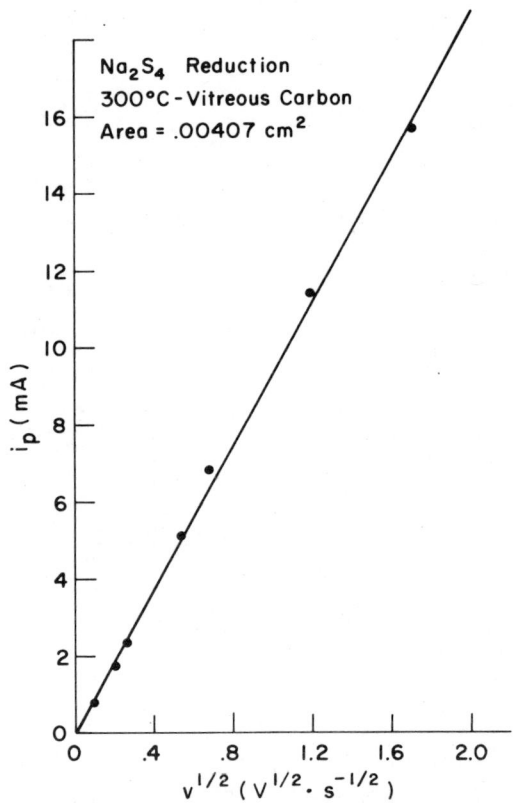

Fig. 19. Peak current versus sweep rate, iR corrected, for wave A, from Ludwig (52).

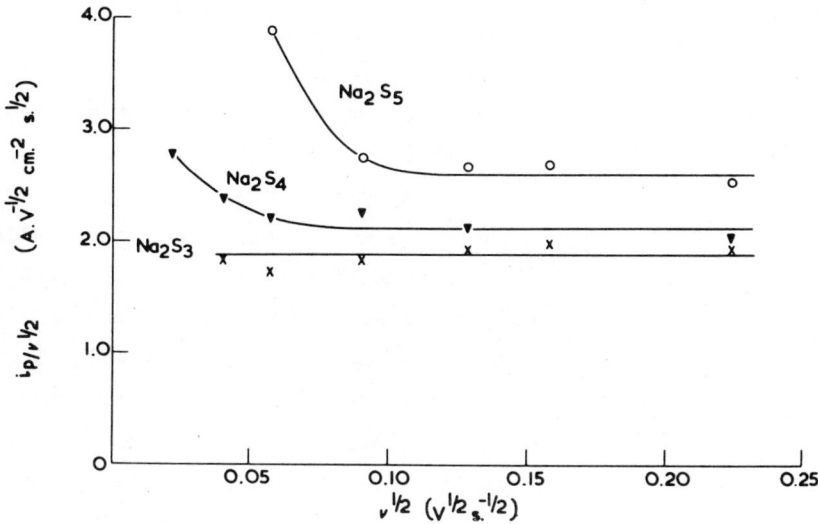

Fig. 20. Current function, $i_p/v^{1/2}$ versus $v^{1/2}$ for wave A in different melt compositions at 350°C, from South et al. (50).

Na_2S_2, to reform the original polysulfide [for the sequence $O + e^- \rightarrow R$, $R + Z \rightarrow O$, Nicholson and Shain (42) term the chemical reaction "catalytic" since the rate of reaction is enhanced because of the regeneration of O]. Our data in Fig. 19 do not indicate an upturn at low sweep rates. Indeed, South et al. also mention that below 0.01 V/sec the major portion of the cathodic curve no longer decreases in current. This strongly suggests that at these slow sweep rates, the onset of convection could easily account for the constant current and for the apparent increase in current function for Na_2S_4. The steeper rise for Na_2S_5, we feel, is partly due to convection but also in accord with the criteria for a catalytic step (42). We ascribe the catalytic reaction to $Na_2S_2 + 2Na_2S_5 \rightarrow 3Na_2S_4$. As is shown later, both Na_2S_4 and Na_2S_5 give very similar reduction waves. In other words, Na_2S_5 can react with Na_2S_2 to regenerate more reactant, but Na_2S_2 could only give Na_2S_3, whose existence in the melt is doubtful; therefore, we would not expect a catalytic step for Na_2S_4. South et al. (50) conclude that Na_2S_2 reacts with elemental sulfur because they measured sulfur activities in the melt of 0.9 in Na_2S_5, 0.2 in Na_2S_4, and 0.04 in Na_2S_3, based on unit activity for pure sulfur, which is in equilibrium with a composition of approximately $Na_2S_{5.2}$ at 350°C. However, these activities do not necessarily mean that the sulfur exists as a free element in the melt. According to the previously mentioned transference experiments of Cleaver and Davies (31), it does not.

South et al. also suggest that the reduction wave represents the reduction of polysulfide rather than sulfur, because the heights of the waves in the three melts are nearly the same, whereas the activities of sulfur vary significantly.

The i_L^{-1} versus $\omega^{-1/2}$ linearity of the rotating disk data shown in Fig. 21 is specific to a first-order reaction. The zero intercept is confirmation that the charge-transfer rate is very high. The data in Figs. 18, 19, and 21 have been shown to indicate the close experimental agreement with theory and the apparently high experimental accuracy that can be attained in the measurements. Thus, even though the $CD^{1/2}$ versus $CD^{2/3}$ comparison involves a $\frac{1}{6}$-power dependence, the data seem reliable enough to yield a value of D and C that should not exceed the error implied earlier. With constant use of the vitreous carbon electrode for either anodic or cathodic sweeps, a slow change in the availability of surface sites for reaction is obtained. This available area is very easily ascertained by conductance measurements (as illustrated in

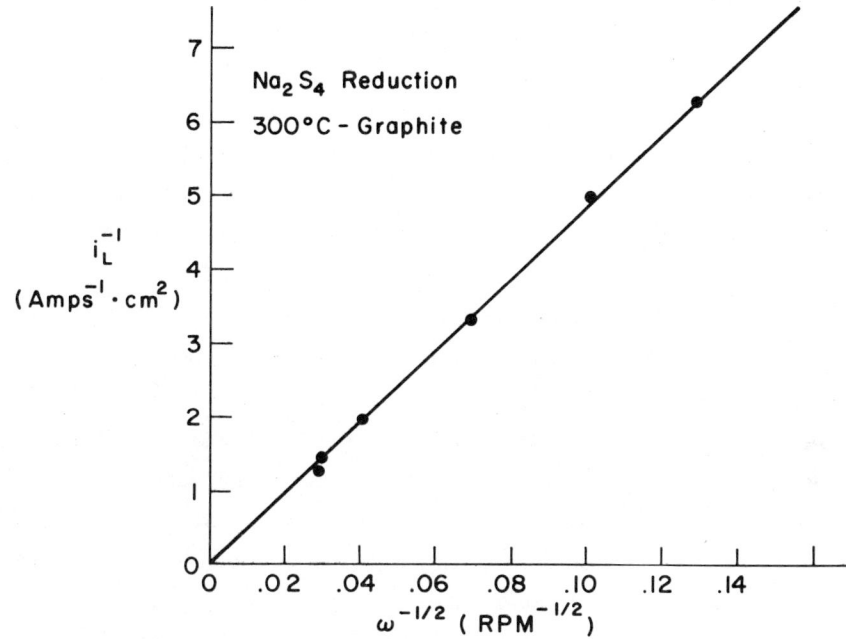

Fig. 21. Rotating disk diffusion limited current versus rotation rate, from Dzieciuch (53).

Fig. 17) at the open-circuit potential (OCV). The original conductance can be regained by holding the potential at voltages more negative than 0.9 V versus the OCV for approximately 1 min. During this time, the current increases rather than decreases. The reduction at these more negative voltages in KCNS solutions is shown in Section 3.4 to be reduction to Na_2S. Probably the tetrasulfide is reduced in steps to $S^=$ with $S_2^=$ existing as an intermediate. South et al. (50) have also found a decrease in film resistance at these negative potentials; they attribute this to a more rapid dissolution of Na_2S than Na_2S_2 in the bulk melt (however, see text accompanying Fig. 25). If quantitative kinetic data are being sought, great care must be taken to renew the surface periodically as described, and to keep the conductance between 85 and 95% of the fresh surface value.

So far, it has been proposed that wave A in Fig. 17 is due to a reversible, two-electron reduction of first order, uncomplicated by chemical reactions. At 300°C, $E_p - E_{1/2}$ for a two-electron reaction should be only 0.027 V (42). Instead an $E_p - E_{1/2}$ of 0.059 V is observed. This strongly suggests a two-step charge-transfer process with superposition of two reversible waves (58). The reverse oxidation wave labeled D in Fig. 17 extends over the same voltage range (-0.25 to 0 V) and is of the same peak height as wave A. If the potential is held constant for 30 to 60 sec at values between -0.25 and -0.35 V, the peak current i_p of wave D is not altered; thus the oxidation process is not coupled with a chemical reaction. Double-pulse galvanostatic (59) and current impulse relaxation techniques (60) gave very high values for the exchange current for the first reduction wave (61) (Fig. 22). Chronocoulometric experiments gave a value of 1 A/cm^2 for the total two-step charge transfer at 300°C (57). The results for the first reduction wave were based on potential excursions of a few millivolts, whereas the result for the total two-step charge transfer came from cathodic potential steps up to 400 mV.

As shown in Fig. 17, the differential conductance remained quite constant on the cathodic sweep until after peak A. The abrupt drop is interpreted as resulting from the formation of an insoluble, partially blocking layer of Na_2S_2. Calculations based on experimental values of the faradaic capacitance and faradaic resistance show that neither the Warburg, nor the charge-transfer resistance change abruptly enough with potential to account for the sudden decrease in conductance (52). There is no evidence to date from any of the experimental techniques of the extent of monolayer adsorption, and therefore the nature of the blocking

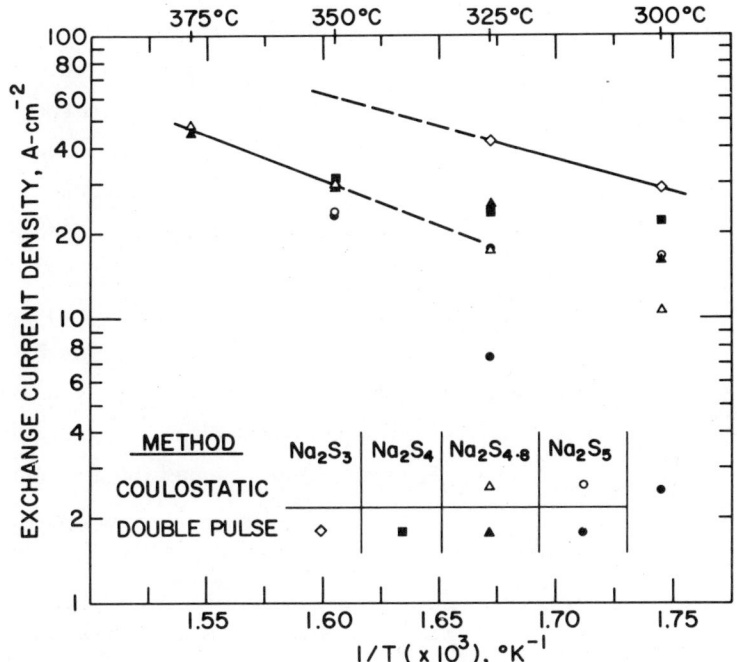

Fig. 22. Exchange currents for the first reduction wave as a function of temperature, determined by two different methods. The wide scatter at lower temperature shows the difficulty of measuring exchange currents of very fast reactions in a viscous medium with low diffusion coefficient and correspondingly fast buildup of diffusion polarization, from Gupta and Tischer (61).

layer cannot be specified. As shown in Fig. 17, when the cathodic sweep is reversed, a potential is reached at which the Na_2S_2 redissolves, the conductance increases, and a cathodic spike occurs due to reduction of Na_2S_4 at the more negative potential suddenly imposed on the electrode by the absence of the blocking layer. As the sweep continues in the positive direction, a rapid change from reduction to oxidation occurs as the material that was precipitated is oxidized. A current akin to a stripping current is obtained. The potential at which the Na_2S_2 begins to be oxidized coincides with the equilibrium potential for the saturated melt, further identifying the solid phase as Na_2S_2. This can be clearly seen by comparing data from Figs. 23 and 24. The potential difference between (solid Na_2S_2/liquid melt) and (liquid $S/Na_2S_{5.2}$ liquid melt) is 0.32 V at 340°C, as shown by the potential difference between the horizontal lines in Fig. 23 (62). An iR corrected plot of sweep curves for compositions $Na_2S_{3.0}$, $Na_2S_{4.0}$, and $Na_2S_{5.2}$ is shown in Fig. 24 (51). The three curves all superpose on the reverse sweeps at -0.3 V versus $Na_2S_{5.2}$, indicating that the precipitated Na_2S_2 from the three reduction processes

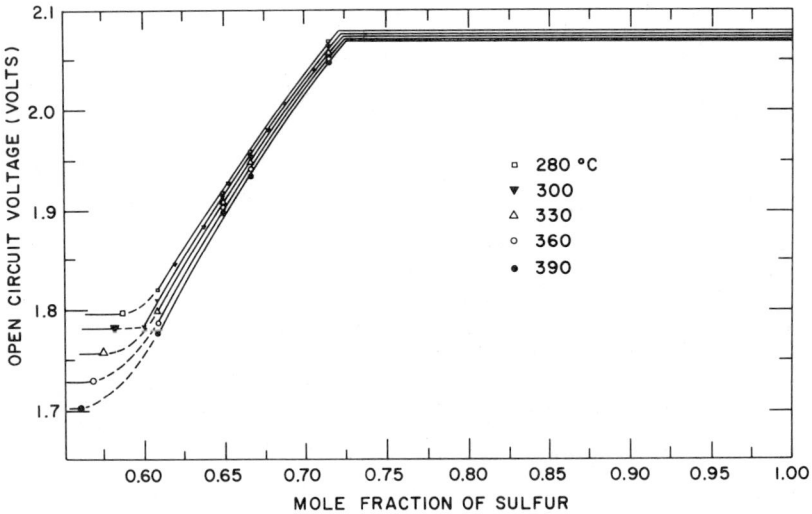

Fig. 23. Equilibrium potentials (versus Na/Na^+) versus sodium polysulfide composition, from Gupta and Tischer (62).

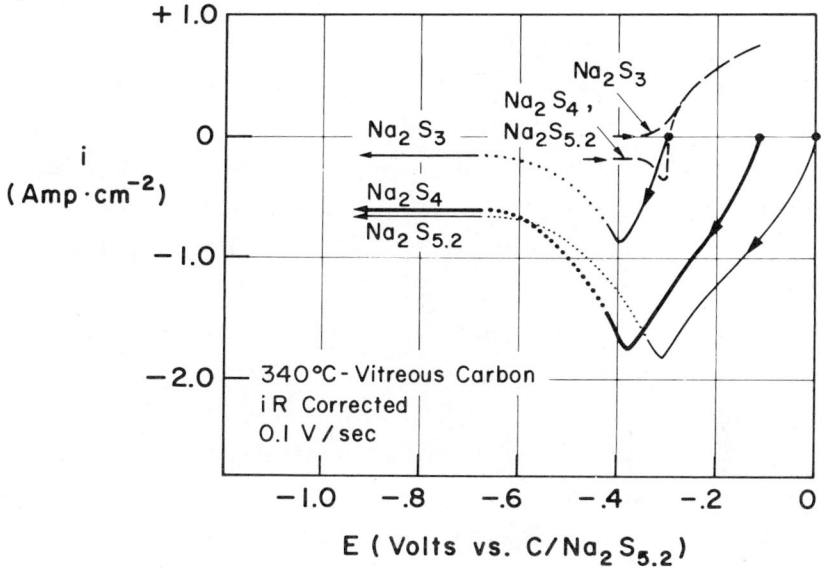

Fig. 24. Cyclic sweeps cathodic of rest potentials. Reverse sweeps partially shown, from Ludwig et al. (51).

redissolves at the potential of the Na_2S_2-saturated melt. Peak A occurs at about the same potential in the three melts, suggesting a common reduction process. The $Na_2S_{3.0}$ wave occurs over a narrower potential range. The rest potential in Na_2S_3 is negative enough so that only the second wave of the two-step reduction occurs, and, in fact, the current function for a two-electron reversible $Na_2S_{3.0}$ reduction agrees with the theoretic value within a few percent, as shown by South et al. (50). It now becomes clear why the experimental values of the current functions for Na_2S_4 and Na_2S_5 found by South et al. are less than the theoretic value for a two-electron wave; namely, the theoretic current function for superposed waves for two consecutive one-electron steps separated by −90 mV is 0.75 [sign according to Polcyn and Shain (58)]. This value compares with the 1.26 value for a two-electron single-step wave with the $n^{3/2}$ factor included. The experimental value of the separation for the two waves for Na_2S_2 and Na_2S_5 is about −90 mV (52) when reduced to 25°C, and therefore the current function should be about 0.75. On the basis of the data of South et al. (50) the value for Na_2S_5 is 0.67 and for Na_2S_4, 0.87. The fact that the Na_2S_3 is proposed to be an equimolar mixture of Na_2S_2 and Na_2S_4 is well supported by the data in Fig. 24. The wave occurs at the same peak potential as the Na_2S_4 wave, and its peak current is less.

It is because of the two-step nature of the reduction wave that we can be quite certain that the chemical dissociation, followed by the one step reduction of S_2^- to $S_2^=$, is not the reaction mechanism. In the mechanism proposed below, $S_4^=$ is written as $^-S_2 - S_2^-$ for convenience only. The reader is referred to Giggenbach (63) and Allen (64) for a discussion of the various resonance forms, electronic configurations, and bond energies of the tetrasulfide ion. The reaction proposed may proceed as a concerted reaction as written, or the bond may break forming $S_2^=$ and S_2^- with the second electron then adding to the radical anion:

For Na_2S_4 and Na_2S_3

$$^-S_2 - S_2^- + e^- \rightleftarrows {}^-S_2 \cdots S_2^=$$

$$^=S_2 \cdots S_2^- + e^- \rightleftarrows 2S_2^=$$

For Na_2S_5

$$^-S_3 - S_2^- + e^- \rightleftarrows {}^-S_3 \cdots S_2^=$$

$$S_2^= \cdots S_3^- + e^- \rightleftarrows S_2^= + S_3^=$$

$$2S_3^= \rightarrow S_2^= + S_4^=$$

The cathodic peak B in Fig. 17 does not seem to result from a separate reaction. If the cathodic sweep is stopped in the trough after peak A and before peak B, the current rises to the level of peak B and then slowly falls. Apparently, wave B is produced by a reorganization in the blocking Na_2S_2 layer, which causes a temporary decrease in resistance. This is not apparent in the slower response of the autobalancing technique used with the a.c. bridge differential conductancé measurements shown in Fig. 17, but becomes quite clear with the faster response obtained when a lock-in amplifier was used to make differential conductance measurements, as shown in Fig. 25 and 26 (51). The rise in conductance at -1.0 V represents the reduction to Na_2S discussed previously.

Occasionally at 0.01 V/sec sweep rate a wave C would begin at -1.0 V, in agreement with the data of South et al. (50). It obviously is involved with the reduction to Na_2S. In Fig. 25 the conductance, but not the sweep data, show the effect of the reduction. Apparently, the reduction is occurring without the formation of a wave; instead of a current decrease in the cathodic direction because the thickening Na_2S_2 layer blocks the access of Na_2S_4, the current remains fairly constant due to the reduction to Na_2S. The expected chemical reaction, $2S^= + S_4^= \rightarrow 3S_2^=$, apparently results in a precipitation of Na_2S_2. The chemically produced Na_2S_2 probably does not block the electrode as effectively as the electrochemically produced Na_2S_2.

The data in Fig. 26 for $Na_2S_{5.2}$ are similar to those of Fig. 25, except that the differential conductance does not decrease very much after the reduction peak, probably because of the reaction

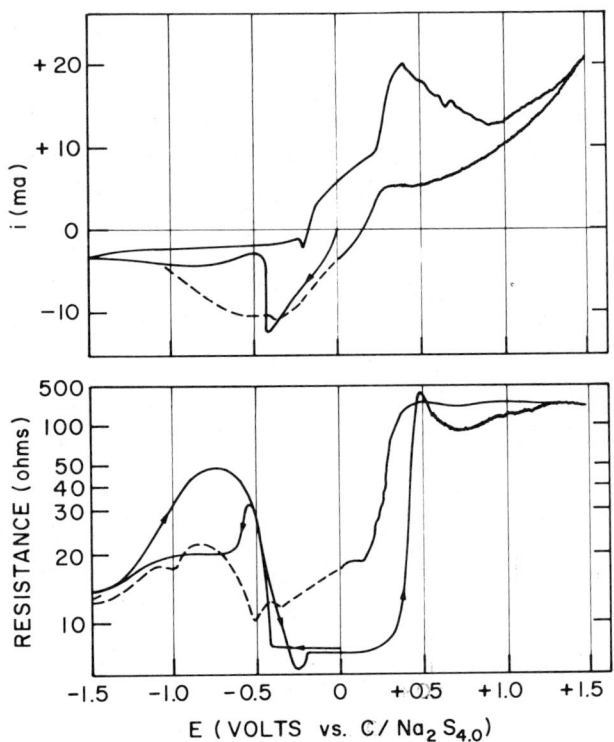

Fig. 25. Cyclic sweep, 0.1 V/sec, and melt resistance in $Na_2S_{4.0}$ at 340°C. Resistance measured with 5 kHz, 5 mV p-p signal. Sweep data are iR uncorrected. From Ludwig et al. (51).

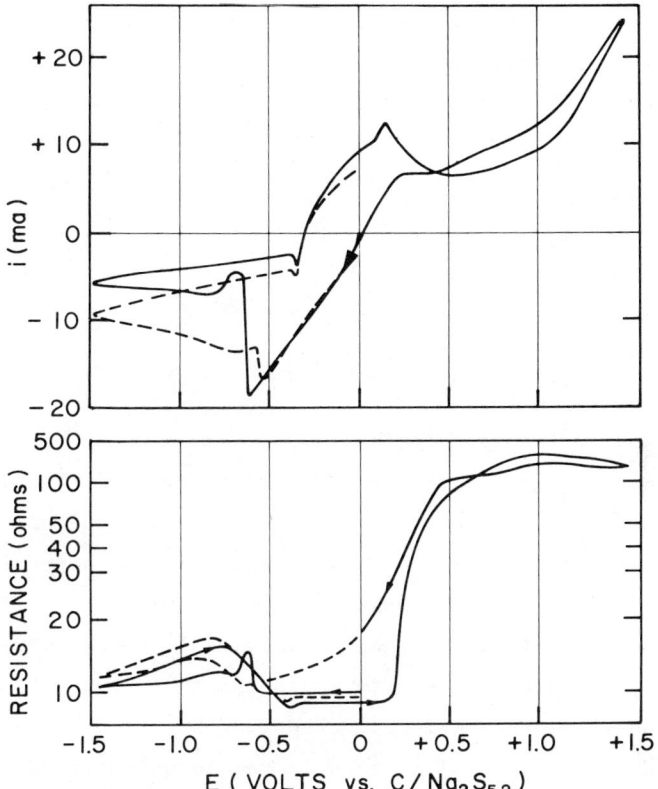

Fig. 26. Cyclic sweep, 0.1 V/sec, and melt resistance in $Na_2S_{5.2}$ at 340°C. Resistance measured with 5 kHz, 5 mV p-p signal. Sweep data are iR uncorrected. From Ludwig et al. (51).

$S_2^= + 2S_5^= \rightarrow 3S_4^=$ discussed previously.

The anodic wave producing peak E in Fig. 17 represents the oxidation of tetrasulfide to sulfur. The differential conductance data indicate the formation of a blocking sulfur layer (see also Fig. 25 and 26). The height of the peak compared to peak A is in excess of that expected for a two-electron wave with a normal current function. However, the value of the current function (i.e., $i_p/v^{1/2}$) falls with increasing sweep rate. This suggests a following chemical reaction based on the removal of sulfur from the electrode. The net electrochemical and chemical reactions are postulated to be (51)

$$S_4^= \rightleftarrows S_4 + 2e^-$$

$$S_4 + 2S_4^= \rightarrow 2S_6^=$$

$$S_4 + 4S_4^= \rightarrow 4S_5^=$$

The $S_5^=$ can be further electro-oxidized, and in this respect the reaction is partially catalytic (42). However, the two chemical reactions and the electro-oxidation of $S_5^=$ can proceed only to an overall composition of $Na_2S_{5.2}$, the composition of polysulfide phase in equilibrium with a separate liquid sulfur phase in the temperature range of interest. The species S_4 and $S_6^=$ were chosen because of their proposed existence in liquid sulfur (65) and polysulfides (66), respectively. The length of the sulfur and polysulfide chains could just as well remain unspecified (e.g., S_2 is known to exist in the liquid state) without altering the kinetic interpretations.

The important aspect of the anodic reaction is not the catalytic part, but the removal of the sulfur by chemical reaction. At faster sweep rates, the slower chemical reactions cannot remove the sulfur and therefore $i_p/v^{1/2}$ decreases with sweep rate. Thus the peak current does not represent a diffusion-controlled process but a process of a chemical reaction removing a blocking layer. This is why peak E shows erratic current flow. Furthermore, at fast sweep rates, where the chemical reaction is too slow to remove the sulfur and form soluble ions, little diffusion can occur. The reaction product, sulfur, is immiscible in the melt, and the product cannot dilute the reactant. Therefore the current increases with little concentration polarization, the slope of the i/E curve being nearly equal to the conductance of the melt. The current peak is determined by the blocking of the electrode by sulfur. On the reverse sweep, the sulfur layer disappears at +0.4 V and the trace nearly superimposes over the forward trace (51) (Fig. 27). A possible explanation of the slight separation of the forward and reverse trace is that some diffusion is taking place. This diffusion may be due to a small amount of chemical reaction with the sulfur, as discussed previously.

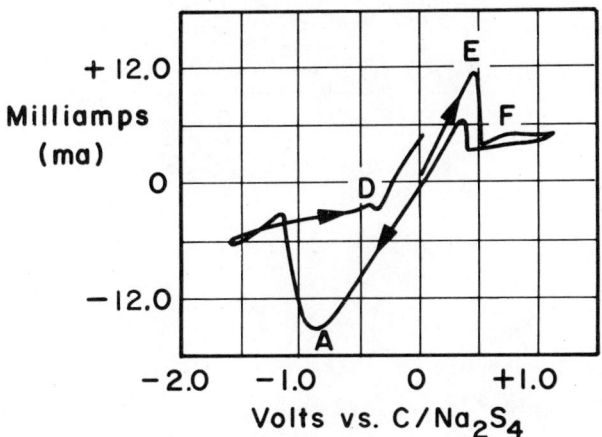

Fig. 27. Cyclic sweep, 11.7 V/sec, iR uncorrected, $Na_2S_{4.0}$ at 300°C, vitreous carbon electrode area is 0.00407 cm^2, from Ludwig (34).

In Fig. 28 (51), a partial preoxidation of lower polysulfides to higher polysulfides is observed before the oxidation to sulfur. The quantity of charge should increase from Na_2S_5 to Na_2S_3, as is evident in Fig. 28. The difference in the peak currents at the slower sweep rate shown in Fig. 28 is an indication of the difference in chemical reaction rate of sulfur with $S_2^=$ (in Na_2S_3), $S_4^=$,

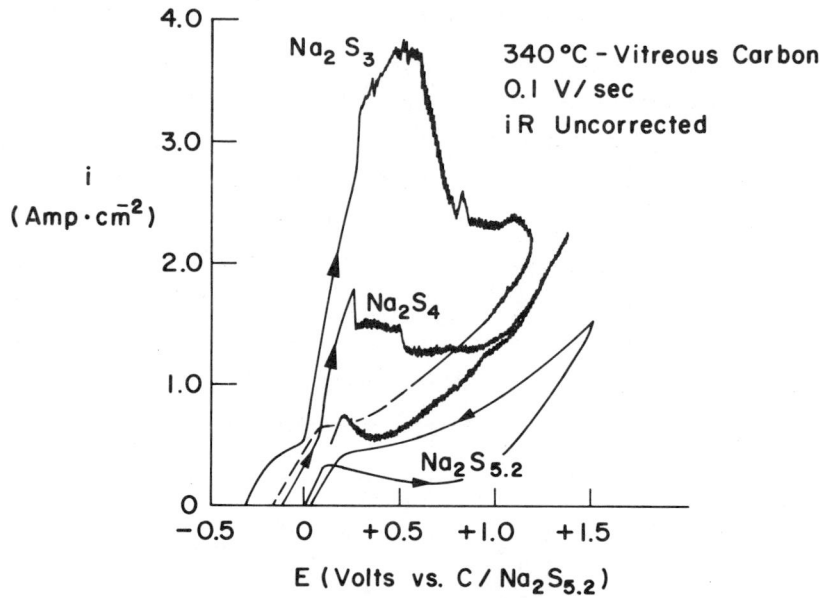

Fig. 28. Cyclic sweeps anodic of rest potentials, from Ludwig et al. (51).

and $S_5^=$ (51). In support of this rate sequence is the observation that if the sweep is stopped at the anodic limit, the sulfur layer covering the electrode is removed only slowly in Na_2S_5 (determined by the slow drift of the OCV back to the original OCV). The drift rate back to the original OCV varies in the order $S_2^= > S_4^= > S_5^=$. Note also that the current in the return sweep in the sulfur-saturated Na_2S_5 melt exceeds the current in the forward sweep in Fig. 28. This is probably due to convective flow, the density of sulfur being less than the density of Na_2S_3, Na_2S_4, or Na_2S_5. Finally, at the intermediate sweep rate of 0.1 V/sec of Fig. 28, the conductance data clearly show that when the electrode is suddenly blocked by sulfur, the current falls (Figs. 25 and 26); the current for Na_2S_3 remains high for a long time (Fig. 28) and the conductance does not fall until the current begins to drop. At high sweep rates when the sweep is in the cathodic direction first or the potential is held at the cathodic potential of peak A before the anodic sweep, a new peak F appears (51,52) shown in Fig. 29 (compare with Fig. 27). The oxidation peak of Na_2S_2 is absent, probably because the sweep rate is in excess of the dissolution rate of Na_2S_2 in the melt. However, in keeping with the same sequence of chemical reaction rates noted previously for sulfur with the polysulfides, we would expect solid Na_2S_2 or Na_2S on the electrode surface to react very rapidly with the sulfur formed on the electrode. Thus peaks E and F may be postulated to resemble an ECE mechanism, with $S_4^=$ remaining adsorbed on the electrode:

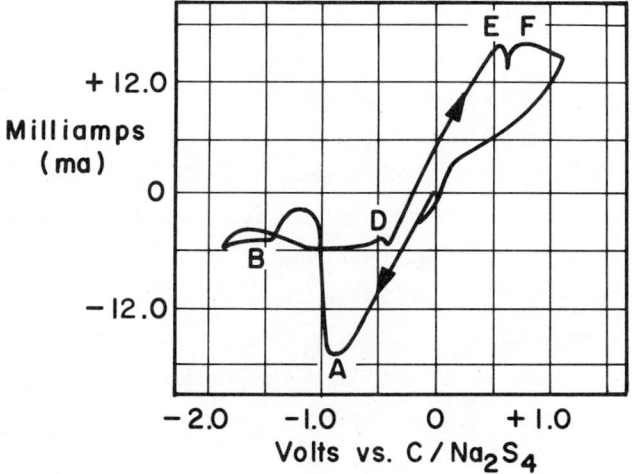

Fig. 29. Cyclic sweep, 13.2 V/sec, iR uncorrected, $Na_2S_{4.0}$ at 300°C, vitreous carbon electrode area is 0.00407 cm^2, from Ludwig et al. (34).

$$S_4^= \rightleftarrows S_4 + 2e^-$$
$$2S_2^= \text{ (ads)} + S_4 \rightarrow 2S_4^= \text{ (ads)}$$
$$S_4^= \text{ (ads)} \rightarrow S_4 + 2e^-$$

South et al. (50) state the $i_p/v^{1/2}$ is constant with sweep rate for their anodic peaks, whereas we found that $i_p/v^{1/2}$ decreases. They state that in Na_2S_4 $i_p/v^{1/2}$ was more than twice the value obtained for the cathodic results, and in Na_2S_5 it was less than half the corresponding cathodic value. We suggest that their sweep rate range, 10 to 50 mV/sec, was too limited for a good conclusion concerning the constancy of $i_p/v^{1/2}$ with sweep rate. Also, their observations on the $i_p/v^{1/2}$ comparisons to the cathodic values are a proof of the chemical reaction rates we proposed, since if the anodic process were diffusion controlled, the $i_p/v^{1/2}$ values would be in closer agreement for Na_2S_4 and Na_2S_5 and would also agree more closely with the cathodic values. Finally an electrode surface that was not restored to its virgin state and remained coated with Na_2S_2, for example, would show the effect of the more rapid chemical reaction of sulfur with Na_2S_2, and $i_p/v^{1/2}$ would not drop as rapidly with increasing sweep rate. The only high sweep rate data obtained by South et al. were a few cyclic sweeps. These data were not used in their $i_p/v^{1/2}$ plots, but do show a large decrease in $i_p/v^{1/2}$ for the anodic peak (50). They also mention that their anodic chronopotentiometric results were not in accord with the Sand equation.

If sulfur is deposited on the electrode anodically, in the following cathodic sweep (Fig. 30) a wave suggesting a catalytic mechanism is obtained (34,51):

$$S_4^= + 2e^- \rightleftarrows 2S_2^=$$
$$2S_2^= + S_4 \rightarrow 2S_4^=$$

where the first reaction is the two-step charge transfer discussed previously. The second step, identical to that discussed for the anodic peaks, is a pseudo-first-order reaction in which the activity of sulfur is constant because of its presence as a separate phase. The cathodic wave in Fig. 30 is in accord with the Nicholson-Shain criteria (42) for a catalytic reaction. The main

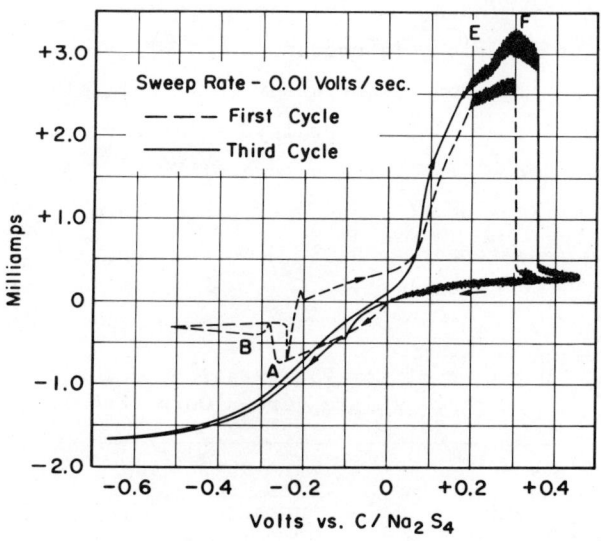

Fig. 30. Cyclic sweeps in $Na_2S_{4.0}$ at 300°C, iR uncorrected, vitreous carbon electrode area is 0.00407 cm^2, from Ludwig et al. (51).

features are the increased cathodic current, the presence of a limiting current rather than a current peak, the lack of hysteresis on the reverse sweep, and the return to the normal peak A at fast sweep rates. The breakup of the blocking sulfur layer can be seen as a small blip at -0.1 V (confirmed also by a rise in conductance to about half the value at the virgin electrode). The first-order rate constant is 0.2 sec^{-1}, calculated on the basis of the ratio of the limiting kinetic current to the pure diffusion current (42) for peak A. The value of this rate constant can be of direct importance to the understanding and designing of the sulfur electrode in the sodium-sulfur cell. Sixty percent of the discharge capacity of the cell involves discharge from a two-phase $S/Na_2S_{5.2}$ melt; therefore the catalytic reduction mechanism just described is operative.

To determine the conditions under which cell electrodes might become blocked with sulfur layers, a vertical graphite rod electrode suspended in a large volume of $Na_2S_{5.2}$ melt at 320°C was operated at various anodic constant current levels (52). The experiment was repeated in an $Na_2S_{4.7}$ melt. The results are shown in Table 1.

Table 1. Duration of Low Polarization. Constant Current at 320°C on Graphite (Ultra Carbon Corp. UF4S)

$Na_2S_{5.2}$ + S		$Na_2S_{4.7}$	
Current (mA/cm^2)	Time (min)	Current (mA/cm^2)	Time (min)
5.0	1230	17.5	1188
10.0	25	25.0	84
15.0	12	63.5	4

The polarization was negligible at the graphite electrodes for the time periods shown. Then within seconds the polarization increased to 15 V, the limit of the constant current power supply. On some of the voltage traces, the final period before the blocking layer formed showed small irregular blips, indicating convective removal of sulfur. The initial negligible polarization in both melts and the higher current levels in the $Na_2S_{4.7}$ as compared with those in $Na_2S_{5.2}$ are in accord with the kinetic observations made with respect to Fig. 28. The sudden blocking of the electrode is in accord with the observation previously mentioned that the electrode surface slowly becomes less active. At OCV the blocking sulfur layer disappeared in minutes, but it took many hours before the graphite recovered its activity. If the constant current experiments were repeated before full recovery of activity, the time period before the sudden blocking was considerably shortened. Polarizations approaching -1.0 V tended to hasten the recovery of the electrode, but the effect was not nearly as dramatic as that at the vitreous carbon microelectrode discussed previously.

A vertical stainless steel rod electrode under the same experimental conditions in $Na_2S_{5.2}$ at 320°C was operated at 500 mA/cm^2 for 20 hr at negligible electrode polarization (52). No blocking layer developed, nor did the electrode corrode. Stainless steel appears to be anodically protected in the polysulfide melts rich in sulfur; just as stainless steels can be anodically protected in aqueous media.

A preliminary explanation for the anodic results at stainless steel compared with graphite in the two-phase region may be found in the different wicking and wetting characteristics of the two materials. Contact angles were measured for molten sulfur and Na_2S_4 on graphite, vitreous carbon, and AISI 446 stainless steel

(52). The results on the metal surface were independent of the surface preparation of the metal (i.e., etched in HCl, oxidized in air at 800°C, untreated or freshly abraded). In a helium atmosphere at 318°C, the contact angles for Na_2S_4 were approximately as follows: on graphite, 100°C; on vitreous carbon, 100°C; on stainless steel, 0 to 5°C; in helium at 282°C, the contact angle for sulfur on all three materials was 20° to 30°. Over the period of 45 min, the contact angle for Na_2S_4 changed only on the vitreous carbon, going from 100° to 70°. These data are preliminary estimates, since the molten materials were not in equilibrium with their vapors and since no attempt was made to clearly define the difference between advancing and receding contact angles. Furthermore, the effect of potential on contact angle was not obtained. However, the difference in wetting of the metal and the graphite by Na_2S_4 is so extreme that it is likely that the metal does not become blocked by sulfur at anodic potentials because it is not preferentially wetted by sulfur.

More work is needed on the nature of the sulfur/carbon interaction. It is hoped that further experimental work will yield information on the extent of surface coverage θ by sulfur on graphite and vitreous carbon electrodes. The observed time-dependent blocking of the electrode is reminiscent of passivation effects observed in other systems. The adsorption-desorption kinetics of sulfur on various electrode surfaces are undoubtedly complex. The wetting observations made above provide only a preliminary approach to the problem.

3.4 Polysulfides in Molten KCNS

The reduction of polysulfides in KCNS or LiCl-KCl supporting electrolyte/solvent involves overlapping voltammetric waves that cannot be separated except for the polysulfide/sulfur and the $S^=/S_2^=$ couples. The oxidation of polysulfide to sulfur, however, has been clearly analyzed and allows us to draw some very interesting comparisons between the kinetics in KCNS and in the pure melt. The oxidation to sulfur is therefore discussed before the reduction of the polysulfides. Cleaver et al. (67) have made a thorough study in dilute (10^{-3} to 10^{-2} molar) solutions. Aikens et al. (68) have examined the entire concentration range from 10^{-3} M to 6 M in an attempt to bridge the gap between the KCNS melts and the pure polysulfide melts.

The first problem encountered in the KCNS melts is that vitreous carbon (VC) is attacked whereas Pt is not (67). In polysulfide melts the reverse situation is found: metals are attacked but not VC. Therefore, all of the work of Cleaver et al. was done with a Pt electrode. The work of Aikens et al. was done with Pt and W electrodes except for one experiment comparing a W and a VC electrode in a NaCNS-KCNS melt at 310°C, 0.56 molar in Na_2S_4. The sweep curves in this one experiment at 0.1 V/sec were practically identical. This one experiment does give us some justification for comparing work in the dilute melts at metals with work in the pure polysulfides at VC.

The sweep data of Cleaver et al. (67) shown in Fig. 31 show an anodic sulfur deposition peak followed by a cathodic sulfur stripping peak. The deposition of sulfur was confirmed by observation of the deposit by the scanning electron microscope in conjunction with analysis by x-ray microprobe. The sulfur deposits in globular form, suggesting that sulfur does not wet the electrode under the experimental conditions. This is in accord with an observation by Ludwig (52) that, with no applied potential difference across the electrode, Na_2S_4 wets the metal (stainless steel) more strongly than does sulfur; however, Na_2S_4 wets

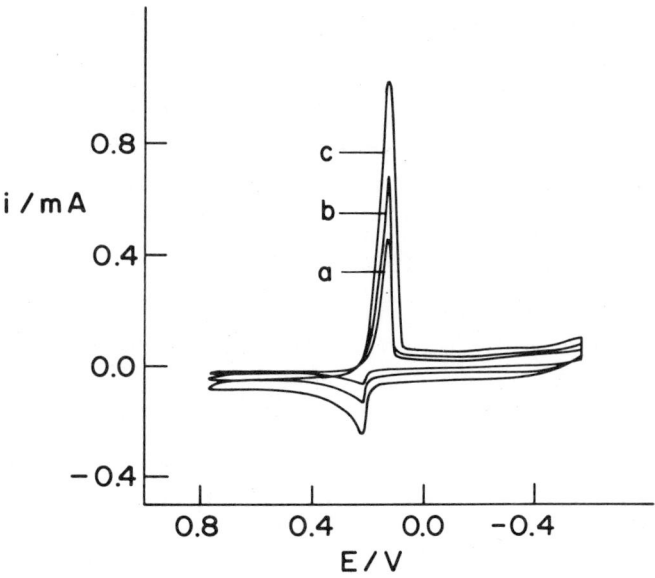

Fig. 31. Cyclic sweeps for 12.6 mM $Na_2S_{4.3}$ at 210°C in KCNS, using a platinum working electrode. Sweep rates: a, 0.03 V/sec; b, 0.1 V/sec; c, 0.3 V/sec. Potentials were measured versus the Ag, Ag_2S|KCNS reference electrode. From Cleaver, et al. (67).

graphite and VC less strongly than does sulfur (see the effects of metal electrodes on the kinetics in the preceding section). The nonwetting of the metal explains why the tail of the sulfur deposition curve is typical of a diffusion wave; that is, the electrode is not being blocked even though sulfur is being deposited.

Both potential step and linear sweep data indicate a diffusion-controlled sulfur deposition process (67). Cleaver et al. (67) have found good agreement between results of voltage sweep and potential step experiments for sulfur deposition on metal with the theory of diffusion-controlled metal deposition (69): E_p is independent of sweep rate to 100 V/sec, indicating a reversible charge-transfer reaction; $i_p/v^{1/2}$ is a constant; i_p is proportional to the nominal concentration of polysulfide; the number of electrons (n) involved in the electrode process, is one, calculated from the concentration dependence of the anodic peak potential; the deposited sulfur does not redissolve for times as long as 240 sec -- the coulombs deposited were completely recovered in sulfur stripping peaks. A potential step experiment also indicated simple diffusion control, as did an experiment relating the charge deposited to the sweep rate. In all of these experiments, solutions of three polysulfide compositions were tested, $K_2S_{3.6}$, $Na_2S_{4.6}$, and $K_2S_{5.9}$, at 210°C over a range of concentration of 1.3×10^{-2} to 1.0×10^{-3} moles/liter in KCNS. The sweep rates ranged from 0.03 to -100 V/sec. A composite plot of the sweep data is shown in Fig. 32. Cleaver et al. reason that, since n = 1, the anodic reaction is most likely

$$S_2^- \rightleftarrows 2S + e^-$$

or

$$S_3^- \rightleftarrows 3S + e^-$$

They do not believe the charge-transfer step is preceded by a rapid dissociation of polysulfide, for example,

$$S_4^= \rightarrow 2S_2^-$$

since their freezing point depression studies of polysulfides in KCNS (45) indicate extensive dissociation into radical ions, independent of charge transfer, that is,

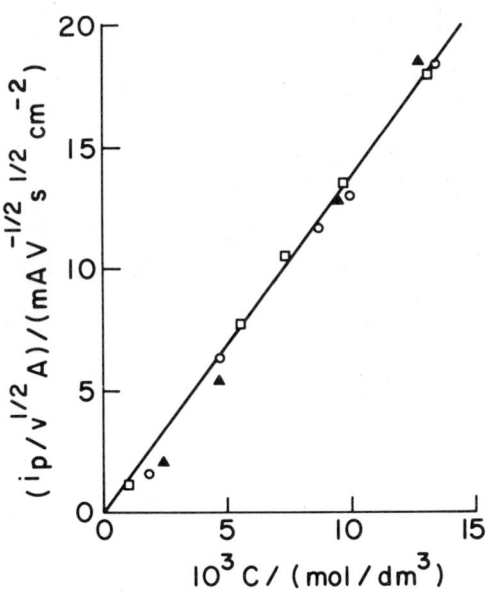

Fig. 32. Plot of $i_p/v^{1/2}$ A, versus concentration, C, at 210°C for $K_2S_{3.6}$ (squares) $Na_2S_{4.6}$ (triangles), and $K_2S_{5.9}$ (circles) in KCNS. From Cleaver, et al. (67).

$$S_6^= \rightleftarrows 2S_3^-$$
$$S_5^= \rightleftarrows S_3^- + S_2^-$$
$$S_4^= \rightleftarrows 2S_2^-$$

Furthermore, the green and blue colors of these solutions have been shown to derive from the radical anions (30,45,46). The fact that all of the points in Fig. 32 fall approximately onto one line suggests that the polysulfides are all extensively dissociated and that the diffusion coefficients of S_2^- and S_3^- are the same. Based on the nominal polysulfide compositions the diffusion coefficient calculated by Cleaver et al. was 1 to 2×10^{-5}

cm^2/sec. Assuming complete dissociation of polysulfide, the value falls in the range of 2 to 5 × 10^{-6} cm^2/sec. The viscosity of the KCNS melt is roughly 10 times less than the viscosity of the pure polysulfide melt (54,70). Through the Stokes-Einstein equation the value of 3 to 9 × 10^{-7} cm^2/sec found for the diffusion coefficient of $S_4^=$ in pure Na_2S_4 (52) (see Section 3.3) is in good agreement with the above range of 2 to 5 × 10^{-6} cm^2/sec.

In the pure melt the dissociation of the polysulfides is suppressed, and the electrode reactions do not involve the radical anions. The tendency toward suppression of dissociation as the concentration of the polysulfides is increased is supported by the work of Aikens et al. (68). They found the same sweep rate relationships for the deposition of sulfur at low polysulfide concentrations, but when i_p was plotted against the nominal polysulfide concentration, the relationships shown in Fig. 33 and 34 were obtained. The data of Fig. 33 are in close agreement with those of Fig. 32 to concentrations of 10^{-1} M, but at the higher concentrations shown in Fig. 34 a decrease in slope suggests that the dissociation of the polysulfides to the radical anions is suppressed. This is confirmed by the respective yellow-green, green, and blue color of the dilute solutions of $Na_2S_{3.2}$, $Na_2S_{4.0}$, and $Na_2S_{5.0}$; the concentrated solutions of these three salts are all red-brown (68). Furthermore, for $Na_2S_{3.2}$ at 240°C a plot of the concentration dependence of the anodic peak potential over the range 10^{-2} mole/liter to about 0.6 mole/liter indicates that n remains equal to one. Therefore, the mechanism has not changed; the one-electron oxidation of the radical anion is still the primary oxidation process. Evidently, the concentration of the polysulfide must become larger before the radical anion concentration is suppressed to the extent that the two-electron mechanism takes over. In Na_2S_3 and Na_2S_5 the competition between the one-electron

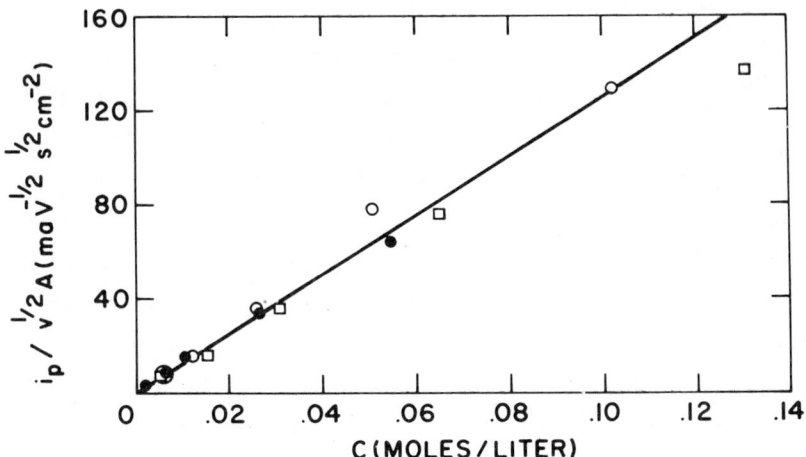

Fig. 33. Current function, $i_p/v^{1/2}$ A, versus concentration, C, for $Na_2S_{3.2}$ (solid circles), $Na_2S_{4.0}$ (open square), $Na_2S_{5.0}$ (open circle) in KCNS at 340°C for peak E. From Aikens et al. (68).

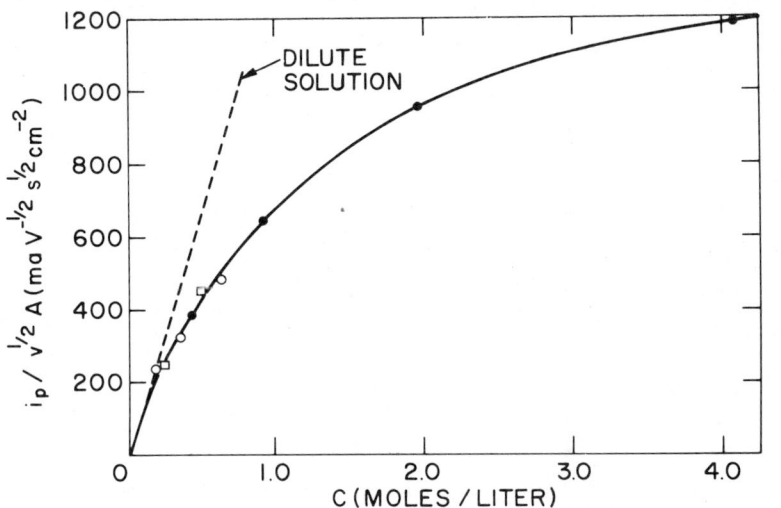

Fig. 34. Current function, $i_p/v^{1/2}$ A, versus concentration, C, $Na_2S_{3.2}$ (solid circle), $Na_2S_{4.0}$ (open square), $Na_2S_{5.0}$ (open circle) in KCNS at 340°C for peak E. From Aikens et al. (68).

and two-electron oxidation becomes noticeable at about 0.5 molar
concentration, but the waves do not interfere significantly with
each other. In $Na_2S_{4.0}$ the wave splitting is more severe and
begins at lower concentrations. The sequence of Figs. 35 and 36
(68) serves to illustrate the effects and is also representative
of Na_2S_3 and Na_2S_5 sweep curves (except, as mentioned, the splitting is less involved in these polysulfides). The potentials in
these figures are given with respect to an $Ag/0.01$ molal Ag^+ electrode. Figure 35b best exemplifies the splitting, shown as the
splitting of peak E into E_1 and E_2 and the splitting of the stripping peak S' into S_1' and S_2'. As the concentration increases from
Fig. 35a to Fig. 35b, the height of the new peaks E_2 and S_2' increases. Peaks C and C' will be shown (cf. text discussing Fig.
37) to definitely represent the reaction $S_2^= + 2e^- \rightleftarrows 2S^=$. Peak
C' is a stripping peak for the solid deposit of Na_2S. Note that,
as the concentration increases from Fig. 35a to 35b, C' becomes
less pronounced. As sweep rate increases (Fig. 35c) peaks E_1 and
S_1' and C' are again enhanced. The suggested interpretation is
that E_1 and S_1' are the one-electron sulfur deposition and stripping peaks, respectively. As the concentration increases, E_1 and
S_1' are suppressed and E_2 and S_2' are enhanced. The wave splitting
in dilute solution may be surmised to be related to an ECE mechanism, such as

$$S_2^= \rightleftarrows S_2^- + e^- \quad E_1$$
$$S_2 + S_2^- \rightarrow S_4^-$$
$$S_4^- \rightleftarrows S_4 + e^- \quad E_2 \text{ (dilute solution)}$$

This would explain the relative enhancement of peak E_1 at increased
sweep rate (71). Finally, as the concentration increases further,

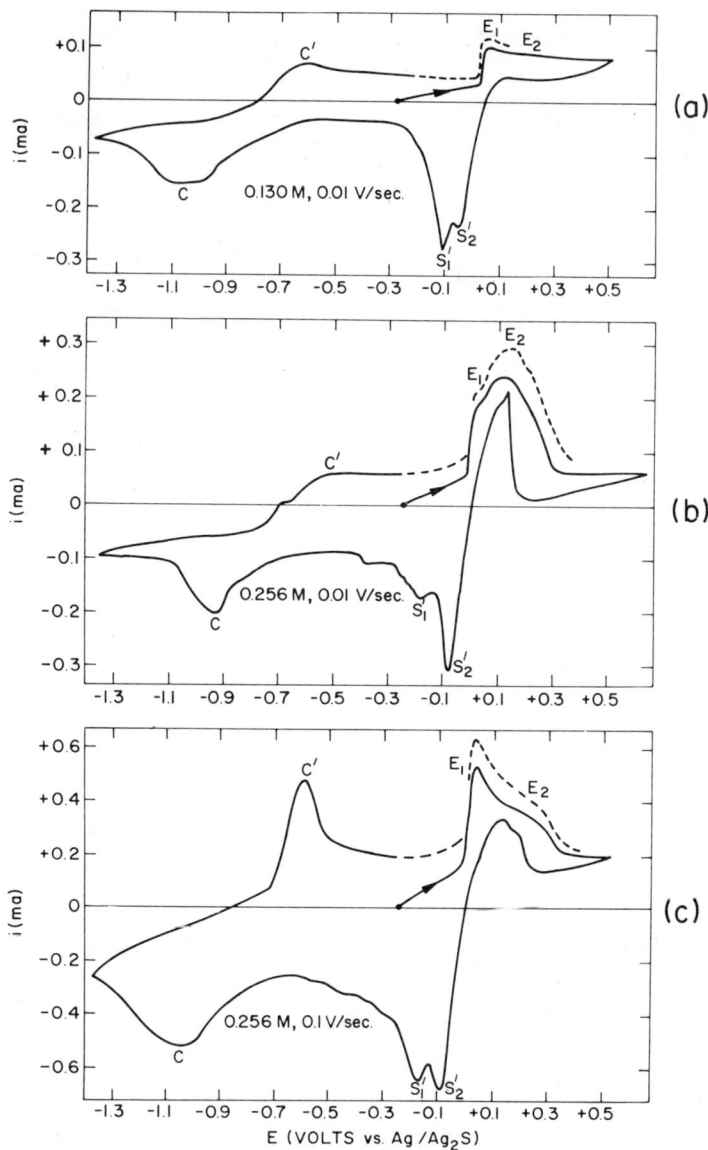

Fig. 35. Cyclic sweeps, iR uncorrected, $Na_2S_{4.0}$ in KCNS at 240°C at a tungsten electrode, area is 0.0044 cm^2, from Aikens et al (68).

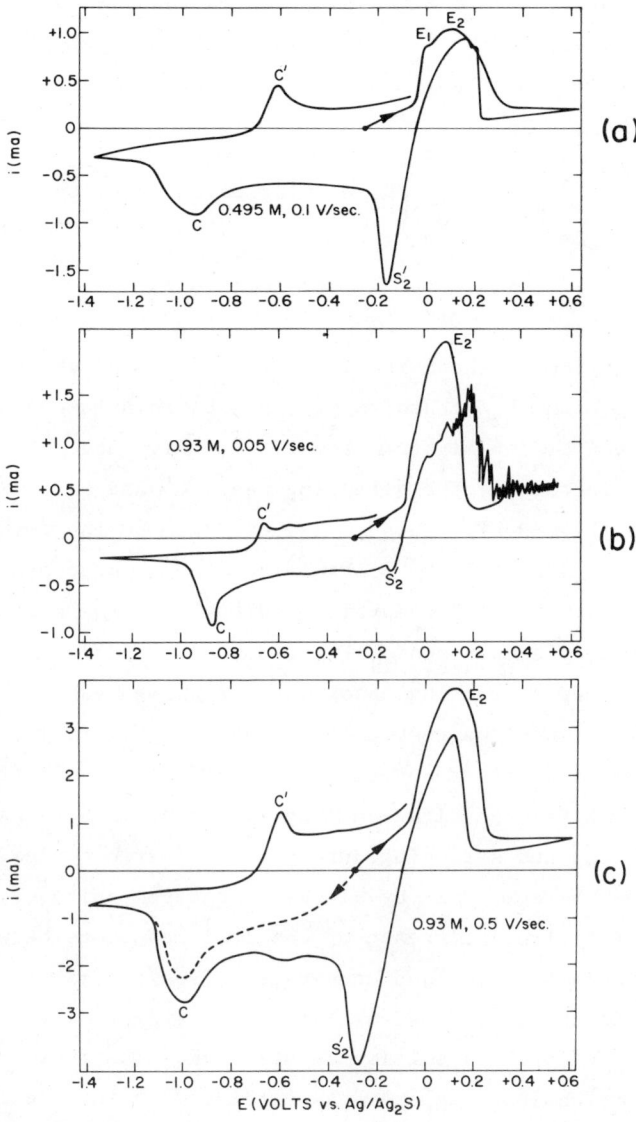

Fig. 36. Cyclic sweeps, iR uncorrected, $Na_2S_{4.0}$ in KCNS at 240°C at a tungsten electrode, area is 0.004 cm^2, from Aikens et al. (68).

the S_2^- concentration is highly suppressed and the reaction, now entirely represented by E_2 and S_2', is the two-electron process of the pure melt:

$$\left.\begin{array}{l} S_4^= \rightleftarrows S_4^- + e^- \\ S_4^- \rightleftarrows S_4 + e^- \end{array}\right\} E_2 \text{ (concentrated solution)}$$

$$\overline{S_4^= \rightleftarrows S_4 + 2e^-} \quad \text{(net reaction)}$$

This progression is shown in Figs. 36a, 36b, and 36c where the changeover to the E_2 mechanism at the 0.93 molar concentration is complete even at the higher sweep rate of Fig. 36c. As the concentration increases, the stripping peaks C' and S_2' diminish in importance (Fig. 36b). This is suggested to be the result of chemical reaction with the polysulfide melt as proposed in Section 3.3. The slow chemical removal of sulfur from the electrode is demonstrated by comparison of Figs. 36b and 36c, in which an increase in sweep rate brings back the stripping peaks. A notable difference between the results of Cleaver et al. (67) and Aikens et al. (68) is in the recovery of charge during the sulfur stripping peak. Cleaver claims equality of charge is obtained for sulfur deposition and stripping during voltammetric cycles. A feature of the voltammograms by Aikens of the polysulfides in KCNS that indicates the occurrence of chemical reactions between the polysulfide in solution and condensed phases deposited on the electrode is the dependence of the relative peak height on the composition of the polysulfide in solution. The ratio of coulombs recovered to coulombs expended as estimated by integration of the peaks for deposition and stripping of sulfur is 0.2 for $Na_2S_{3.2}$, 0.4 for Na_2S_4, and 0.9 for Na_2S_5. The same sequence was found in the pure melt studies (51). The ratio of coulombs recovered in stripping of sodium sulfide to coulombs expended in deposition of sodium sulfide, estimated in the same manner, using the areas under the peaks is 0.5 for $Na_2S_{3.2}$ and 0.3 for Na_2S_5.

The sweep data of Cleaver et al. (67) shown in Fig. 31 do not have as much fine structure as shown in Figs. 35 and 36, nor do the C, C' peaks show. The excursion to negative potentials is not far enough to yield the C and C' peaks. When Cleaver added Na_2S_2 to KCNS, some fine structure similar to peaks E_1, E_2 and S_1', S_2' in Figs. 35 and 36 did begin to show. However, the peaks did not appear to be split, but were more widely separated. Cleaver interpreted these peaks in terms of the reversible reactions.

$$S_2^= \rightleftarrows S_2^- + e^-$$
$$S_2^- \rightleftarrows 2S + e^-$$

These reactions are probably correct. For the data shown in Figs. 35 and 36 for the Na_2S_4 melt, the concentration of $S_2^=$ in the initial anodic sweep is not high enough to account for the E_1 peak. The second cycle E_1 peak is not enhanced, although $S_2^=$ was generated at peak C'. Also, the oxidation of $S_2^=$ has been shown by Ludwig et al. (51) to occur at potentials cathodic relative to the E_1, E_2 waves. This preoxidation is seen in the enhanced currents observed between peaks C' and E_1 on the second cycle sweeps in Figs. 35 and 36. Thus it appears that $S_2^=$, S_2^- are the principal anodic reactants in an Na_2S_2/KCNS melt, whereas S_2^- and S_4^- are the principal anodic reactants in an Na_2S_4/KCNS melt. It can be seen from Figs. 35 and 36 that the sulfur chemical stripping rate increases with the concentration S_2^- (or $S_4^=$). But for the electrochemical stripping we propose $S_2^=$ as the intermediate. If S_4^- + $e^- \rightarrow 2S_2^-$ were the charge-transfer step, the coulombs stripped in

Fig. 35 would be different. Therefore, the S_2' reactions are, in Na_2S_2,

$$S_2^- + e^- \rightleftarrows S_2^=$$
$$S_2^= + 2S \rightleftarrows 2S_2^-$$
$$\overline{}$$
$$2S + e^- \rightleftarrows S_2^- \quad \text{(net reaction)}$$

and in dilute Na_2S_4,

$$S_2^- + e^- \rightleftarrows S_2^=$$
$$2S_2^= + S_4 \rightleftarrows 4S_2^-$$
$$\overline{}$$
$$S_4 + 2e^- \rightleftarrows 2S_2^- \quad \text{(net reaction)}$$

The second cathodic reaction suggested for Na_2S_2 is

$$S_2^- + e^- \rightleftarrows S_2^=$$

and for dilute Na_2S_4 (the S_1' reduction),

$$S_2^- + e^- \rightleftarrows S_2^=$$
$$S_2^= + S_2 \rightleftarrows 2S_2^-$$
$$\overline{}$$
$$S_2 + e^- \rightleftarrows S_2^- \quad \text{(net reaction)}$$

The slight differences in the potential of the S_1' and S_2' reactions in dilute Na_2S_4 are postulated to be the result of the differences in equilibrium constants or reaction rates for the two postulated following chemical reactions (42). Circumstantial evidence has been presented in this chapter for the participation of sulfur polymeric species S_2, S_4, and S_8 in the reactions discussed. Obviously, considerably more work would be required to firmly establish the polymer chain length of reacting sulfur species.

Note that an inverse anodic spike is present at all concentrations greater than 0.2 molar in Figs. 35 and 36 because of the sudden removal of a partially blocking sulfur layer (confirmed by conductance measurements). In Na_2S_3 the blocking layer does not form an inverse spike except at much higher concentrations, but the blocking effect, as determined by conductance measurements, is there nonetheless. The sweeps in Figs. 35 and 36 begin from the rest potential and proceed in the anodic direction. They show a prewave before wave E. The number of coulombs passed in this prewave region depends on the sulfur content of the solute in a manner similar to that in the pure melt. In Fig. 36c the dashed curve shows the behavior when the sweep begins in the cathodic direction from the rest potential. In KCNS the separation of the $S_4^=$ reduction wave A and the $S_2^=$ reduction wave C is not as well defined as in the pure melt.

Sodium sulfide (Na_2S) is not soluble in KCNS, but it dissolves slightly in NaCNS-KCNS. Its cyclic sweep, starting from the rest potential, is shown in Fig. 37 (68). Peak C is absent in the cathodic sweep, and C' is present in the anodic sweep. Peak C is obtained in the reduction of an Na_2S_2 containing melt, whereas C' is not obtained in the oxidation. This is further corroboration that waves C and C' represent the Na_2S_2/Na_2S couple.

3.5 Practical Sulfur Electrodes

The large surface graphite felt electrode used in the sodium-sulfur cell is necessary for best utilization of the sulfur in the

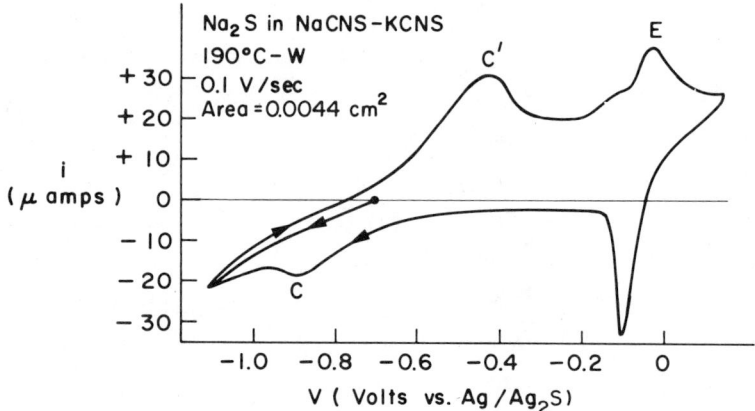

Fig. 37. Cyclic sweep, from Aikens et al. (68).

cell. A theoretic model for this complicated system is being developed by Wayner, but has not yet reached the state where it can be tested conclusively by measurements in a cell (73,74).

Two methods have been applied by Minck to following the progress of the reaction through the thickness of the felt electrode (73,75).

In one approach two kinds of small graphite fiber probes are inserted into the felt:

1. in this probe the fiber protrudes out of its glass sheath and touches the electrode fibers, making it possible to measure the potential of the electrode fiber at the place of contact (electronic probe);
2. the fiber in this probe is slightly recessed in its glass sheath and acts as an $S/S_x^=$ electrode determining the potential in the melt (ionic probe).

A major difficulty in this setup lies in designing and placing these probes in such a way as to cause minimum interference with the flow of the melt. A sodium-sulfur tube cell was built having the usual concentric arrangement: Na/β"-alumina/graphite felt in sulfur-polysulfide/graphite current collector with three electronic and two ionic probes. Experimental results on discharge indicate that up to 0.06 A-hr/cm^2 (relative to the conductive ceramic surface), all the electrode processes are occurring near the ceramic surface. The current flows electronically through the more conductive graphite rather than ionically through the melt. As this region is depleted of uncharged sulfur and polarization at the graphite fiber surface increases, the reaction zone moves away from the ceramic. From 0.6 to 0.8 A-hr/cm^2 this movement proceeds toward the current collector. Eventually, above 0.8 A-hr/cm^2, the current collector becomes part of the active electrode. The actual figures have only relative meaning, because the available volume of sulfur in this cell was not well defined. Although this method needs further development, it has proved its usefulness. The application of a sodium/glass membrane probe in combination with an ionic probe is being considered for determining the local polysulfide composition.

The other method for following the progress of the reaction through the depth of the sulfur electrode uses a long graphite felt electrode precisely fitted into an α-alumina cylinder terminated at one end by a β"-alumina disk and at the other by a graphite current collector. After passage of a definite amount of charge, the cell is quenched and sliced. The slices are analyzed for composition of the sulfur-sulfide melt. This experiment was carried out with various current densities and times of discharge. Although there may be some fingering (protrusion of polysulfide fingers into sulfur) and some redistribution of melt during quenching, the results permit definite conclusions about the progress of the discharge process in the sulfur electrode (Fig. 38). At low discharge rates (22 mA/cm^2), the concentration has time to equalize, which results in a low gradient. At higher current densities an increasingly steeper gradient of concentration is observed. As the reaction progresses, the cell resistance increases, because the current travels farther and farther in the melt correspondingly shortening the conducting path in the lower

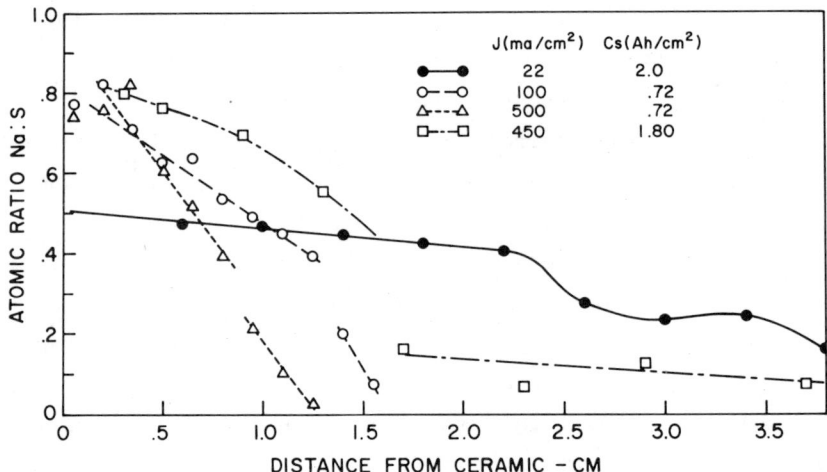

Fig. 38. Variation of polysulfide composition within the porous electrode, from Minck (73,75).

resistance graphite felt. Analysis shows compositions essentially below the liquidus line in the reacted region spreading from the ceramic electrolyte, indicating deposition of Na_2S_2 in significant amounts. These deposits do not seem to be a serious hindrance to mass transport, but they must contribute to the electrical resistance. This is in agreement with the kinetic investigations of the reaction at a smooth surface, where the Na_2S_2 deposition never causes blocking, as does sulfur during charging at a carbon surface. One may envision the Na_2S_2 deposit as something like a crystalline powder. The maximum concentration of sodium found in these experiments is independent of the depth of discharge, once saturation has been reached near the ceramic electrolyte.

An important consideration in designing practical electrodes is the ratio of felt resistance to melt resistance. A very conductive felt makes the current flow preferentially through the felt to the vicinity of the ceramic electrolyte during charging and may lead to accumulation of sulfur in that area, once the melt is saturated there. The insulating sulfur blocks off the rest of the melt from further reaction.

The results of these studies have been utilized in the design of electrodes for sodium-sulfur cells. The porous electrode must provide sufficient surface area so that the interfacial reaction does not become rate determining, even when part of the surface is covered (especially in the charging mode). It must also provide sufficient open space for diffusion and convection to ease transport processes. Wicking must also be considered as a means of transport. For the charging process the advantage of a metal surface in rejecting sulfur, as it is preferentially wetted by polysulfide, can only be used if a material with these properties can be found that does not corrode or that corrodes sufficiently slowly. Corrosion leads to increased resistance by loss of material; corrosion products increase contact resistance, and their accumulation hinders transport processes. Accumulation near the ceramic electrolyte can cause uneven current distribution; this and the direct interaction of some corrosion products with the ceramic contribute to ceramic degradation.

Figure 39 shows the performance of a metal-free tube cell (inner compartment Na, outer compartment S/Na_2S_x) with an open graphite felt structure (filling only part of the sulfur compartment but covering the ceramic electrolyte) at various current densities (relative to the outer ceramic surface) and at two

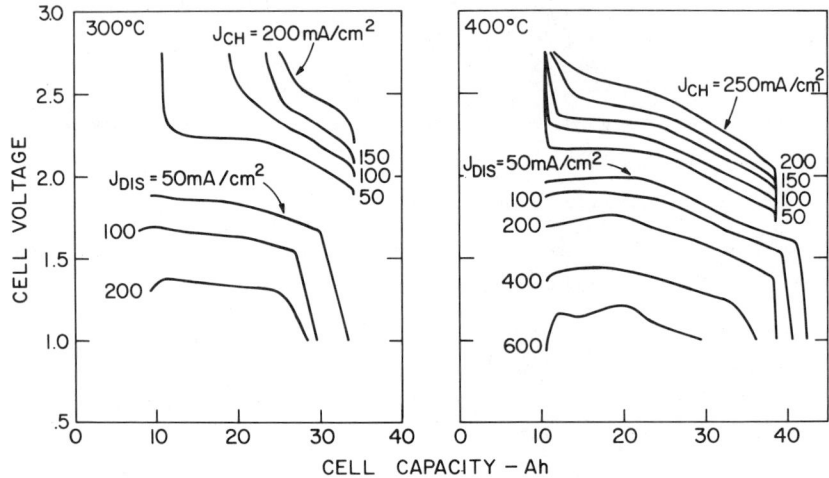

Fig. 39. Charge and discharge characteristics for cell with shaped electrode, from Minck and Chang (73,76).

temperatures (73,76). The operation at the higher temperature results in a better performance, but corrosion problems may not permit continuous operation at 400°C. In Fig. 40 the influence of the cut-off voltages for charge and discharge on the capacity is depicted. Finally, Fig. 41 shows the power output in various modes of operation versus the energy delivered. The efficiency shown is related only to discharge (i.e., relative to OCV rather than charge voltage).

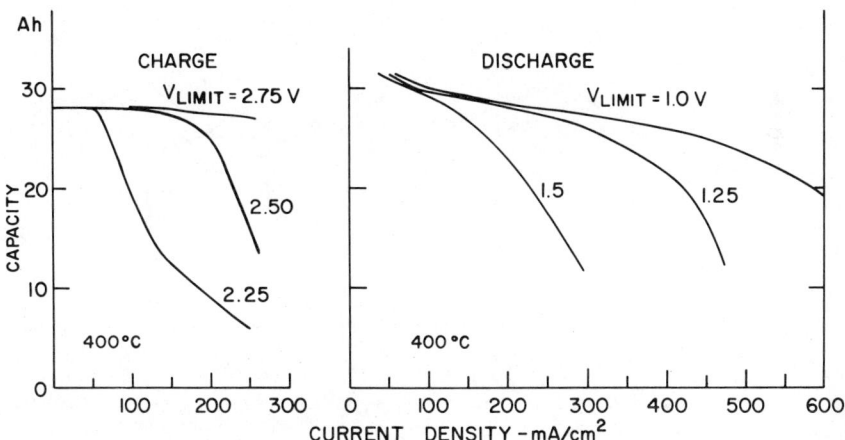

Fig. 40. Effect of cell voltage limits on charge and discharge capacity, from Minck and Chang (73,76).

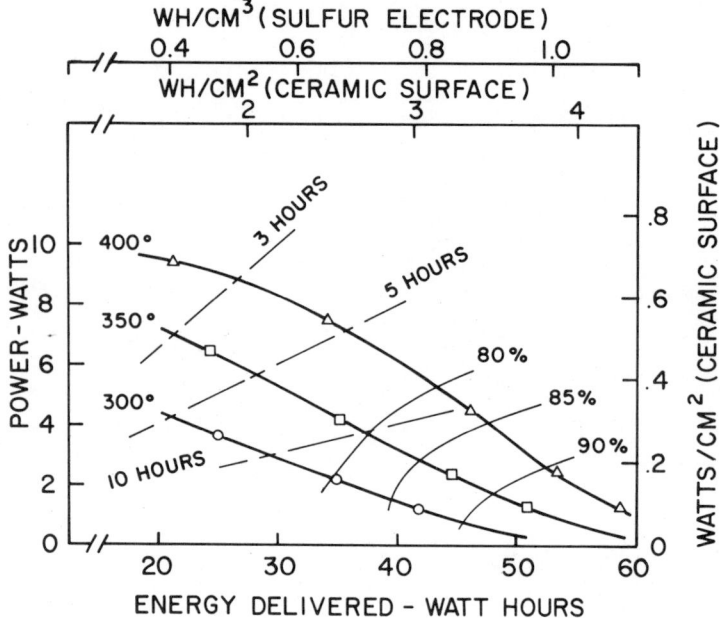

Fig. 41. Performance characteristics for cell with shaped electrode. The loci of constant discharge times and constant voltage efficiencies are superimposed on the power versus energy curves, from Minck and Chang (73,76).

The problems still to be solved in designing a practical sulfur electrode are predominantly related to materials. The ideal electrode material would combine the good conductivity and wetting properties of a metal with the corrosion resistance of carbon materials. Another problem is to ensure good and lasting electrical contact between the porous electrode and its backing (in most cases the container) that serves as a lead out.

4 SULFUR CATIONS

Recent studies (cf. 23) have proved the existence of positive sulfur ions in the presence of strong Lewis acids.

Adhami and Herlem (77) found that sulfur dissolved in fluosulfonic acid gives two anodic polarographic waves. Controlled potential coulometry (of a fresh solution) at the foot of the first wave makes both waves disappear completely after passage of three electrons per sulfur atom. After the electrolysis, there is no reduction wave. From their results the authors conclude that sulfur is dissolved as S_2^{2+} and oxidized in the following two steps:

$$S_2^{2+} = 2S^{2+} + 2e^-$$
$$S^{2+} = S^{4+} + 2e^-$$

S^{2+} is unstable and dismutes into S^{4+} and S_2^{2+}. The absence of a reduction wave is tentatively attributed to reaction with the solvent forming SF_4, which escapes,

$$S^{4+} + 4FSO_3^- \rightarrow SF_4 + 4SO_3$$

It seems there are media in which sulfur cations are more stable. Bjerrum (78) found S(II) and S(IV) in various media including chloroaluminate melts. The use of the system S_x/S_y^{n+} (n = 2,4) as the cathode in a battery with an $AlCl_3$ - NaCl melt electrolyte has been proposed by Mamantov et al. (79). In combination with a suitable anode (e.g., Al or Na separated by β alumina), this system forms a cell with an emf above 2.5 V. Only exploratory experiments with an aluminimum-sulfur cell have thus far been reported. They look promising enough to warrant further study.

The authors wish to thank Ford Motor Company for permission to publish this review, and gratefully acknowledge the support of NSF-RANN for part of this work.

REFERENCES

1. G. Tammann and H. O. v. Samson-Himmelstjerna, Z. Anorg. Allg. Chem. 216, 288 (1934).
2. G. Delarue, Soc. Chim. France, Bull. 1960, 906, 1654.
3. H. A. Laitinen and C. H. Liu, J. Amer. Chem. Soc. 80, 1015 (1958); H. A. Laitinen and J. W. Pankey, J. Amer. Chem. Soc. 81, 1053 (1959).
4. K. Kiukkola and C. Wagner, J. Electrochem. Soc. 104, 379 (1957).
5. H. Reinhold, Z. Elektrochem. 40, 361 (1934).
6. L. S. Darken, J. Amer. Chem. Soc. 72, 2909 (1950).
7. A. G. Verduch and C. Wagner, J. Phys. Chem. 61, 559 (1957).
8. C. Wagner, Thermodynamics of Alloys, Addison-Wesley, Reading, Mass., 1952, p. 14.
9. R. Schenck, I. Hoffman, W. Knepper, and H. Vögler, Z. Anorg. Allg. Chem. 240, 173 (1939).
10. N. K. Gupta and R. P. Tischer, J. Electrochem. Soc. 119, 1033 (1972).
11. T. G. Pearson and P. L. Robinson, J. Chem. Soc. 1930, 1473.
12. D. G. Oei, Inorg. Chem. 12, 435 (1973).
13. E. Rosen and R. Tegman, Chem. Scripta 2, 221 (1972).
14. W. T. Thompson and S. N. Flengas, Can. J. Chem. 46, 1611 (1968); J. Electrochem. Soc. 118, 419 (1971); 119, 399 (1972).
15. J. Berkowitz and J. R. Marquart, J. Chem. Phys. 39, 275 (1963); J. Berkowitz and W. A. Chupka, ibid. 40, 287 (1964).

16. D. Detry, J. Drowart, P. Goldfinger, H. Keller, and H. Rickert, Z. Physik. Chem., N.F. 55, 314 (1967).
17. M. V. Merritt and D. T. Sawyer, Inorg. Chem. 9, 211 (1970); R. P. Martin, W. H. Doub, Jr., J. L. Roberts, Jr., and D. T. Sawyer, Inorg. Chem. 12, 1921 (1973).
18. R. Bonnaterre and G. Cauquis, J. Chem. Soc., Chem. Commun. 1972, 293.
19. T. Chivers and I. Drummond, Inorg. Chem. 11, 2525 (1972); J. Chem. Soc. Dalton 1974, 631.
20. F. Seel and H. J. Güttler, Angew. Chem. 85, 416 (1973); Angew. Chem. Int. Ed. 12, 420 (1973).
21. W. Giggenbach, J. Inorg. Nucl. Chem. 30, 3189 (1968); J. Chem. Soc. Dalton 1973, 729.
22. T. Chivers, Nature 252, 32 (1974).
23. T. Chivers and I. Drummond, Chem. Soc. Rev. 2, 233 (1973).
24. J. R. Coleman and M. W. Bates, in D. H. Collins (Ed.), Power Sources 2, Proc. 6th Internat. Symp. Brighton 1968, Pergamon New York, 1968, p. 289.
25. G. A. Toporischev, O. A. Esin, and V. N. Kalugin, Fiz. Khim. Rasplav. Shlakov 1970, 34.
26. S. I. Rempel' and E. M. Malkova, Zhur. Prikl. Khim. 25, 558 (1951).
27. F. G. Bodewig and J. A. Plambeck, J. Electrochem. Soc. 116, 607 (1969).
28. J. H. Kennedy and F. Adamo, J. Electrochem. Soc. 119, 1518 (1972).
29. J. P. Bernard, A. de Haan, and H. Van der Poorten, Compt. Read. Ser. C 276, 587 (1973).
30. J. Greenberg, B. R. Sundheim, and D. M. Gruen, J. Chem. Phys. 29, 461 (1958); cf. also D. M. Gruen, R. L. McBeth, and A. J. Zielen, J. Amer. Chem. Soc. 93, 6691 (1971).

31. B. Cleaver and A. J. Davies, Electrochim. Acta 18, 733 (1973).
32. J. R. Birk and R. K. Steunenberg, Advances in Chemistry Series, American Chemical Society, Washington, D.C., No. 140, p. 186 (1975).
33. M. J. Weaver, Electrochemical Studies of Sulfur Electrodes in Molten Salts, Ph.D. thesis, Imperial College, London, England, 1972.
34. Research on Electrodes and Electrolyte for the Ford Sodium-Sulfur Battery, Annual Report for Period June 30, 1973 to June 29, 1974 (July, 1974), National Science Foundation, Contract No. NSF C-805 (AER-73-07199).
35. B. Cleaver, A. J. Davies, and D. J. Schiffrin, Electrochim. Acta 18, 747 (1973).
36. B. R. Sundheim, Fused Salts, Mc-Graw Hill, New York, 1964.
37. J. Crank, The Mathematics of Diffusion, Oxford University Press, New York, 1956.
38. F. A. L. Dullien and L. W. Shemilt, Trans. Faraday Soc. 58, 244 (1962).
39. R. W. Laity, J. Phys. Chem. 67, 671 (1963).
40. C. Wagner, in W. Jost (Ed.), Diffusion in Solids, Liquids, Gases, Academic Press, New York, 1952.
41. W. H. Reinmuth, Anal. Chem. 34, 1446 (1962).
42. R. S. Nicholson and I. Shain, Anal. Chem. 36, 706 (1964).
43. D. G. Oei, Inorg. Chem. 12, 438 (1973).
44. G. J. Janz, in Research on Electrodes and Electrolyte for the Ford Sodium-Sulfur Battery, Annual Report for Period June 30, 1974 to June 29, 1975 (July 1975), National Science Foundation, Contract No. NSF C-805; to be published in Inorg. Chem.
45. B. Cleaver and A. J. Davies, Electrochim. Acta 18, 741 (1973).

46. J. Gerlock, Ford Motor Co. personal communication. See also W. Giggenbach, Inorg. Chem. 10, 1308 (1971); T. Chivers, and I. Drummond, Chem. Soc. Rev. 2, 239, 240, 242, 244 (1973).
47. E. Rosén and R. Tegman, Chem. Scripta 2, 63 (1972).
48. N. Weber and J. T. Kummer, Advances in Energy Conversion Engineering, Amer. Soc. Mech. Eng., New York, 1967.
49. S. M. Selis, Electrochim. Acta 15, 1285 (1970).
50. K. D. South, J. L. Sudworth, and J. G. Gibson, J. Electrochem. Soc. 119, 554 (1972).
51. F. A. Ludwig, R. P. Tischer, D. A. Aikens, and K. W. Fung, Extended Abstracts No. 386, 147th Meeting, The Electrochem. Soc., May 1975.
52. F. A. Ludwig, unpublished data.
53. M. A. Dzieciuch, Ford Motor Co., personal communication.
54. B. Cleaver and A. J. Davies, Electrochim. Acta 18, 727 (1973).
55. S. C. Abrahams and J. L. Bernstein, Acta, Cryst. B25, 2365 (1969).
56. B. Cleaver, A. J. Davies, and M. D. Hames, Electrochim. Acta 18, 719 (1973).
57. K. W. Fung, D. A. Aikens, F. A. Ludwig, and R. P. Tischer, unpublished data.
58. D. S. Polcyn and I. Shain, Anal. Chem. 38, 370 (1966).
59. H. Gerischer and M. O. Krause, Z. Phys. Chem. N.F. 10, 264 (1957); 14, 184 (1958); H. Matsuda, S. Oka, and P. Delahay, J. Amer. Chem. Soc. 81, 5077 (1959).
60. W. D. Weir and C. G. Enke, J. Phys. Chem. 71, 275 (1967).
61. N. K. Gupta and R. P. Tischer, unpublished data.
62. N. K. Gupta and R. P. Tischer, J. Electrochem. Soc. 119, 1033 (1972).
63. W. Giggenbach, J. Inorg. Nucl. Chem. 30, 3189 (1968); J. Chem. Soc. Dalton, 729 (1973).

64. T. L. Allen, *J. Chem. Phys.* 31, 1039 (1959).
65. B. Meyer, T. Strayer-Hansen, D. Jensen, and T. V. Oommen, *J. Amer. Chem. Soc.* 93, 1034 (1971); B. Meyer, T. V. Oommen, and D. Jensen, *J. Phys. Chem.* 75, 912 (1971).
66. R. Bonnaterre and G. Cauquis, *J. Chem. Soc. Chem. Commun.* 293 (1972).
67. B. Cleaver, A. J. Davies, and D. J. Schiffrin, *Electrochim. Acta* 18, 747 (1973).
68. D. A. Aikens, K. W. Fung, F. A. Ludwig, and R. P. Tischer, unpublished data.
69. T. Berzins and P. Delahay, *J. Amer. Chem. Soc.* 75, 555 (1953).
70. G. J. Janz, *Molten Salts Handbook*, Academic, New York, 1967.
71. R. S. Nicholson and I. Shain, *Anal. Chem.* 37, 178 (1965).
72. L. W. Kao, K. W. Fung, D. A. Aikens, and F. A. Ludwig, *Anal. Chem.* 47, 1121 (1975).
73. *Research on Electrodes and Electrolyte for the Ford Sodium-Sulfur Battery*, Annual Report for Period June 30, 1974 to June 29, 1975 (July 1975), National Science Foundation, Contract No. NSF C-805 (AER-73-07199).
74. P. C. Wayner, Jr., unpublished.
75. R. W. Minck, unpublished.
76. R. W. Minck and Y. C. Chang, unpublished.
77. G. Adhami and M. Herlem, *J. Electroanal. Chem. Interfac. Electrochem.* 26, 363 (1970).
78. N. J. Bjerrum, personal communication.
79. G. Mamantov, R. Marassi, and J. Q. Chambers, Electrochem. Soc., Fall Meeting, New York 1974, *Extended Abstr.* 74-2, No. 8; *Research and Development Tech. Rep.* ECOM-0600-F, April 1974.

Index

A

A.C. impedance technique, 434
Activation energy, 281, 292, 430
Activity
  of graphite, 455
  of sulfur, 393, 407, 408, 409, 438, 453
Adsorption, 105, 112, 130, 141, 142, 157, 167, 172, 173, 175, 185, 187, 251, 255, 293, 297
  from the gas phase, 262, 277
  of water, 262, 263, 266, 277, 301
  rare gas, 263
  strength for H_2, 298
Air/fuel ratio, 10, 73, 76, 77, 80, 82
Alcohols, 71
Alumina, 71
Aluminium silicates, 28, 42
Ammonia, 71
Amplification, 93, 95
Anharmonic vibration, 335, 352, 373
Anion, 409
  sulfur, 409
  polysulfide, 409
  radical, 444, 460, 461
Anisotropic temperature factor, 334
Antifoggants, 104, 174
Antioxidants, 160, 185, 186
Aqueous solution, 410
Asbestos, 64
Autocatalysis, 175, 185

B

Band, gap, 108, 112, 137, 153
  model, 142, 146, 164, 155
Batteries, 391, 477
  organic solvent primary, 414
  secondary, 391
  sodium-sulfur, 425
  solid electrolyte, 326
  storage, 392
Benzoic acid, 58
Beta alumina, 326, 327, 336, 355, 471, 473, 477
  copper, 349
  gallium analogs of, 355
  iron analogs of, 355
  lithium, 361
  potassium, 355
  sodium, 336
  thallium, 345
Bismuth oxide
  Bi_2O_3, 365, 366
Biomedical technology, 62
Bleaching, 132, 140, 152, 163, 189, 191
Blocking layer, see surface film
Blood, 62
Blue color (of sulfide solutions), 411, 412, 413, 416, 418, 420, 421, 460, 461
Bond energy, see (polysulfide) ion
Born-Mayer model, 374
Bridging layers, 338, 351

C

Cadmium oxide, 71
Calcium fluoride
  CaF_2, 331, 363
Carbon monoxide, 35, 36, 60, 66, 71, 72
Catalysts, 73, 74
  dual-bed, 74
  three-way, 75, 80
Catalytic
  converter, 73
  mechanism, catalytic step, 418, 437, 438, 448, 452, 453, 454
Cell, with transference, 409, 438

silver sulfide,398
   formation,398, 408
   sodium-sulfur,407,454,469,473
   solid state,395,398
Ceramic electrolyte,see beta-alumina
Ceric oxide,CeO_2,364
Chalcogenide,47
Characteristic curve,115,122, 123,186
Charge transfer,425,427,440,453, 459
   rate,438(see also exchange current)
   resistance,440
   step,467
Chemical potential,241,245
   of electrons,219,221
Chemical reaction,419,425,440, 445,448,449,452,466
   coupled,following,419,422,434
   rate,434,449,451,452,468
Chemisorption,71,72
   of water,273
Chloride melt,414,416
Chloroaluminate melt,477
Chronocoulometry,440
Chronopotentiometry, chronopotentiogram,421, 424,430,452
Cobalt complexes,189
Cobalt oxide,68,69,72
   phthalocyanine,53,56
Color development,95,161,169,188, 189
Complex plane analysis,376
Concentration,gradient,472
   polarization,449
Condensation,127
Conductance,23,28,68,71,72
   measurements,67,433,440,445, 447
   equivalent,431
   in solids,12,18,38

Conductivity,ionic,99,101,107, 110,171
   electronic,95,110,111,118, 143
Contact,angle(sulfur, polysulfide),445,456
   resistance(electrode-current collector),473
Controlled potential electrolysis,416,419
Convection,414,437,450,455,473
Copper compounds
   Cu_2Se,359
   CuBr,332,334
   CuI,332,333,334,359
   CuO,71
   Cu_2HgI_4,333,359
   $KFe_{11}O_{17}$,355
   Cu_2S,333,359
   KCu_4I_5,360
Corrosion,(role of - in Na-S cell), 445,473,474, 476
   products,473
Coulometry,410,422,477
Coulometric,analysis,63,64,65
   cell,63,64
   sensors,65
Cross section,121,125
Current function,431,437,443, 444,447
Current impulse relaxation technique,440
Current interruption technique,434
Current voltage curve,$S^=$ in chloride melt,416
Cyanine dyes,141
Cyclic sweep,see voltammetry

D

d metals,230,294,298
d and sp metals,252,257
Density of states,141,144,146, 289

Desensitization,134,140,155,157
Desensitizer,135,150,151,152,175
Desorption of water,297
Developer,161,162,163,171,181,187 188
Development,photographic,center, 130,133,166,176
  chemical,95,160,163,167,168, 172,175,178
  color,95,161,169,189
  ideal,168,169,172
  lith,185
  postfixation physical,95,160, 167,172,175
  praefixation physical,168,170
  superadditive,186
Differential thermal analysis, 405
Diffusion,106,117,119,120,121,125, 127,173,177,192,473
  coefficient,6,63,405,426,427, 430,431,460
  control,426,434,448,449,452, 458,459
  counter,426,427
  current,6,7,454
  equations,426
  layer,6,50
  limited process,427
  transfer process,161
  species,430,431
Dimethyl formamide,413
Dimethyl sulfoxide,409,411,412, 414
Dislocation,103,113,114,172
Disorder,332
  electronic,108,109,163
  ionic,97,98,170
  structural,103
  type(groups),328,331
Dissociation of polysulfide ions, see ion (radicals)
Dissolution,complex,167,168,170, 192
  rate(of Na_2S_2 in polysulfide melt),451

reaction(of sulfur),405
Doping,99,103,105,111,127,128
Double layer,248,258,276,282, 283,285,286,290
Dye,see sensitizer, desensitizer

E

ECE mechanism,418,451,463
Electrocatalysis(development), 161,166,167,187,188,190
Electrochemical,kinetics,279
  potential of electrons,243
  work function,226,246,249 260,300
Electrode,battery,392
  counter,52,57,63,64,65
  gas-diffusion,8,48,49
  graphite felt,469,470,474
  kinetics of the sulfur,392, 414,424,425,428,430, 445,457,473
  lead sulfide,416
  material,429,457
  metal sulfide,392,393
  palladium,66
  platinum,30,44,46,49,63,65, 66,77
  potential,absolute,239,242, 243
  silver,reference,463
  sodium,404,409
  sulfur,391,392,395,398,401, 404,412,414,415,454
  sulfur-potential,392,409
  stainless steel,455,456,457
  theory(development),166,167, 175,178,182
Electrolyte,supporting,425,456
  ceramic,see beta-alumina
  solid,325
Electromotive force,394
Electron,see mobility,trap, injection

Electron affinity,116,143,147
    (see sensitizing dye)
Electron transfer,see injection,
    sensitization, optical,
    development
Electronegativity,266
    of elements,228
Electroneutrality,427
Electronic configuration,see
    (polysulfide) ion
    states,284
    transference number,348
Electrons in the solution,244,
    245
Emf,395,396,400,401,404,405,408,
    409,477
    series,392
Emission,control system,76
    decontamination,75
    values,73
Energy,distribution,141,144,146
    gap midpoint,144,145,151,154
    of activation,280
    level,see sensitizing dye
    transfer,see sensitization,
        optical
Entropy,effects,293,295
    and enthalpy,7,405
Equilibrium,chemical,9
    constant,468
    potential,16,17,18,19,21,23,25,
        27,28,30,31,32,34,46
    preceding chemical,9
    thermodynamic,11,30,32,36,44
    voltage,17
ESR,428
Exchange current,494,434,440
Exciton,93,142
Exposure,96,122,127,130,132,135,
    175

F

Faradaic impedance,433,440
Fermi-energy,109,11,112,139,142,
    144,145,164

level,215,218,220,258,267,
    284,301
Ferric oxide,Fe_2O_3,71,72
Fick's law,427
Film,see surface film
Filamentary silver,167,168,169,
    171,172,173,185,193
Fixation,95,192
Fluorite structure,331,363
Free energy,395,396,397,405
Freezing point depression,
    429,459
Frenkel disorder,97,98,100
Fused salts,411,414,425,428,431

G

Galvanostatic charging,425
    double-pulse galvanostatic
        technique,440
Gelatin,99,128,129,138,150,
    160,171,174
Gibbs-Duhem equation,395,405,
    407
Gold,58
Graphite felt,see electrode

H

Haven ratio,345
Half-wave potentials,60
Heat of adsorption,230
Hole,see mobility, trap,
    photohole, injection
Hollandites,362
Hydrogen,33,35,44,293,297
    adsorption,297
    evolution,292,298,301
    m metals,369-370
    sulfide,71
Hydrophilicity,268,269,270,287
    scale,270,273,301

I

Identity card,216

Induction period,164,175,185
Inhibition,65,66
Injection,electron,145,163,166
  hole,132,140,152,163,166
Inner layer capacity,270
Intensification,163
Intercalation by water,351
Interdiffusion,426
Interstitialcy mechanism,345,348
Interstitials in metals,366
Ion,polysulfide,392,409,412,413
    428,see also solvation
    shell
  resonance forms of polysulfide,
    444
  dissociation into-radicals,
    411,412,428,444,459,460,
    461
  exchange,341
  positive sulfur,429,477
  electronic configuration of
    polysulfide,444
  bond energy of polysulfide,444
Ionic conductivity,temperature
    dependence,331
  mobility,size effect,348
Ionization energy,116,143,144,
    see sensitizing dye

K

Kinetic energy of electrons,218
Kinetics,see electrode
Kink sites,97,98,102,103,105,192

L

λ-values,73,75,80,82,83
Lanthanum fluoride,LaF_3,365
Latent image,93,94,111,122,123,
    160
  growth,122,123,124,126,133
  inner,127,163,171
  size,122,130,131,132
  stability,128,129,130
  surface,125,127,166

Latent preimage,122
Lattice defects,97,98
Layer structures,362,365
Leveling(silver),173,174,193
Lewis acids,477
Life time,electrons,117,119,
    121,134
  holes,119,121,134
Liquid junction,see potential
Liquidus line,472
Lith development,186
Lithium compounds
  LiCl-KCl eutectic melt,393,
    416,418,419,420,421,
    422,424,456
  Li_2SO_4,359
  Li_4SiO_4,361
  Li_2WO_4,359
  Li_4GeO_4,361
Localized motion,rattling,376
Limiting current,7,50,52,53,
    57,60,62,64,454
  diffusion,6,9,59
  measurement,54
  sensors,51,60

M

Mass transport(transfer),414,
    427,472,473
Membrane,50,51,57,58,59,60
Metal-hydrogen bond strength,
    292
Metal sulfide,393,see also
    electrode
M-H bond,297
Migration,427
Minimum energy path,374
Minimum energy positions,335
Mixed potential,11,46,47,48,49
Mobility,electronic,117,118,
    119,120
  ionic,101,106
Model for water,276
  at interfaces,277
Molten salts,see fused salts

Molten sublattice,332
Monobath processes,95,161,167, 193
Mossbauer spectroscopy,376
Mullite phase,28,42

N

Nernst,equation,5,16,33,34,35, 37,38,39,41,44,78
  formula,401
  glower,324
  slope,422
  Einstein equation,431
Nicholson-Shain criteria,453
Nitrate melt(NaNO$_3$-KNO$_3$ eutectic), 425
Nickel oxide,71
Nitric oxide,NO,60,71
Nitrogen dioxide,concentration, 47,48,53,54,56,71
Nitrous oxide,71
Nonpolarizable electrodes,243
Non-silver photography,96,113
Nucleation,122,123,130

O

Open-circuit voltage decay,425, see also potential
Operative potential,241,245,274
Optical absorption,reflectance, 376
Organic,electrolytes,58
  iodide compounds,362
  substances,71
Orientation,of dipoles,276,277, 236
  of the solvent,239
  of water,247,261,265,266,267, 268,271,273,274,278,289, 299,300,301
Oxide films,283
Oxidizable gases,37
  in the presence of excess oxygen,36

Oxygen,concentration,32,37,38, 41,49,60,62,63,71,79
  dissolved in water,65

P

Partial pressure,16,17,24,27, 46
  hydrogen,34,45
  oxygen,14,15,16,19,20,22,23, 24,27,29,30,31,32,33, 34,36,39,40,42,43,44, 68,69,78,79
  sulfur,30
Passivation,456
Peaking splitting,463
Phase,diagram (Na$_2$S-S),394, 401,405
  transformation,329
Photocurrent,135,138,157
Photoelectron,93,96,122,125, 127
Photographic materials,96
Photohole,96,122,125,127
Photolysis,127,138
photomoment,135,136,146
Physical measurements on solid electrolytes,357
Platinum,see electrodes
  sulfide,425
Polarizable electrodes,243
Polarographic,analysis,7
  gas detectors,8
Polarography,477
Polycrystalline surfaces,224, 249,261,274
Polysulfide,anions,427,428
  melt,409,411,414,425,428, 429,455,457
  phase,448
  solutions,412
  species,410,412
  absorption spectra of,412
  reaction mechanism of,412
  pre-oxidation of,449,467
  viscosity of,431,461
  see also ion,dissociation

Potassium, antimonate, $KSbO_3$, 361
  hydroxide, 64
  sulfate, K_2SO_4, 361
  thiocyanate, 414, 418, 429, 439, 456, 457, 459, 467, 469
Potential, energy, 220
  of zero charge, 258, 264, 273, 276, 278, 282, 287
  profiles, 373, 374
  separation, 418
  step, 440, 458, 459
  decomposition, 416
  equilibrium, 421, 442
  half wave, 416, 421
  liquid junction, 409
  open circuit, 439
  rest, 432, 443, 469
  see also electrode
Potentiometry, 424
Pre-oxidation, see polysulfide
Pre-exponential factor, 283, 293
Propane, C_3H_8, 60
Propylene carbonate, 58

Q

Quantum yield, 95, 119, 144, 156
Quenching, 139, 141, 145, 152, 158

R

Radical, 138, 148, 162, 179, 181, 186
Ramsdellite, 362
Randles-Sevcik analysis, 432
Rate constant, 454
Reaction rate, see chemical reaction
Recombination, 93, 95, 96, 113, 123, 124, 127
Redox reactions, 282
Redox system, 425
Reduced absolute potential, 243 244
Relaxation time, 106, 107, 121, see also lifetime
Reorientation, 273

  of the solvent, 250
  of water, 269
Resonance forms, see ions
Rotating disk technique, 430, 431, 438
Rutherford back-scattering, 376
Rutile, 362

S

Sand equation, 452
Scattering, light, 376
  Neutron, 376
  Raman, 376
  X-ray diffuse, 376
Semiconductors, 68, 70, 71, 72
  fast diffusion in, 370372
Sensitivity, photographic, 96, 115, 123, 156
Sensitization, chemical, 96, 115, 125, 126, 127, 129, 130
Sensitization, optical, 94, 134
  energy transfer, 136, 137
  electron transfer, 136, 137, 140, 142, 155
Sensitizing dyes, 141, 142, 149, 151, 152, 175
  energy levels, 139, 146, 147, 149, 150
  oxidation potentials, 147, 148, 149, 153, 154, 155, 158
  reduction potentials, 136, 144, 147, 148, 149, 153, 154, 155, 156, 158
  regeneration, 138, 140, 158
Sensors, alcohol, 49
  limiting current, 51, 60
  membrane, 51
  $-\lambda$, 77, 79, 81
  semiconductor, 68, 72, 83
  titanium dioxide, 82
  zirconium dioxide, 82
Silicates, 362
Silver
  cathode, 65

compounds
  MAg_4I_5 materials, 360
  $RbAg_4I_5$, 327, 332
  AgI, 330, 332, 333, 357, 360
  $Ag_{19}I_{15}P_2O_7$, 361
  $Ag_7I_4PO_4$, 361
  $Ag_6I_4WO_4$, 361
  Ag_3SX materials, 357
  Ag_2Te, 359
  Ag_2SO_4, 361
  Ag_2HgI_4, 331, 332, 33, 335, 359
electron affinity, 116, 164
iodides, 327
ionization potential, 116, 164
filaments, see also filamentary
    silver
  see also halides
  nucleus, 94, 123, 126, 128, 130, 133,
    161, 164, 176
  redox potential, 128, 155
Silver halides, 93, 96, 97, 108, 109
  electronic disorder, 97, 100,
    101, 104
  electronic conductivity, 110
  ionic conductivity, 101
  ionic disorder, 101
  ionic mobility, 101
  space charge, see also space
    charge
Single crystal faces, 261, 273,
  274
Slag, 414, 415
Sodium-sulfur, see cell, battery
Solid anion conductors, 363
  carbon conductor, 356, 363
  state, 395, 348
Solid electrolytes, 5, 12, 13, 16,
  18, 25, 30, 33, 41, 42, 43, 58,
  62, 63, 76, 77, 80
Solvated electrons, 237
Solvation energy, 104, 147, 148
  shell of polysulfide ion, 412
Solvent, aprotic, 410, 418, 428
  organic, 409, 410
Sommerfeld model, 217, 221

sp metals, 234, 260, 266, 267, 274,
  290, 291, 294, 298
sp and d metals, 225, 300
space charge, 99, 102, 104, 105,
  126, 164
Spectra, absorption, 412
  laser Raman, 428
Spectrophotometry, 410
Splitting, 463, 467
Stainless steel, see electrode
Steel manufacture, 37
Stokes-Einstein equation, 461
Stripping current, peak, 442,
  457, 463, 466, 467
Sulfide, in slags, 415
  double-systems, 396
  see also electrode, metal
    sulfide, thermodynamic
    data
Sulfur, chloride, 393, 422
  density of, 450
  dioxide, 71
  liquid, 401, 448
  homonuclear-species, 414
  see also activity, electrode,
    ion, thermodynamic data,
    vapor
Superadditivity, 186
Supersensitization, 140, 156
Supersensitizer, 157, 159
Surface, charge, 105, 113, 126, 164
  concentration, 98
  film, 431, 440, 441, 445, 448, 449,
    454, 455, 456, 458, 469, 473
  resistance, 432, 439, 440
  potential, 98, 101, 102
    of water, 236, 262, 264, 265,
      271
  state, 166
  tension, 130, 131, 185

T

Ternary system, 395
Thermal properties, 331

Thermodynamic,data(sulfur,
    sulfide),392-401
  potential,242
Thermodynamics of irreversible
    processes,409,426
Theoretical models,372
Thorium oxide,18,28,30,42,364
Tin oxide,71,72
Titanium dioxide,68,69,71,80,82
Transference,428,438
  see also cell
  number,110,120
Transient methods,415,429,434
Transition metal oxides,68
Transition,290
  metals,250,265,267,273
  metal oxides,68
Transpiration method,408
Transport processes,see mass
    transport
Traps,electron,112,113,116,121,
    123,125,133,134,140
  hole,112,114,115,116,118,121,
    125,128,134,140,157
Tungsten bronzes,48,362
Tunnel structures,361
Tunneling,143,158
Tysonite structure,365

U

Ultrasonic attenuation,376
Underpotential deposition,251,255
Utilization,see electrode

V

Vanadium oxides,362

Vapor(sulfur),397,398,416
  pressure,396,398,401,407,409
  molecular composition of,397,
    398,408
V-centers,114,125,127
Viscosity,see polysulfide
Volcano curves,231
Voltage sweep,see voltammetry
Voltammetry,409
  cyclic(sweep),410,411,418,
    419,421,422,425,429,
    432,452,456,463,466,467
  linear sweep,431,432,458
Voltammogram,see voltammetry

W

Warburg resistance,440
Wetting properties(of sulfur
    and polysulfide),414,
    455,456,457,473,476
Whisker,172
Wicking properties(for sulfur
    and polysulfide),455,
    473
Work function,217,219,223,235,
    238,248,258,262,266,
    274,279,280,282,284,
    286,290,292,293,297,300

Z

Zirconium oxide,13,14,15,19,
    20-28,30,39,43,58,62,
    77,82,324,363,364
Zinc oxide,71,72